Dust in the Galactic Environment

The Graduate Series in Astronomy

Series Editors: **R J Tayler**, University of Sussex
R E White, University of Arizona

The Graduate Series in Astronomy includes books on all aspects of theoretical and experimental astronomy and astrophysics. The books are written at a level suitable for senior undergraduate and graduate students, and will also be useful to practising astronomers who wish to refresh their knowledge of a particular field of research.

The Graduate Series in Astronomy

Dust in the Galactic Environment

D C B Whittet

Associate Professor, Rensselaer Polytechnic Institute, New York, USA

Institute of Physics Publishing
Bristol, Philadelphia and New York

British Library Cataloguing-in-Publication Data

A catalogue record for this book is available from the British Library.

ISBN 0 7503 0204 6(hbk)
    0 7503 0209 7(pbk)

Library of Congress Cataloging-in-Publication Data are available

Series Editors: **R J Tayler**, University of Sussex
                    R E White, University of Arizona

Published by IOP Publishing Ltd, a company wholly owned by The Institute of Physics, London.

IOP Publishing Ltd
Techno House, Redcliffe Way, Bristol BS1 6NX, UK
335 East 45th Street, New York, NY 10017-3483, USA

US Editorial Office: IOP Publishing Inc., The Public Ledger Building, Suite 1035, Independence Square, Philadelphia, PA 19106

Typeset by TeX
Printed in the UK by Cambridge University Press

To

Clair and James

with love

Ever the dim beginning,
Ever the growth, the rounding of the circle,
Ever the summit and the merge at last, (to surely start again,)

All space, all time,
The stars, the terrible perturbations of the suns,
Swelling, collapsing, ending, serving their longer, shorter use,

Ever the mutable,
Ever materials, changing, crumbling, recohering....

Walt Whitman

Contents

Preface   **ix**

1.   Introduction   **1**

1.1   Dust in the Galaxy—our view from within   2
1.2   Interstellar environments   15
1.3   The significance of dust   22
1.4   The problem of grain composition   25

2.   Element abundances and depletions   **31**

2.1   The origin and evolution of the chemical elements   32
2.2   The Solar System abundances—a standard reference   39
2.3   Abundance trends in the Galaxy   43
2.4   The observed depletions   47

3.   Interstellar extinction and scattering   **57**

3.1   Theory and methods   58
3.2   Average properties   65
3.3   Spatial variations in the extinction curve   73
3.4   Models for interstellar extinction   80

4.   Interstellar polarization and grain alignment   **83**

4.1   Extinction by non-spherical particles   83
4.2   Visual polarimetry of reddened stars   86
4.3   The spectral dependence of polarization   92
4.4   Grain alignment   106

5.   Spectral absorption features   **115**

5.1   The 2175 Å feature   116
5.2   The optical diffuse bands   129
5.3   The infrared absorption features   144

6.   Continuum and line emission   **165**

6.1   Theoretical considerations   166
6.2   Continuum emission from the galactic disk   174

6.3   Spectral emission features   180

7.   The origin and evolution of interstellar grains   193

7.1   Dust in the ejecta of evolved stars   194
7.2   The evolution of dust in the interstellar medium   215
7.3   Dust in the environments of young stars   238

8.   Towards a unified model for interstellar dust   257

References   265

Index   287

Preface

Dust is a ubiquitous feature of the cosmos, impinging directly or indirectly on most fields of modern astronomy. Dust grains composed of small (submicron-sized) solid particles pervade interstellar space in the Milky Way and other galaxies: they occur in a wide variety of astrophysical environments, ranging from comets to giant molecular clouds, from circumstellar shells to galactic nuclei, and the study of this phenomenon is a highly active and topical area of current research. This book aims to provide an overview of the subject, covering general concepts, methods of investigation, important results and their significance, relevant literature, and some suggestions for promising avenues of future research. It is aimed at a level suitable for those embarking upon post-graduate research, but will also be of more general interest to researchers, teachers and students as a review of a significant area of astrophysics. As a formal text for taught courses, it will be particularly useful to final-year undergraduates and M.Sc. students studying the interstellar medium. My aim throughout has been to produce a compact, coherent text which will stimulate the reader to investigate the subject further. It would be unproductive to consider dust grains in isolation. Our concept of interstellar space has changed from a passive 'medium' to an active 'environment', and the underlying theme of the book is the significance of dust in astrophysics with particular reference to the interaction of the solid particles with their gas phase environment.

It would be impossible to treat all aspects of this subject in depth in a single book of reasonable size. The scope and sheer volume of research on cosmic dust and related topics can be judged from the proceedings of two recent international symposia on the subject (Bailey and Williams 1988; Allamandola and Tielens 1989), which contain, between them, some 50 review papers as well as numerous shorter contributions. In order to keep the present text unified and reasonably compact, I have focussed the discussion on interstellar dust in our own Galaxy. The Galaxy is both the environment of planetary systems and the most accessible example of the building blocks of the Universe, and thus provides the link between planetary and extragalactic astronomy. Solid particles within our Solar System are considered where directly relevant to the galactic context, e.g. as a pointer to the nature of presolar interstellar grains. Similarly, dust in external galaxies is discussed as an integral part of the text rather than as a distinct topic requiring separate chapters.

The text is divided into eight chapters, the first of which provides a historical perspective for current research and a review of interstellar environments. The observed properties of interstellar grains are considered in Chapters 2 to 6, beginning, in Chapter 2, with their influence on gas phase abundances. Chapters 3, 4 and 5 examine the extinction, polarization and spectral absorption characteristics of the grains, respectively, and Chapter 6 considers continuum and line emission. In Chapter 7, we discuss the origin and evolution of the dust, tracing its lifecycle in a succession of environments from circumstellar shells to diffuse interstellar clouds, molecular clouds, protostars and protoplanetary disks. The final chapter attempts to provide an overview of the subject. It is assumed throughout that the reader is familiar with basic concepts in stellar astronomy, such as magnitude and distance scales and the spectral classification sequence, and has a qualitative familiarity with the stellar evolutionary cycle according to current models. The reader with little or no background in astronomy may find the introductory texts by Shu (1982) and Kutner (1987) helpful.

Système Internationale (SI) units are used in addition to the units of astronomy, but the unsuspecting reader should be aware that the cgs system is still widespread in the astronomical literature. One exception to SI in the present text is that the Ångstrom ($1\,\text{Å} = 10^{-10}\,\text{m}$) is generally used to denote wavelengths in the visible and ultraviolet spectral regions. This has the minor but convenient advantage that all dust-related spectral features in this part of the spectrum can be specified uniquely by the wavelength to four significant figures without a decimal point, hence (e.g.) '$\lambda 6284$' for the diffuse absorption feature centred at $628.4\,\text{nm}$. The Ångstrom enjoys such common usage in the astronomical literature that it might be regarded as an adoptive unit of astronomy.

The author has found astrophysical dust to be a challenging and rewarding topic of study. An important reason for this is the wide variety of techniques involved, embracing observational astronomy over much of the electromagnetic spectrum, theoretical modelling, and laboratory astrophysics. Interpretation and modelling of observational data may lead the investigator into such diverse fields as solid-state physics, scattering theory, mineralogy, organic chemistry, surface chemistry, and small-particle magnetism. Moreover, in spite of much activity and considerable progress in recent years, there is no shortage of challenging problems: the nature of the grain alignment mechanism and the origin of the optical diffuse bands are just two examples. If this book attracts students of physical sciences to study cosmic dust, it will have succeeded in its primary aim. It is also my hope that physicists with interest and expertise in small-particle systems may be encouraged to consider grains in the laboratory of space. As Huffman (1977) has remarked, "it is a difficult experimental task to produce particles a few hundred Ångstroms in size, keep them completely isolated from one another and all other solids, maintain them in ultra-high vacuum at low temperature, and study photon interactions with the particles at remote wavelengths ranging from the far infrared to the extreme ultraviolet. This is the opportunity we have in the case of interstellar dust."

I am indebted to my wife and family for their love, support, patience and understanding during the preparation of this book. Many colleagues and friends have contributed over the years to the development of my knowledge and ideas on interstellar dust, and my research has benefitted immeasurably from interactions with others attracted to this strangely fascinating topic. I should like to record my thanks, particularly, to Kashi Nandy for stimulating my early interest in the subject, to Andrew Adamson, Michael Bode and David Williams for helping to sustain it, and to Walter Duley and Peter Martin for providing hospitality, intellectual stimulus, and practical support during a period of sabbatical leave in Toronto at the Physics Department, York University, and the Canadian Institute for Theoretical Astrophysics. The text was completed whilst I was on leave at the Laboratory for Space Research, Groningen. I am indebted to Roger Tayler for his thoughtful and constructive comments on the manuscript, and to Mark Bailey for his advice and assistance in the selection of a number of illustrations. The lines from Whitman's poem 'Eidólons' prefacing the book are based on those selected by Vagn Holmboe to inscribe the score of his Tenth Symphony. The script was produced by the author using the Adam Hilger macro package for the TEX typesetting system, "intended for the creation of beautiful books—and especially for books that contain a lot of mathematics" (Knuth 1986). I leave the reader to assess the irrelevance of this quotation.

Finally, I should like to thank Wolfgang Amadeus Mozart, Jean Sibelius, and Wilhelm Stenhammar for writing music which, I have found, induces a state of mind ideally suited to relaxation after long nights at the word-processor.

D C B Whittet

Rensselaer Polytechnic Institute

August 1991

1

Introduction

"The discovery of spiral arms and—later—of molecular clouds in our Galaxy, combined with a rapidly growing understanding of the birth and decay processes of stars, changed interstellar space from a stationary 'medium' into an 'environment' with great variations in space and in time."

H C van de Hulst (1989)

Interstellar space is, by terrestrial standards, a near-perfect vacuum: the average particle density in the solar neighbourhood of our Galaxy is approximately $10^6 \, \text{m}^{-3}$ (one atom per cubic centimetre), a factor of about 10^{19} less than in the terrestrial atmosphere at sea level. However, dense objects such as stars and planets occupy a tiny fraction of the total volume of the Galaxy, and the tenuous interstellar medium† contributes roughly a third of the mass of the galactic disk. The stellar and interstellar components are continually interacting and exchanging material: new stars condense from interstellar clouds, and, as they evolve, they bathe the surrounding ISM with radiation; ultimately, many stars return a substantial fraction of their mass to the ISM, which is thus continuously enriched with heavier elements fused from the primordial hydrogen and helium by nuclear processes occurring in stars. A major proportion of these heavier atoms are locked up in submicron-sized solid particles (dust grains), which account for roughly 1% of the mass of the ISM and are almost exclusively responsible for its obscuring effect at visible wavelengths. Despite their relatively small contribution to the total mass, the remarkable efficiency with which such particles scatter, absorb and re-radiate starlight ensures that they have a very significant impact on our view of the Universe. For example, the attenuation between us and the centre of the Galaxy is such that, in the visual waveband, only one photon in every 10^{12} reaches our telescopes.

---

† For convenience, the term 'interstellar medium' (ISM) is used to refer, collectively, to interstellar matter over all levels of cloud density, embracing a wide range of environments (§1.2).

1

The energy absorbed by the grains is re-emitted in the infrared, accounting for some 20% of the total bolometric luminosity of the Galaxy.

This chapter aims to provide an introductory overview of the phenomenon of cosmic dust and its rôle in interstellar processes. We begin (§1.1) by reviewing the early development of knowledge on interstellar dust, assessing the impact of its obscuring properties and spatial distribution on our view of the Universe, whilst simultaneously introducing some basic concepts and definitions. We then examine the environments to which the grains are exposed (§1.2), and discuss the importance of dust as a significant chemical and physical constituent of interstellar matter (§1.3). A summary of current models for interstellar grains is included in the final section (§1.4).

1.1   DUST IN THE GALAXY—OUR VIEW FROM WITHIN

The influence of interstellar dust may be discerned with the unaided eye on a dark, moonless night at a time of year when the Milky Way is well-placed for observation. In the Northern hemisphere, the background light from our Galaxy splits into two sections in Aquila and Cygnus; Southern observers are better placed to view such irregularities: the dark patches and rifts were seen by Aborigine observers as a 'dark constellation' resembling an emu, with the Coal Sack as its head, the dark lane passing through Centaurus, Ara and Norma as its long, slender neck, and the complex system of dark clouds towards Sagittarius as its body and wings. Discoveries in the twentieth century enable us to recognize the Milky Way in Sagittarius as the nuclear bulge of a dusty spiral galaxy, seen from within the disk at a distance of a few kiloparsecs. Our view of our home Galaxy is impressively illustrated by wide-angle, long-exposure photographs of the night sky, such as that shown in figure 1.1. The Milky Way is a fairly typical spiral of Hubble type Sb, with a nucleus and disk surrounded by a spheroidal halo containing globular clusters (see Mihalas and Binney 1981 for a wide-ranging review of the stucture and dynamics of the Galaxy). There is a striking resemblence between figure 1.1 and photographs of external spiral galaxies of similar morphological type seen edge-on, such as NGC 891, illustrated in figure 1.2. The visual appearance of such galaxies tends to be dominated the equatorial dark lane which bisects the nuclear bulge. Obscuration is less evident (but invariably present) in spirals inclined by more than a few degrees to the line of sight. These results indicate that dark absorbing material is a common characteristic of such galaxies, and that this matter is concentrated into disks which are thin in comparison to their radii. Contours of infrared intensity superposed on the optical photograph in figure 1.2 illustrate that emission from dust is also aligned with the disk of the galaxy.

Figure 1.1   A wide-angle photograph of the sky, illustrating the Milky Way from Vulpecula (left) to Carina (right). The nuclear bulge in Sagittarius is below centre. Photograph courtesy of W Schlosser and Th Schmidt-Kaler, Ruhr Universität, Bochum, taken with the Bochum super wide-angle camera at the European Southern Observatory, La Silla, Chile. The secondary mirror of the camera system and its support are seen in silhouette.

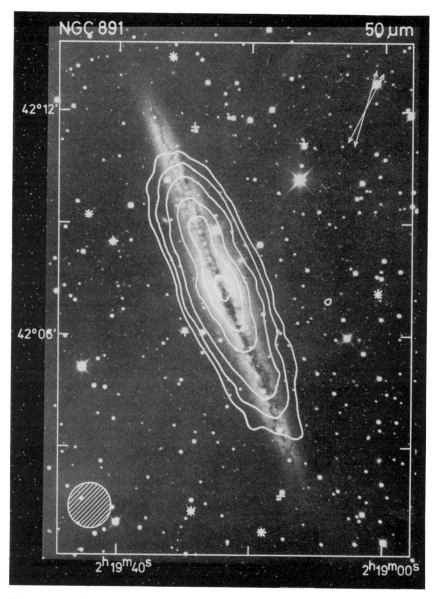

Figure 1.2   An optical photograph of the edge-on spiral galaxy NGC 891, illustrating high dust opacity concentrated in the disk. Contours of 50 μm infrared emission measured with the Infrared Astronomical Satellite (IRAS) are superposed (Wainscoat *et al* 1987). Photograph courtesy of P R Wesselius, Laboratory for Space Research, Groningen.

1.1.1 Early Investigations of Extinction and Reddening

The study of extinction by interstellar dust can perhaps be said to have begun with Wilhelm Struve's analysis of star counts (Struve 1847), which demonstrated that the apparent number of stars per unit volume of space declines in all directions with distance from the Sun (see Batten 1988 for a modern account of this work). This led Struve to hypothesize that starlight suffers absorption in proportion to the distance travelled, and on this basis he deduced a value for its amplitude which is in remarkably good agreement with current estimates. His proposal did not gain acceptance, however, and no further progress was made until the beginning of the 20th century, when Kapteyn (1909) recognized the potential significance of extinction:

> *"Undoubtedly one of the greatest difficulties, if not the greatest of all, in the way of obtaining an understanding of the real distribution of the stars in space, lies in our uncertainty about the amount of loss suffered by the light on its way to the observer."*

Both Struve and Kapteyn envisaged uniform absorption, but Barnard's photographic survey of dark 'nebulae' provided evidence for spatial variations (Barnard 1910, 1913, 1919, 1927). The existence of dark regions in the Milky Way had been known for many years: William Herschel regarded them as true voids in the distribution of stars ('holes in the sky'), a view which still prevailed in the early 20th century. However, detailed morphological studies convinced Barnard that at least some of the 'holes' contain interstellar clouds which absorb and scatter starlight. For example, the association of dark and bright nebulosities in regions such as the well-known complex near ρ Ophiuchi strongly supported this view (e.g. Barnard 1919; see Seeley and Berendzen 1972a, b for an in-depth historical review). It was also suggested at about this time (Slipher 1912) that the diffuse radiation surrounding the Pleiades cluster might be due to scattering by particulate matter.

Confirmation that the interstellar extinction hypothesis is correct came some years later as the result of two distinct lines of investigation by the Lick Observatory astronomer R J Trumpler (1930a, b, c). If dust is present in the interstellar medium, its obscuring effect will clearly influence stellar distance determinations, introducing another degree of freedom in addition to apparent brightness and intrinsic luminosity Trumpler sought to determine the distances of open clusters by means of photometry and spectroscopy of individual member stars. Spectral classification provides an estimate of the luminosity, and the distance modulus is obtained by comparing apparent and absolute magnitudes. In the Johnson (1963) notation†, the standard distance equation may be written:

$$V - M_V = 5\log d' - 5, \tag{1.1}$$

---

† Trumpler used an early magnitude system, but we adopt modern usage.

where V and M_V are the apparent and absolute visual magnitudes, respectively, and d' is the apparent mean cluster distance in parsecs. Having evaluated d', Trumpler then deduced the linear diameter of each cluster geometrically from the measured angular diameter. When this had been done for many clusters, a remarkable trend became apparent: the deduced cluster diameters appeared to increase with distance from the Solar System. From this, Trumpler inferred the presence of a systematic error in his results due to obscuration in the interstellar medium, and concluded that a distance-dependent correction must be applied to the left-hand side of equation (1.1) in order to restore normality and render the cluster diameters independent of distance:

$$V - M_V - A_V = 5\log d - 5, \tag{1.2}$$

where d is now the true distance. The quantity A_V represents interstellar 'absorption' at visual wavelengths in the early literature, but should correctly be termed the 'extinction' (the combined effect of absorption and scattering). A_V tends to increase linearly with distance in directions close to the galactic plane; for the open clusters, a mean rate of $\sim 1\,\mathrm{mag\ kpc^{-1}}$ is required.

Trumpler then considered the implications of his discovery for the colours of stars. If interstellar extinction is produced by submicron-sized particles then a reddening effect is to be expected, exactly analogous to the reddening of the Sun at sunset by particles in the terrestrial atmosphere. Considering radiation from a distant star seen through a dusty intervening medium, an absorbed photon is completely removed from the beam and its energy converted into internal energy of the particle, whereas a scattered photon is deflected from the line of sight. Reddening occurs because absorption and scattering are, in general, more efficient for shorter wavelengths in the visible: thus red light is less extinguished than blue light in the transmitted beam, whereas the scattered component is predominantly blue. A problem which had puzzled stellar astronomers in the 1920s was the fact that many stars close to the galactic plane appear redder than expected on the basis of their spectral types. In effect, there was a discrepancy in stellar temperatures deduced by spectroscopy and photometry. Spectral classification gives an estimate of temperature based predominantly on the presence and relative intensities of spectral lines produced in the stellar photosphere. Colour indices such as $(B - V)$ in the Johnson system are indicators of temperature based on the slope of the stellar continuum and its equivalent blackbody temperature. A number of stars which show early-type spectral characteristics, indicating high surface temperatures, were known to have colour indices more appropriate to cool late-type stars. This anomaly is explained if the stars are reddened by the intervening medium: the appearance of a stellar spectrum over a limited spectral range is not drastically altered by reddening, in the sense that the wavelengths and relative strengths of characteristic lines are essentially unchanged; spectral classification therefore gives a good indication of the temperature of the underlying star independent of reddening. However, colour indices depend on both temperature and reddening, and this information cannot be separated easily until the spectral type of the star is known. By comparing the apparent brightnesses over a range of wavelengths of intrinsically similar stars with different degrees of reddening, Trumpler showed that interstellar extinction is a roughly linear function of wavenumber ($A_\lambda \propto \lambda^{-1}$) in the visible region of the spectrum. This important result, subsequently verified by more detailed studies (e.g. Stebbins, Huffer and Whitford 1939), implies the presence of solid particles with dimensions comparable to the wavelength of visible light†. Such particles may be expected to contain $\sim 10^9$ atoms if their densities are comparable with those of terrestrial solids.

In order to study in detail the observed properties of interstellar extinction and reddening, it is essential to use background stars with known spectral characteristics as probes. The degree of reddening or 'selective extinction' of a star in the Johnson B, V system is quantified as the colour excess

$$E_{B-V} = (B - V) - (B - V)_0 \qquad (1.3)$$

where $(B - V)$ and $(B - V)_0$ are observed and 'intrinsic' values of the colour index. As the extinction is always greater in the B-filter (central wavelength $0.44\,\mu$m) than in V ($0.55\,\mu$m), E_{B-V} is a positive quantity for reddened stars and zero (to within observational error) for unreddened stars. Intrinsic colours are determined as a function of spectral type by studying nearby stars and stars at high galactic latitudes which have little or no reddening. Colour excesses may be defined for any chosen pair of photometric passbands by analogy with equation (1.3): another commonly used measure of reddening in the blue–yellow region is the colour excess E_{b-y} based on the Strömgren (1966) intermediate passband system ($E_{b-y} \simeq 0.74 E_{B-V}$). The relationship between total extinction at a given wavelength and a corresponding colour excess depends on the wavelength dependence of extinction, or extinction curve. In the Johnson system, the extinction in the visual passband may be related to E_{B-V} by

$$A_V = R_V E_{B-V} \qquad (1.4)$$

---

† The term 'smoke' was often used to describe these particles in the early literature. 'Smoke' implies the product of combustion, whereas 'dust' implies finely powdered matter resulting from the abrasion of solids. The former is arguably more appropriate, at least as a description of the particles condensing in stellar atmospheres, now regarded as a primary source of interstellar grains. However, 'dust' has become firmly established in modern usage.

where R_V is termed the ratio of total to selective visual extinction. The quantity E_{B-V} is directly measureable, whereas A_V is known only if R_V can be determined. Following the pioneering work of Trumpler, much effort was devoted in subsequent years to the empirical evalution of R_V (e.g. Whitford 1958, Johnson 1968, and references therein), in view of its fundamental significance in photometric distance determinations (equations (1.2) and (1.4)). Theoretically, R_V is expected to depend on the composition and size distribution of the grains. However, in the low-density interstellar medium, R_V is found to be virtually constant, and a value of

$$R_V \simeq 3.05 \pm 0.15 \qquad (1.5)$$

may be assumed for most lines of sight. The origin of this result and its limits of applicability are discussed in Chapter 3.

Extinction by dust renders interstellar space a polarizing as well as an attenuating medium. This was first demonstrated by Hall (1949) and Hiltner (1949), who showed that the light of reddened stars is partially plane polarized, typically at the 1–5% level. The origin of this effect is widely accepted to be the directional extinction of elongated grains which are aligned in some way, i.e., their long axes have some preferred direction. A model which produces alignment by means of an interaction between the spin of the particles and the galactic magnetic field was proposed by Davis and Greenstein (1951). These authors assumed that the grains are paramagnetic, and are set spinning by collisions with atoms in the interstellar gas. Paramagnetic relaxation then results in the grains tending to be orientated with their angular momenta parallel (and hence their long axes perpendicular) to the magnetic field lines. Although the strength of the galactic magnetic field was subsequently shown to be too weak to give alignment in precisely the manner suggested by Davis and Greenstein, it seems nevertheless highly probable that an analogous process is occurring in the interstellar medium.

1.1.2 The Distribution of Reddening Material

Studies of obscuration within external systems give us a qualitative picture of the large-scale distribution of dust in typical spiral galaxies: it is strongly confined to galactic disks, producing prominent equatorial dust lanes in edge-on systems. In contrast, there is a general sparcity of dust in elliptical galaxies. In terms of association with stellar populations, dust is clearly a population I phenomenon. As discussed below, this picture is entirely consistent with the distribution of reddening material within our own Galaxy and with the strong spatial correlation between dust and newly formed stars.

Detailed quantitative investigations of the variation of E_{B-V} with galactic longitude, latitude (ℓ, b), and distance in the Galaxy require two-colour photometry and spectroscopy at optical wavelengths for large numbers of stars, making use of equations (1.2), (1.3) and (1.4), or equivalent forms, and the absolute magnitude versus spectral-type calibration (e.g. Gottlieb 1978). Such investigations have been carried out by FitzGerald (1968), Lucke (1978) and Perry and Johnston (1982). Complementary data on the total extinction may be obtained statistically by means of star counts in individual dark clouds (Bok 1956). Analogous techniques have also been used to study the galactic reddening and extinction of extragalactic objects at intermediate and high galactic latitude (e.g. de Vaucouleurs and Buta 1983). The results of these studies confirm that the particles responsible for reddening are closely constrained to the disk of the Galaxy, essentially within a layer no more than $\sim 200\,pc$ thick in the solar neighbourhood. For example, FitzGerald determines the scale height of $\langle E_{B-V}/d \rangle$, averaged for different longitude zones, to be in the range 40–100 pc. Figure 1.3 shows the distribution of colour excess in galactic coordinates for stars out to 2 kpc. Comparison with studies of the interstellar gas distribution show that dust and gas are generally well-mixed (e.g. Dame and Thaddeus 1985).

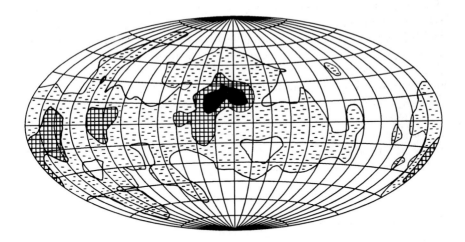

Figure 1.3 The distribution of reddening material within 2 kpc of the Sun, displayed in galactic coordinates. The galactic centre is at the centre of the diagram, with longitude increasing to the left. Lines of constant longitude and latitude are spaced 15° apart. Successive contours represent 0.2, 0.6 and 1.0 mag kpc^{-1} (E_{B-V}/L). (Adapted from Lucke 1978.)

The small-scale distribution of dust in the galactic plane is extremely uneven. Little or no reddening occurs close to the Sun ($E_{B-V} < 0.03$ for almost all stars studied within 100 pc), suggesting that we are currently situated in a comparatively clear ('intercloud') region of the interstellar medium. This is consistent with the negative result of an experiment to detect gravitationally focussed interstellar grains entering the Solar System by means of their predicted contribution to the zodiacal light (Bertaux and Blamont 1976). At distances beyond 100 pc, the dust shows a strong tendency to concentrate into clouds, well-known examples including the Coalsack and the star-forming regions in Taurus and Ophiuchus. The distribution of clouds within a 500 pc radius of the Sun is dominated by the local stellar group known as Gould's Belt, a disk-like system of young stars tilted at about 18° to the galactic plane, which includes the Orion and Scorpio-Centaurus OB associations. This effect may be seen in figure 1.3: there is a tendency for reddening material to extend above the galactic plane towards the centre ($\ell = 0°$), and below towards the anti-centre ($\ell = 180°$). Remarkably, the Taurus and Ophiuchus dark clouds are almost symmetrical about the Sun, and may be part of the same macroscopic cloud structure (Turon and Mennessier 1975). These clouds are amongst the nearest regions of recent star formation to the Solar System, and are invaluable for studies of the nature and evolution of interstellar grain material in a relatively dense environment. The general distribution of dust in the solar neighbourhood illustrates the spatial association between interstellar material and young stars.

The extinction in directions away from the galactic plane, although generally small, is of considerable significance, as corrections for the dimming of primary distance indicators (such as Cepheids, novae, and supernovae) in external systems by dust in the Galaxy influence the extragalactic distance scale. The reddening of high-latitude stars ($|b| > 20°$) is independent of distance beyond a few hundred parsecs. If the disk of the Galaxy is treated as a flat, uniform slab with the Sun in the central plane, a systematic dependence of extinction on b is expected; it may easily be shown that this takes the form of a cosecant law†:

$$A_V(b) = A_P \operatorname{cosec} |b| \qquad (1.6)$$

where A_P is the visual extinction at the galactic poles. The general applicability of this idealized representation is open to considerable doubt. The degree of extinction near the galactic poles has proved particularly controversial: some authors (e.g. McClure and Crawford 1971) argue in favour of polar 'windows', with $A_V(b) \leq 0.05$ for $b > 50°$, whereas de Vaucouleurs and Buta (1983) deduce a dependence compatible with equation (1.6) with

---

† This relation is exactly equivalent to Bouguer's law for extinction in a plane-parallel planetary atmosphere, used to correct for telluric extinction in astronomical photometry.

$A_P \simeq 0.15$, on the basis of galaxy counts and reddenings. The reddenings of intermediate-luminosity stars indicate that the polar obscuration is patchy (e.g. Hilditch *et al* 1983), which is consistent with the uneven distribution of $100\,\mu m$ emission from dust at high $|b|$ (§1.1.3). Extinctions based on equation (1.6) should thus be treated with caution: the only reliable method of determining the galactic reddening of extragalactic objects is to investigate the distribution of E_{B-V} in galactic field stars close to each individual line of sight. Burstein and Heiles (1982) have produced maps of galactic reddening which cover almost the entire galactic sphere for $|b| > 10°$, which are invaluable for this purpose.

The general correlation between extinction and distance in the galactic plane depends statistically on the number of clouds which happen to lie along each given line of sight. On average, a column $L = 1\,kpc$ long intersects several (~ 5) diffuse clouds which produce a combined reddening $E_{B-V} \simeq 0.6$. Making use of equations (1.4) and (1.5) to express this in terms of total extinction, the mean rate of visual extinction with respect to distance is

$$\left\langle \frac{A_V}{L} \right\rangle \simeq 1.8 \text{ mag kpc}^{-1}. \tag{1.7}$$

This result is based on studies of stars with distances up to a few kiloparsecs from the Sun, and is applicable only in the galactic plane. At greater distances, $\langle A_V/L \rangle$ is more difficult to estimate, as even luminous OB stars and supergiants become too faint to observe at visible wavelengths. The visual magnitude of a typical supergiant would exceed 20 for distances greater than $6.5\,kpc$ and average reddening. Photometry at infrared wavelengths may be used to penetrate to greater distances if a sufficiently luminous background source is available; assumptions regarding the wavelength dependence of extinction and its spatial uniformity then allow visual extinctions to be calculated. The extinction towards the infrared cluster at the galactic centre is estimated to be $A_V \sim 30$ mag over the $\sim 8\,kpc$ path (Roche 1988), a result which suggests an increase in the rate of extinction, compared with the solar neighbourhood, as we approach the nucleus.

As is evident from the above discussion, the concentration of dust in the galactic plane can seriously hinder investigation of the structure and kinematics of the Galaxy using visually luminous spiral arm tracers such as early-type stars and supergiants. Models for spiral structure beyond $\sim 3\,kpc$ are based largely on long-wavelength astronomy (radio and infrared observations), although a few 'windows' in the dust distribution, where the rate of extinction is unusually low, allow studies at visible wavelengths to distances $\sim 10\,kpc$. However, in general, the morphological structure of our own Galaxy is less well explored than that of our nearest neighbours. A further implication of some significance is that it is extremely difficult to detect novae and supernovae in the disk of the Galaxy. Studies of external galaxies suggest that the expected mean supernova rate in spirals of similar

Hubble type to our own is ~ 1 per 25 years (e.g. Tammann 1974), with an uncertainty of approximately a factor of 2, but historical records suggest that only 5 visible galactic supernovae have been seen in the past 1000 years, the last of which was 'Kepler's star' in 1604 (Clark and Stephenson 1977). The apparent discrepancy is attributed to the presence of extinction.

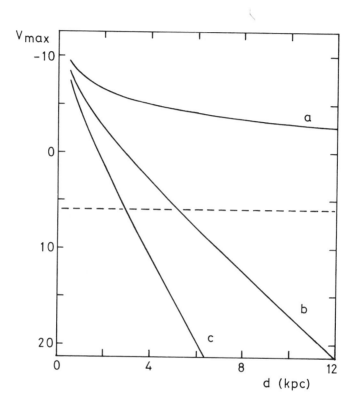

Figure 1.4   Plot of maximum visual magnitude V_{max} against distance d for a typical type II supernova for three values of the rate of visual extinction: (a) zero, (b) 2 mag kpc^{-1}, and (c) 4 mag kpc^{-1}.

Figure 1.4 plots the apparent visual magnitude at maximum luminosity (V_{max}) against distance (equation (1.2)) for a typical type II supernova with an absolute visual magnitude of -18 at maximum. Curves are shown for three values of the rate of extinction in mag kpc^{-1}: (a) $\langle A_V/L \rangle = 0$, a good approximation for high galactic latitude; (b) $\langle A_V/L \rangle = 2$, which is close to the mean rate within a few kiloparsecs of the Sun (equation (1.7)), and (c) $\langle A_V/L \rangle = 4$, the mean value estimated for the line of sight to the galactic centre. The dashed line indicates the threshold of naked-eye visibility, and an image at the threshold of detection on a typical photographic

Schmidt survey plate would lie close to the faintest end of the magnitude range shown. We deduce that for cases (b) and (c), supernovae further than 5 and 3 kpc from us, respectively, would never reach naked-eye visibility. A supernova explosion occuring at the galactic centre ($d \sim 8$ kpc) would have $V_{max} \sim 28$, and would thus be undetectable even if we were fortunate enough to observe this line of sight with the most sensitive optical detectors at precisely the right time. The situation would be much improved in the infrared: for example, in the K-band at $2.2\,\mu$m, the extinction in magnitudes is approximately a factor of 10 less than in V. Assuming that the absolute magnitudes of the supernova are comparable in the K and V bands, the equivalent form of equation (1.2) for K gives $K_{max} \sim -0.5$. Thus, the same supernova would be comparable in brightness with, say, Vega in the near infrared!

1.1.3   Diffuse Radiation from the Galactic Disk

The energy removed from the transmitted beam when starlight passes through a dusty medium is either scattered from the line of sight or absorbed as heat. Interstellar grains redistribute the spectrum of the interstellar radiation field, attenuating starlight most efficiently at visible and ultraviolet wavelengths, and re-emitting in the infrared. Both scattering and re-emission contribute to the diffuse background radiation from the galactic disk.

Scattering is the principal cause of continuum extinction at wavelengths below $\sim 1\,\mu$m (e.g. Witt 1988). Direct observational evidence for scattered light in the interstellar medium is provided by the presence of blue reflection nebulae surrounding individual dust-embedded stars or clusters, bright filamentary nebulae and halos around externally heated dark clouds, and, on the macroscopic scale, by weak ultraviolet background radiation from the disk of the Milky Way, the diffuse galactic light (DGL). Faint reflection nebulosity at high galactic latitude (Sandage 1976) provided the first evidence for 'cirrus' clouds subsequently studied in detail in the infrared. Observations of scattered light provide important diagnostic tests for grain models, constraining the optical properties of the grains through determination of albedo and phase function (§3.1.2 and §3.2.2). Observations of the DGL are particularly valuable, but also extremely difficult, in view of its intrinsic weakness and the problem of separating the DGL component of the sky brightness from the stellar background, the zodiacal light, and, in the case of ground-based studies, telluric airglow.

An absorbing dust grain must re-emit a power equal to that absorbed to maintain thermal equilibrium. Grain temperatures ~ 10–50 K are predicted for submicron-sized grains under typical interstellar conditions (van de Hulst 1946), implying peak emission at wavelengths ~ 60–$300\,\mu$m in the infrared. The diffuse emission from the disk of our own Galaxy and other spirals has been studied in detail with the Infrared Astronomical Satellite (IRAS). Figure 1.2 compares $50\,\mu$m contours with the optical image of NGC 891. Figure 1.5 maps the distribution in the Galaxy of cool, compact interstellar clouds detected by IRAS at $100\,\mu$m. Comparison of figure 1.5 with figure 1.3 (the corresponding map of optical reddening) shows the expected similarity between the distributions of absorbing and emitting grains: both are broadly confined to the galactic plane, with the addition of local structure associated with Gould's Belt. The scale height for $100\,\mu$m emission is comparable with those of reddening and H I, and somewhat greater than that characteristic of CO (Beichman 1987). At high galactic latitudes, the smooth component of the $100\,\mu$m emission tends to follow a cosec $|b|$ law analogous to equation (1.6). Superposed upon this is the patchy emission (cirrus) associated with individual high-latitude clouds.

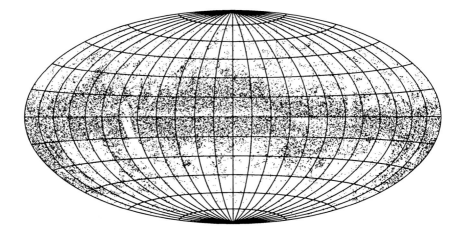

Figure 1.5   Distribution with respect to galactic coordinates of IRAS sources detected at $100\,\mu$m only, displayed in the same format as figure 1.3. Diagram courtesy of T Prusti, Laboratory for Space Research, Groningen.

The emission at $100\,\mu$m is attributed to 'classical' grains typically 0.1–$0.2\,\mu$m in radius, which are known to exist from studies of interstellar extinction in the visible (§1.1.1). IRAS also showed the presence of substantial diffuse emission at shorter wavelengths (12 and $25\,\mu$m) which requires the presence of a hotter component ($T \sim 100$–500 K) previously expected only in circumstellar environments. Such grain temperatures cannot be reached in the interstellar medium by classical grains which absorb ambient starlight and re-emit under thermal equilibrium. However, very small grains (VSGs) of low heat capacity may be subject to large, transient increases in grain temperature due to the absorption of individual energetic photons, a phenomenon discussed previously by (e.g.) Greenberg and Hong (1974). A population of VSGs with dimensions $< 0.01 \, \mu$m may explain a number of the observed properties of interstellar dust.

1.2   INTERSTELLAR ENVIRONMENTS

Interstellar gas is composed predominantly of hydrogen, which may be in one of three physical states or 'phases', molecular, atomic or ionized. The gas is acted upon by cosmic rays, electromagnetic radiation, gravity, shock waves and magnetic fields. The physical processes involved lead to the presence of a vast range of environments, from tenuous, hot plasmas to cold, dense clouds. In this section, we estimate the average density of the ISM in the galactic disk near the Sun, discuss briefly the factors which determine the physical state of the gas, and identify conditions of density, temperature and phase which characterize typical environments. For more detailed and wide-ranging reviews of physical processes in interstellar clouds, the reader is referred to Spitzer (1978), Hollenbach and Thronson (1987), and Morfill and Scholer (1987).

1.2.1   The Gas-to-Reddening Ratio and Mean Density in the Galactic Disk

Column densities of atomic hydrogen (H I) have been measured in the lines of sight to 100 reddened stars within ~ 1 kpc of the Sun by Bohlin, Savage and Drake (1978), by means of Lyman-α absorption-line spectroscopy with the Copernicus satellite. Results were combined with corresponding Copernicus data for molecular hydrogen (Savage et al 1977) to determine the total (nucleon) column density N_{H} (m^{-2}) in atomic and molecular form, i.e.

$$N_{\mathrm{H}} = N(\mathrm{H\,I}) + 2N(\mathrm{H_2}). \tag{1.8}$$

The ionized component of the gas contributes only a tiny fraction of the total mass of interstellar material in the disk of the Galaxy (see §1.2.2 below) and may be neglected here. Bohlin et al demonstrated that N_{H} and E_{B-V} are well-correlated, strengthening the view that gas and dust are generally well-mixed in the ISM. The mean ratio of gas column density to reddening deduced from the correlation is

$$\left\langle \frac{N_{\mathrm{H}}}{E_{B-V}} \right\rangle \simeq 5.8 \times 10^{25} \ \mathrm{m^{-2} \ mag^{-1}} \tag{1.9}$$

with scatter for individual stars typically less than 50%. Converting reddening in equation (1.9) to extinction via equations (1.4) and (1.5), we have

$$\left\langle \frac{N_\mathrm{H}}{A_V} \right\rangle \simeq 1.9 \times 10^{25} \ \mathrm{m}^{-2} \ \mathrm{mag}^{-1}. \tag{1.10}$$

Equations (1.10) and (1.7) may then be combined to eliminate the extinction, giving a value for the mean hydrogen number density:

$$\langle n_\mathrm{H} \rangle = \left\langle \frac{N_\mathrm{H}}{L} \right\rangle \simeq 1.1 \times 10^6 \ \mathrm{m}^{-3} \tag{1.11}$$

or about one atom per cm^3. This is a good macroscopic average for $\langle n_\mathrm{H} \rangle$ in the solar neighbourhood of the galactic plane, but large departures from this result occur in individual regions: as discussed below (§1.2.2), the ISM tends to divide into clouds with $n_\mathrm{H} \gg \langle n_\mathrm{H} \rangle$ and intercloud gas with $n_\mathrm{H} \ll \langle n_\mathrm{H} \rangle$.

Expressing equation (1.11) as a mass density, we have

$$\langle \rho_\mathrm{H} \rangle = m_\mathrm{H} \langle n_\mathrm{H} \rangle \simeq 1.8 \times 10^{-21} \ \mathrm{kg} \ \mathrm{m}^{-3}$$

or, more conveniently,

$$\langle \rho_\mathrm{H} \rangle \simeq 0.027 \ \mathrm{M}_\odot \ \mathrm{pc}^{-3}. \tag{1.12}$$

Assuming that the abundances of the elements in the Solar System are applicable to the interstellar medium (§2.2), the result in equation (1.12) above should be multiplied by a factor of 1.4 to obtain the average density summed over all elements:

$$\langle \rho_\mathrm{ism} \rangle \simeq 0.038 \ \mathrm{M}_\odot \ \mathrm{pc}^{-3}. \tag{1.13}$$

About $\sim 20\%$ of this gas is molecular (see Dame *et al* 1987). For comparison, the observed density of matter in stars in the disk of the Galaxy is

$$\langle \rho_* \rangle \simeq 0.065 \ \mathrm{M}_\odot \ \mathrm{pc}^{-3} \tag{1.14}$$

(Mihalas and Binney 1981, p 224). Thus, the interstellar medium contributes roughly 30–40% of the total observed mass.

The density of material in the disk of the Galaxy may be estimated independently by investigating the motions of stars perpendicular to the galactic plane (z-motions), a technique pioneered by Oort (1932). Recent determinations of ρ_0, the kinematical mass density at $z = 0$, are reviewed by Kuijken and Gilmore (1989); the best current value appears to be

$$\rho_0 \simeq 0.18 \ \mathrm{M}_\odot \ \mathrm{pc}^{-3} \tag{1.15}$$

with scatter of $\sim 50\%$. Thus, the total density of observed material, given by $\langle \rho_* \rangle + \langle \rho_{ism} \rangle \simeq 0.10\,M_\odot\,pc^{-3}$, is consistent with the kinematical value to within considerable uncertainty: there is no convincing evidence for 'missing mass' in the solar neighbourhood of the galactic disk.

1.2.2   The Physical State of the Interstellar Medium

The physical properties of interstellar gas are described by its density (n), temperature (T), and phase (molecular, atomic or ionized). For pressure equilibrium between regions with different temperatures, the product nT should be constant. The temperature of a cloud depends on the balance of heating and cooling mechanisms, as discussed in detail by Spitzer (1978, pp 131–149). Clouds are heated by the interstellar radiation field (ISRF) and by the cosmic-ray flux, and cool by means of spectral line emission from ions, atoms and molecules and by continuum emission from the dust. The molecular transitions which cool the gas are generally rotational, driven by collisional excitation, such that the cooling rate varies as n^2. A small increase in density can therefore lead to a substantial increase in heat loss; this lowers the temperature and pressure, causing contraction of the region and further increase in density. Thus, the ISM tends naturally to separate into cool, dense regions (clouds) and hot, rarefied regions (the intercloud medium).

The physical state of the gas is determined primarily by the local intensity of the ISRF. In unobscured regions, the ISRF comprises the integrated light of all stars in the Galaxy, and has a mean energy density of $\sim 7 \times 10^{-14}\,J\,m^{-3}$ in the solar neighbourhood (Allen 1973). A small but important fraction of this is carried by photons of sufficient energy $h\nu$ to ionize H I:

$$H\,I + h\nu \rightarrow p + e \quad (h\nu \geq 13.6\,eV) \tag{1.16}$$

and a considerably larger fraction is carried by photons capable of photodissociating H_2:

$$H_2 + h\nu \rightarrow H + H \quad (h\nu \geq 4.48\,eV). \tag{1.17}$$

The resulting phase structure of the ISM is characterized by a number of discrete components in which molecular, atomic or ionized hydrogen gas predominate. Table 1.1 lists the properties of four principal components which represent typical environments, based on the so-called three-phase model of McKee and Ostriker (1977) with the subdivision of the cold gas into atomic and molecular phases. In this model, the structure is regulated by supernova explosions. On average, a given point in the ISM will be swept out by an expanding remnant once every few million years, leading to disruption of existing cloud structure and the establishment of low-density

Table 1.1   A four-component model for the interstellar medium

Description	Phase	T (K)	n (m^{-3})	ϕ
Cold molecular	H$_2$	~ 15	$> 1 \times 10^8$	< 0.01
Cool atomic	H I	~ 80	$\sim 3 \times 10^7$	~ 0.03
Warm	H I, H II	~ 8000	$\sim 3 \times 10^5$	~ 0.2–0.3
Hot ionized	H II	$\sim 5 \times 10^5$	$\sim 5 \times 10^3$	~ 0.7–0.8

bubbles of hot gas. Thus, supernova explosions maintain the intercloud medium in its hot, ionized state†.

Along a given line of sight which intercepts regions typified by each of the discrete physical components listed in table 1.1, the average nucleon density is

$$\langle n \rangle = \sum \phi_i n_i \qquad (1.18)$$

where ϕ_i and n_i, are the volume filling factor and number density, respectively, of the ith component. The filling factors sum to unity ($\sum \phi_i = 1$) if all components of the ISM are accounted for. The values of the filling factors are spatially variable, changing systematically with galactocentric distance R_G (molecular clouds are more common at $R_G \sim 3$–8 kpc) and with distance from the galactic plane (hot coronal gas predominates in the halo). In the current context, we are concerned primarily with material in the galactic disk: the lines of sight in which we measure dust properties are generally either parallel to the plane of the disk or sufficiently short (e.g. to nearby high-latitude clouds) that little or no high-z gas is included. Even in the disk, the values of ϕ_i are very uncertain (see Kulkarni and Heiles 1987 for detailed discussion). However, the contrast in density between the hot and cold components is sufficiently large that a general conclusion may be drawn on the basis of the crude estimates for ϕ in table 1.1: whilst the volume of the interstellar medium is mostly filled by hot, ionized gas, the nucleon density (equation (1.18)), and hence the mass, is dominated by the atomic and molecular phases, i.e., by clouds.

In order to characterize the cloudy state of the ISM in a simple way, consider two distinct types of idealized interstellar cloud (figure 1.6). These are assumed to have no internal sources of luminosity, and to be immersed in an intercloud medium of hot, low-density, ionized gas. The ISRF per-

---

† The intercloud H II component should be distinguished from H II regions, which result from ionization of the ambient interstellar gas by individual OB stars or associations.

meates the intercloud medium virtually unattenuated, as it contains little dust and the plasma is optically thin to ionizing radiation. The clouds within are externally heated by the ISRF. A cloud of moderate density $(n_H \sim 10^7\text{--}10^8\,\text{m}^{-3})$ in which the predominant phase is H I is optically thick to radiation beyond the Lyman limit, but remains relatively transparent to radiation which dissociates H_2 (equation (1.17)), although some gas phase molecules (e.g. CO, OH, CH and CH^+, as well as H_2) have detectable abundances. The H I gas is encased in a shell of warm gas, the outer 'halo' of which is partially ionized (equation (1.16)) by the hard ultraviolet photons in the ISRF, which it strongly absorbs. The warm gas is predominantly neutral, however, and heated by soft X-ray photons $(h\nu \sim 40\text{--}120\,\text{eV})$ emitted by the hot intercloud gas.

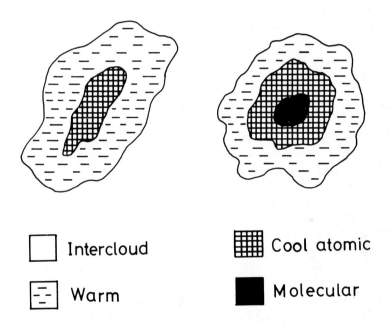

Figure 1.6   Idealized representation of diffuse and dense clouds in the interstellar medium.

If a cloud contains concentrations which are sufficiently dense $(n > 10^8\,\text{m}^{-3})$, virtually all of the H I within these regions is converted to H_2 by grain surface catalysis (§7.2.1) on a timescale short compared with cloud lifetimes. This molecular region is opaque to both ionizing and dissociating radiation, being effectively shielded from the external ISRF by the outer layers of the cloud, which remain predominantly atomic. At such densities, extinction by dust is also important, adding to the screening effect. The boundary between the atomic and molecular zones, in which photodissociating radiation is strongly absorbed, is expected to be quite thin. Observational evidence for a rapid phase change is provided by studies of interstellar H_2 absorption lines (Savage *et al* 1977). Figure 1.7 plots the column density of molecular hydrogen against the total column density (equation (1.8)) for a large number of stars. Individual lines of sight may contain several clouds of different mean density, which introduces scatter into the diagram, but there is a clear trend for $N(H_2)$ to increase steeply with N_H initially, and then to flatten off. The solid curve represents a model for clouds with mean internal density $\langle n_H \rangle = 6 \times 10^7\,m^{-3}$ and differing mean thickness. At large column densities, the conversion of H I to H_2 is almost complete.

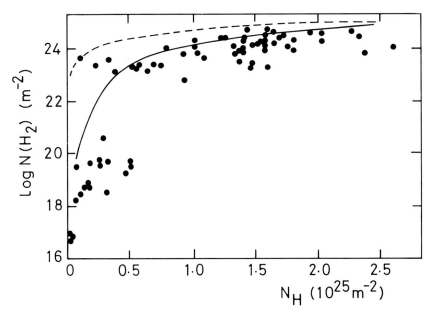

Figure 1.7   Plot of the column density of molecular hydrogen against that of total (atomic + molecular) hydrogen (log $N(H_2)$ versus N_H) for early-type stars (Savage *et al* 1977). The solid curve represents a model for clouds with mean internal density $\langle n_H \rangle = 6 \times 10^7\,m^{-3}$ and differing mean thickness. The dashed line indicates the locus of total atomic to molecular conversion ($N_H = 2N(H_2)$).

Although the external energetic photon field is strongly attenuated in a molecular cloud, it should be noted that significant internal ionization can occur due to penetration of cosmic rays. Gas phase reactions between ions and molecular hydrogen (formed on dust) are believed to provide important pathways for the formation of polyatomic molecules in dense clouds (§7.2.1).

In summary, we adopt the labels 'diffuse' and 'dense' to describe discrete

clouds in which the mass of gas is predominantly atomic and molecular, respectively. Between them, these two cloud types account for almost all the
mass of interstellar matter in the galactic disk. The dense clouds are encased in an outer protective layer of atomic gas, and both dense and diffuse
clouds are suspended in a hot, ionized substrate, the intercloud medium.
Individual clouds may have a large range in size, mass and morphology.
The smallest clouds tend to be destroyed quickly by evaporation into the
surrounding hot medium. A typical diffuse cloud has dimensions $\sim 5\,\mathrm{pc}$
and mass $\sim 30\,M_\odot$, whereas giant molecular clouds may have masses up
to $\sim 10^6\,M_\odot$. The clouds are in a continuous cycle of evolution. A self-
gravitating dense cloud is initially quiescent, but the onset of star formation leads to the appearance of wind-blown cavities and H II regions in the
molecular material which will ultimately dissipate the cloud.

1.2.3   The Interstellar Environment of the Solar System

We noted in §1.1.2 that there is little evidence for reddening within about
100 pc of the Sun, indicating that the Solar System does not currently reside
in the immediate vicinity of an interstellar cloud of significant opacity†.
This is confirmed by studies of the local interstellar gas. On a scale of
several tens of parsecs, the Sun is surrounded by an irregularly shaped,
low-density region with $n \sim 4 \times 10^3\,\mathrm{m}^{-3}$ and $T \sim 10^6\,\mathrm{K}$, known as the
local bubble (Cox and Reynolds 1987). These results are typical of the
intercloud environment (table 1.1). On a more local scale, the gas in the
immediate vicinity of the Solar System ($d < 10\,\mathrm{pc}$) has somewhat higher
density, $n \sim 10^5\,\mathrm{m}^{-3}$, a temperature $T \sim 10^4\,\mathrm{K}$, and a degree of ionization
$\sim 70\%$ (Cowie and Songaila 1986), conditions more typical of the warm
gas associated with the outer layers of clouds. It appears that the Sun lies
near the edge of a small 'cloudlet' $\sim 5\,\mathrm{pc}$ in extent, centred roughly in the
direction of the galactic centre. There is negligible reddening associated
with this structure, and the density is presumably nowhere sufficient for it
to contain a cool phase.

Cox and Reynolds (1987) discuss in detail the question of whether the
local interstellar medium (LISM) represents a typical interstellar environment, or whether it has been disturbed in some way by a specific, recent,
local event such as a supernova explosion or the integrated stellar wind of
an OB association in Gould's Belt. If the model of McKee and Ostriker
(1977) for the general ISM is broadly correct, there appears to be little
difficulty in accounting for the LISM in terms of normal hot, ionized intercloud gas. On the basis of estimates for the filling factors (table 1.1), we
would predict that the probability of finding the Solar System in the cool,

---

† The closest dense cloud appears to be Lynds 1457 at $d \sim 65\,\mathrm{pc}$ (Hobbs
et al 1986).

dense environment of an interstellar cloud at any given time is no more than a few per cent. However, the possibility remains that the local bubble is an anomaly, perhaps representing the cavity blown by a single, active supernova remnant, in which case the LISM may bear little resemblance to typical, average interstellar conditions. This question is unresolved.

1.3   THE SIGNIFICANCE OF DUST

Early studies of interstellar dust (§1.1.1) were motivated primarily by a desire to correct photometric data for its presence rather than by an intrinsic interest in the dust itself or an appreciation of its astrophysical significance. According to Gaustad (1971) it was once the case, as far as the typical optical astronomer was concerned, that "...if you simply tell him the reddening law, particularly the ratio of total to selective extinction, he can unredden his clusters, correct the distance moduli, find the turnoff points, determine the age of the galaxy, and be happy!" Times have changed.

Far beyond having 'nuisance value', dust is now recognized as a vital ingredient of interstellar material in its own right. Historically, this revolution came about largely as a result of the exploration of new regions of the electromagnetic spectrum. Ultraviolet spectroscopy of interstellar gas indicated a short-fall, or depletion, in the abundances of many of the heavier elements with respect to solar values, which is most readily explicable if the missing atoms are tied up in solid particles. Spectral line observations at radio, microwave and ultraviolet wavelengths resulted in the discovery of a variety of interstellar gas phase molecules in previously unsuspected quantities, which led, in turn, to an appreciation of the complexity of interstellar chemistry and the importance of chemical and physical interactions between gas and dust. Observations in the infrared have allowed detailed studies of the dust itself, its chemical composition and structure (which we discuss in §1.4 below), and have elucidated its rôle in regulating the thermal equilibrium of interstellar clouds and the circumstellar shells of both young and evolved stars.

Notwithstanding recent developments, the classical reason for the study of interstellar dust remains vitally important and is assuming prominence in extragalactic as well as galactic astronomy. Some implications of the distribution and degree of interstellar reddening in the Milky Way are discussed in §1.1.2 above. In view of the fundamental significance of the ratio of total to selective extinction to photometric distance determinations, it is important to investigate the sensitivity of this parameter to environment: systematic, widespread variations in the value of R_V would have serious consequences for our understanding of galactic structure (e.g. Johnson 1968) and the extragalactic distance scale. A detailed knowledge of the degree and spectral dependence of extinction is needed in order to recover the intrinsic luminosities and spectral energy distributions of reddened objects from observed fluxes. This procedure is particularly important to studies of newly formed stars, as it is within the dark clouds associated with regions of recent star formation in the Galaxy that the most dramatic variations in dust properties have been found to occur. The soft ultraviolet radiation field in the clouds is regulated principally by extinction, and changes in the optical properties of the grains can thus have important implications for molecular abundances governed by photodissociation rates. On the wider stage, the question of whether cosmologically distant objects are obscured by dust is vital to our understanding of the structure and evolution of the Universe (Ostriker and Heisler 1984). Significant extinction by dust either within or between intervening galaxies would hinder detection of objects at very high redshift and could obscure our view of galaxy formation (Disney 1990). Interstellar matter associated with previously unsuspected degrees of opacity in spiral galaxies could conceivably represent a significant form of 'missing mass' in the Universe (Valentijn 1990).

The ubiquity of molecules in interstellar gas raises a fundamental theoretical problem concerned with their origin: gas phase mechanisms for the production of H_2, the most abundant molecule, are much too slow to account for the observations, and formation on grain surfaces seems to be the only viable alternative. Although gas phase mechanisms may be capable of producing a number of other molecules, given an adequate supply of H_2, the grains must be included in a complete model for interstellar chemistry (Williams 1987). The molecular abundances and atomic depletions are sensitive to interstellar environment. In diffuse H I clouds, grains act as catalysts for the production of H_2: adsorbed atoms trapped at surface defects in the grain structure recombine; the binding energy thus released is partly absorbed into the grain lattice, whilst the resulting molecule is ejected from the surface and returned to the gas. In dense, quiescent clouds, the grains provide nucleation centres for the growth of volatile molecular mantles, thus removing material from the gas, whereas grains exposed to shocks, local heating, or the unattenuated interstellar radiation field in low-density regions become sources of gas phase atoms and molecules. At all levels of density, the exchange of material between interstellar gas and grains is essential to models for the chemical composition and evolution of the ISM as a whole.

The probable detection of polycyclic aromatic hydrocarbons (PAHs) in the interstellar medium, by means of characteristic vibrational modes observed in the infrared, adds another dimension to the problem of understanding interstellar chemistry. These molecules have sizes typically $\sim 1\,\mathrm{nm}$, and they may be considered to represent the transition zone between large molecules and small particles. The observations suggest that PAHs may account for $\sim 5\%$ of the available interstellar carbon, and it seems likely that some or all of the VSGs responsible for the mid-infrared diffuse galactic emission (§1.1.3) are in this form. The origin of these particles is probably closely linked to that of other forms of solid carbon in the interstellar medium.

Dust is a vital ingredient in models for many classes of infrared source, including both compact objects (young stars, evolved stars, and some galactic nuclei) and extended objects (planetary nebulae, H II regions, 'starburst' regions, globules, cirrus clouds, and entire galaxies). Models for these diverse phenomena have in common the presence of dust heated by the absorption of ambient starlight. In many cases, emission from the dust dominates the emergent spectral energy distribution in the infrared and constitutes an important cooling mechanism. Modelling such sources is often difficult, and may involve detailed radiative transfer calculations, as, for example, in the case of an optically thick circumstellar shell. At far-infrared wavelengths, the spectral dependence of the dust emissivity is an important factor in the models. Theoretically, we expect this to take the form $Q_\lambda \propto \lambda^{-\beta}$ (§6.1), where the value of the spectral index β depends on the degree of crystallinity in the grain structure. The grain composition assumed would thus affect, for example, the flux from a source in the sub-millimetre region predicted on the basis of fits to 1–$100\,\mu m$ data. The size distribution of the grains is also important, with the possibility of near- and mid-infrared excess emission arising when the smallest particles are subject to transient heating.

A fundamental problem exists in determining the mass of any dense molecular cloud opaque to starlight because the principal ingredient (H_2) cannot be observed directly: ultraviolet spectroscopy yields meaningful data (§1.2.1) only in relatively unobscured regions (diffuse clouds and the outer layers of dense clouds). It is usual to evaluate cloud masses from millimetre-wave observations of gas phase CO, assuming a constant ratio of integrated CO emission intensity to H_2 column density (van Dishoeck and Black 1987). In reality, this ratio is likely to be sensitive to environmental factors: in cold, quiescent clouds, for example, a significant fraction of the CO may be depleted onto grains where, in solid form, it is undetected by the millimetre observations. However, the infrared and sub-millimetre continuum emission from a cloud provides a direct means of evaluating the mass of *dust* within it (Hildebrand 1983), allowing an independent estimate of the total cloud mass if the dust to gas ratio is known.

Dust is involved intimately in the formation and evolution of population I stars. Stars are born within cocoons of dust and gas, and theoretical considerations suggest that the presence of dust may be essential to the gravitational collapse of massive protostars, providing an efficient radiator of excess heat energy. We have noted that young (pre-main-sequence) stars show a strong spatial association with dense clouds, and the common occurrence of infrared excess emission indicates the presence of cool circumstellar shells or disks. The dust provides raw material for the formation of planetary systems, and may contain molecules relevant to the origin of organic life. It is not yet known whether or not planetary systems are common products of accretion in disks around young stars. In a number of cases, of which Vega is the best-known example, the survival of a diffuse disk into the main-sequence phase has been demonstrated. The effect of radiation pressure on grains may strongly influence the kinematics of stellar envelopes in both young and evolved stars. Grains condensing in the envelopes of late-type giants and supergiants may drive the outflows from such objects: mass-loss rates as high as $10^{-5} \, M_\odot \, yr^{-1}$ have been estimated, which may, in some cases, be sufficient to influence the ultimate fate of a star during the crucial late phases of its evolution. The dusty outflows of evolved stars contribute significantly to the enrichment of the ISM with heavy elements.

1.4   THE PROBLEM OF GRAIN COMPOSITION

Observations of extinction and reradiation imply the existence of particles in interstellar space which range in size from $\sim 1 \, nm$ to $\sim 1 \, \mu m$. In view of the level of research activity devoted to these particles since Trumpler's pioneering work (§1.1.1), particularly in the last 25 years, it is perhaps surprising that their chemical composition has proved extremely difficult to establish, and is, indeed, a matter of continued speculation, controversy and debate. The reason for this is that a number of the observed properties of interstellar dust do not provide very specific compositional information, and there is consequently a lack of uniqueness in the modelling. The 'continuum' effects of extinction, scattering, polarization and emission have much to tell us about various properties of the dust (e.g. particle size, shape, alignment properties), but do not place decisive constraints on its composition. Conversely, this uncertainty does not preclude some important investigations: for example, one can deduce empirical corrections for interstellar reddening without knowledge of the grain composition. The key to the solution of the problem of composition is the detailed study of the dust-related absorption and emission features in astronomical spectra, and the comparison of observed spectra with appropriate laboratory data. In this section, we summarize the main theories and highlight some of the observations which may be used to discriminate between them. These observations, and the modelling processes which lead to the formulation of grain theories, are discussed in greater depth in subsequent chapters of this book.

A fundamental restriction on grain composition is provided by information on the relative abundances of the chemical elements in the interstellar medium. Under interstellar conditions, significant quantities of hydrogen can exist in the solid phase only if the H atoms are chemically bonded to heavier elements, whilst helium is obviously excluded by its extreme volatility and chemical inertia. The bulk of the grain mass is undoubtedly provided by elements with atomic number $z \geq 6$, of which the most abundant (excluding inert gases) are C, N, O, Mg, Si and Fe. The depletions of these elements in the gas phase, most pronounced for the metals, provide indirect but compelling evidence for their inclusion in grains (Chapter 2). The combined mass of all the condensible elements with $z \geq 6$ is 1.7% of the total mass for normal population I abundances. Allowance for hydrogenation leads to an upper limit of $\sim 2\%$ on the dust to gas ratio.

Amongst the first materials to be proposed for interstellar grains were ices (Lindblad 1935) and metals (Schalen 1936). On these, ices are particularly favoured by the high cosmic abundances of the constituent atoms. It was suggested that particles composed of saturated molecules (H_2O, NH_3 and CH_4) would form in interstellar clouds by random accretion, for which indirect support is provided by the presence of radicals such as CH and OH which could be the precursors or destruction products of ice particles. This proposal was developed and extended by Oort and van de Hulst (1946) into a detailed model for ice nucleation and growth in interstellar clouds. Extinction calculations based on optical data for laboratory ices, assuming grains with sizes ranging up to approximately the wavelength of visible light, give an excellent fit to the observed extinction curve in the visible (see also Greenberg 1968).

Although the ice model gained widespread acceptance in the 1950s and early 1960s, Platt (1956) argued that random accretion would tend to produce small particles composed of unsaturated molecules with unfilled energy bands, rather than classical ice grains. These 'Platt particles' were postulated to have radii no more than $a \sim 1\,nm$. They are better described as macromolecules than solid particles, in the sense that their optical properties are determined largely by quantum mechanics rather than solid-state band theory. Although little attention was paid to them at the time, Donn (1968), in a prophetic paper, drew attention to the possible astrophysical significance of very small grains, linking Platt particles with polycyclic aromatic hydrocarbons. These have been the subject of renewed interest and intensive research in the 1980s, as possible carriers of a number of observed spectral features and diffuse continuum emission in the infrared, discussed in detail in Chapter 6.

It seemed in the early 1960s that the nature of interstellar grains was well-established, and that subsequent research would merely clarify some of the details rather than radically change the picture. The ice model of Oort and van de Hulst could explain the observed properties of the dust in the spectral range then accessible, and presented a logical and self-consistent picture of grain formation and evolution. The information explosion which has occurred in recent decades, with the extension of spectral coverage, had two dramatic effects on interstellar grain research: it greatly stimulated interest in the subject, as discussed in §1.3 above, and simultaneously demonstrated that the ice model for interstellar grains is inadequate, or at best incomplete. It is no longer possible to explain all of the observed properties of interstellar grains in terms of a unified model involving a single substance or class of substances condensing under similar conditions; mixtures of materials with diverse origins must be invoked, in which different components account for different aspects of the data.

Vital compositional clues are provided by spectroscopic observations in the ultraviolet and the infrared. Ices have absorption edges in the ultraviolet (Field *et al* 1967), occurring, for example, in the case of water-ice, at $\lambda \sim 1600\,\text{Å}$. The spectra of reddened stars show no evidence for structure near this wavelength, but do show a very strong and ubiquitous absorption feature centred at $2175\,\text{Å}$ which is widely attributed to the presence of small graphite grains with radii $a < 0.02\,\mu m$ (Stecher and Donn 1965). The continuous far-ultraviolet rise in extinction at wavelengths $\lambda < 1500\,\text{Å}$ suggests the presence of a further population of very small particles of unspecified composition.

In the infrared, vibrational transitions provide a means of testing for the presence of grain signatures in regions which are too heavily obscured to permit observation in the ultraviolet with current technology. For example, the stretching vibrations of O–H bonds in H_2O-ice produce a strong feature at $\lambda \sim 3\,\mu m$, and this provides an acid test for the ice model of interstellar grains. The first attempts to observe this feature in the spectra of reddened stars were unsuccessful (Knacke *et al* 1969), confirming that ice is no more than a minor constituent of diffuse clouds. Subsequent investigations have shown that the presence or absence of ice in the ISM is critically dependent on physical conditions along the particular line of sight observed, as well as on the amount of reddening. Stars reddened by long path-lengths of low-density material do not generally show appreciable ice absorption, whilst those obscured by dense molecular clouds generally do (Gillett *et al* 1975b; Willner *et al* 1982; Whittet *et al* 1988). Ices thus require a shielded environment to condense and survive, and cannot account for the optical extinction of reddened stars observed through diffuse clouds; in dense, quiescent molecular clouds, however, they appear to be ubiquitous.

Infrared spectroscopy has demonstrated the presence of other grain signatures. Perhaps the most important of these as a compositional indicator is the broad feature centred at approximately $9.7\,\mu m$, which is seen in the spectra of several different types of source, including reddened stars, H II regions, oxygen-rich circumstellar envelopes, comets, and some galactic nuclei. Significantly, the feature may be present in either absorption or emission, depending on the nature of the source and the degree of foreground extinction, whereas the $3\,\mu m$ ice feature is never seen in emission. Absorption at $9.7\,\mu m$ occurs in normal reddened stars, with a strength that correlates with visual extinction (Roche and Aitken 1984a), and is also prominent in sources with a molecular cloud in the line of sight (Willner *et al* 1982). The corresponding emission feature is frequently observed in the spectra of H II regions, T Tauri stars, and late-type giants and supergiants. Sources with $9.7\,\mu$m emission or absorption generally also show a corresponding weaker feature at $18\,\mu$m. The carrier must be robust, capable of surviving in a wide variety of physical conditions, and oxygen-rich, as the features are observed in circumstellar shells only when the abundance of O exceeds that of C. Both 9.7 and $18\,\mu$m features are widely attributed to silicate dust, which has the required refractory properties and spectral activity at the appropriate wavelengths, produced by Si–O stretching and O–Si–O bending vibrations, respectively. Indirect support for this interpretation is provided by the ubiquity of silicates in meteorites and interplanetary dust. However, the term 'silicate' is rather imprecise, covering, in the geological context, a range of possible chemical compositions and mineral structures based on SiO_4 tetrahedra. The profile of the observed $9.7\,\mu$m feature and the form of the far-infrared emissivity indicate that interstellar silicates are predominantly amorphous rather than crystalline, and abundance considerations suggest that they are likely to be rich in Mg and Fe, but their detailed composition is not specified uniquely by the infrared observations.

Spectroscopic detection of silicates, and, more tentatively, graphite, led to the formulation of models in which the diffuse-cloud extinction is produced by a mixture of refractory grains rather than volatile ices (see Wickramasinghe and Nandy 1971 and references therein for a description of early work in this area). These particles appear to originate in the atmospheres of evolved stars of differing C/O ratio, and so it is natural to assume the existence of distinct oxygen-rich and carbon-rich populations. Mathis, Rumpl and Nordseick (1977), in a key paper widely referred to as 'MRN', showed that the extinction curve from the near infrared to the far ultraviolet could be explained by graphite and silicates. An important factor in the MRN model is the use of a power-law size distribution, subsequently adopted by other investigators, in which the relative numbers of particles according to size (assumed spherical of radius a) within a given range is $n(a) \propto a^{-3.5}$ (see Chapter 3). This is the form of size distribution expected when particles are subject to collisional abrasion, as may occur in red giant winds (Biermann and Harwit 1980). The particles are assumed to remain uncoated in diffuse interstellar clouds, and to provide nucleation centres for the growth of ices when immersed in molecular clouds.

Subsequent research has emphasized that the nature and evolution of the carbon-rich component of the dust is the key issue in the compositional debate. Pure solid carbon may assume a variety of forms ranging from crystals (graphite, diamond) to amorphous (soot-like) material. In a C-rich environment, C atoms naturally tend to become arranged into groups of hexagonal rings, and individual groups containing typically $\sim$ 5–30 hydrogenated rings are essentially PAH molecules (see Robertson and O'Reilly 1987). In amorphous carbon, such groups or 'islands' are assembled randomly, and there is no long-range order. In contrast, graphite is highly ordered, comprising regular stacks of platelets formed from planar groups of rings. Intermediate polycrystalline forms may also occur (e.g. Tielens and Allamandola 1987a). The MRN model postulates that graphite particles contribute significantly to the extinction at all wavelengths, requiring $\sim 60\%$ of the available carbon to be in this form. Unless there is an efficient mechanism for producing graphite in the interstellar medium, this is hard to reconcile with observations of C-rich late-type stars, which imply that the carbon dust in their ejecta is predominantly amorphous rather than graphitic (Whittet 1989). Theoretical considerations also suggest that the carbon grains nucleating in stellar envelopes are amorphous or polycrystalline rather than graphitic (e.g. Gail and Sedlmayr 1984).

Current models for interstellar grains assume that either amorphous carbon or organic materials contribute to the visual extinction. Mathis and Whiffen (1989) extend the 'unmantled refractory' model of MRN to include composite grains which are aggregates of smaller amorphous carbon, silicate and iron particles, similar in structure to interplanetary dust. Alternatives based on composite (core/mantle) grains have also been proposed (Greenberg 1989; Duley et al 1989a). Greenberg and co-workers emphasize that when hot stars form in molecular clouds, irradiation of the icy mantles condensed on mineral grains may result in rearrangement of the chemical bonds to produce complex organic polymers. This process has been demonstrated in the laboratory under simulated interstellar conditions, leading to a grain model in which the visual extinction is produced by silicate grains with organic refractory mantles (with the addition of a separate population of smaller graphite particles which absorb in the ultraviolet). However, the mantles are subject to dissipation by shocks, and it is not clear to what extent irradiation over cosmic timescales will convert the organic mantles to amorphous carbon. Duley et al propose that silicate grains which originate in stellar atmospheres acquire mantles of hydrogenated amorphous carbon in interstellar clouds, as a result of direct depletion of C atoms onto the grain surface. The Greenberg and Duley models are thus in agreement, in that they attribute the visual extinction to silicates coated with a non-graphitic C-rich mantle, although the mantles are assumed to originate in different ways. Both predict significant absorption at $3.4\,\mu$m, as a result of C–H bond vibrations, and this has been observed. A key difference, comparing Duley et al with other models, is the interpretation of the mid-ultraviolet ($\lambda2175$) absorption, which Duley et al attribute to OH$^-$ sites on the surfaces of silicates rather than to graphite.

A departure from grain models based on refractory minerals and ices is provided by Hoyle and Wickramasinghe (1986; see also Jabbir et al 1986 and references therein), who propose that interstellar dust is composed of biological organisms and related materials, including bacteria, viruses, diatoms, proteins and polysaccharides. The optical properties of bacteria are claimed to be consistent with the observed extinction curve in the visible, and graphite spheres, assumed to be the ablation products of organisms, provide the $\lambda 2175$ ultraviolet extinction bump. Identifications for various spectral features have been proposed. There are a number of objections to this model, some of which are outlined briefly here. The spectra of biological materials are, in general, complex, frequently exhibiting features which have no counterpart in astronomical spectra. Strong absorptions are predicted in the ultraviolet, notably in the wavelengths range 2500–3000 Å, but these have not been observed in the ISM[†]. Similarly, the mid-infrared spectra of bio-organics show detailed structure which seems incompatible with (e.g.) the claim that polysaccharides account for the 9.7 μm feature. The 9.7 μm feature is, in any case, observed in O-rich circumstellar shells with temperatures 1000–1500 K, which is incompatible with an origin in organic material. Perhaps the most fundamental test is provided by the fact that biological organisms contain significant quantities of phosphorus: the ratio of P to C by number of atoms is approximately 1:50 in biota, whereas the equivalent cosmic abundance ratio is $1:10^3$. We shall return to this point in Chapter 2. For further critical discussion of the bacterial model, the reader is referred to Butchart and Whittet (1983), Duley (1984), Davies *et al* (1984) and Whittet (1984a, b).

---

† A claimed detection proved to be spurious (see §5.1.1).

2

Element Abundances and Depletions

"Figuratively, when we study depletions, it is as if we were looking at the crumbs left on the plate after the grains have eaten their dinner."

E B Jenkins (1989)

It is well known that hydrogen and helium together comprise some 98% by mass of all observed matter in the Universe. As inhabitants of a terrestrial planet, we are accustomed to situations in which the rarer, heavier elements predominate. The most abundant elements which commonly condense to form solids are C, N, O, Mg, Si and Fe, and these six elements are presumed to contribute most of the mass of interstellar grains. Observational evidence for their presence in grains is provided by measurements of gas phase abundances, which are generally much less (by a factor ~ 100 in the case of Fe) in interstellar clouds, compared with stellar atmospheres where essentially all material is in the gas. The depletions of heavy elements in interstellar gas thus provide qualitative evidence for their inclusion in the dust and, more tentatively, quantitative information on the probable distribution of element abundances in the solid material. The sensitivity of the depletions to environment gives important clues on the nature of the formation and destruction mechanisms for dust grains which operate in the interstellar medium.

In this chapter, we begin (§2.1) by outlining the origin and evolution of the chemical elements, with emphasis on the production of the key heavy elements†. In §2.2, we discuss abundance measurements for the Solar System and assess their applicability as a standard for the interstellar medium. Galactic abundance trends are examined in §2.3, and the depletions observed in the interstellar medium are reviewed in detail in §2.4. The implications of these results for the nature and evolution of interstellar grains are discussed in the final section (§2.4.4).

---

† The term 'heavy elements' is used to refer collectively to those elements with atomic weight ≥ 12.

2.1   THE ORIGIN AND EVOLUTION OF THE CHEMICAL ELEMENTS

A complete discussion of the synthesis of the elements is beyond the scope of this book, but an outline of the processes which lead to the presence of heavy elements in the interstellar medium is pertinent. For detailed discussions and a guide to the relevant literature, the reader is referred to Boesgaard and Steigman (1985), Tayler (1975), and Trimble (1975, 1991).

2.1.1   The Cosmic Cycle: An Overview

The origin and evolution of the chemical elements may be illustrated conveniently by a flow chart as shown in figure 2.1. According to modern cosmological theory, the Universe was created in the promordial fireball or big bang, occurring some 15 billion years (15 Gyr) ago. The abundances of the elements emerging from the big bang were determined by nuclear reactions occurring during the first few minutes of expansion. Denoting the fraction of the total mass contained in hydrogen, helium, and all other elements combined by the symbols X, Y, and Z, respectively, the primordial abundances from the big bang were $X_P \simeq 0.76$, $Y_P \simeq 0.24$, and $Z_P \simeq 0.00$†. This matter was dispersed and subsequently included into galaxies containing interstellar media from which stars formed. The first generations of stars condensed from material with element abundances essentially unchanged from the primordial values, i.e., composed almost exclusively of hydrogen and helium. Subsequent nuclear processing in stars leads to the synthesis of heavier elements. As stars evolve and die, processed material is returned to the interstellar medium: this 'feedback loop' of mass loss in stellar winds, and in supernova explosions and other cataclysmic events, leads to a continuous increase in the heavy-element endowment of successive generations of stars, up to the present mass fraction of $Z \simeq 0.02$.

As the temperatures and densities required for nucleosynthesis in the post-big-bang era are normally reached only in stellar interiors, it follows that the bulk of the heavy elements produced by nucleosynthesis will not be recycled to the interstellar medium but effectively lost by inclusion in collapsed stellar remnants, whilst the ejected material often tends to remain relatively hydrogen-rich. This is confirmed by observational evidence which shows that the majority of evolved stars show little or no direct evidence for the presence of intrinsic heavy-element enhancement in their photospheric spectra. The cosmic cycle is therefore rather inefficient. Enrichment may occur as a result of episodic mixing of core and envelope material in stellar interiors, producing, for example, carbon-rich outflows

---

† Z_P is not identically zero because trace amounts of lithium are believed to be produced in the big bang.

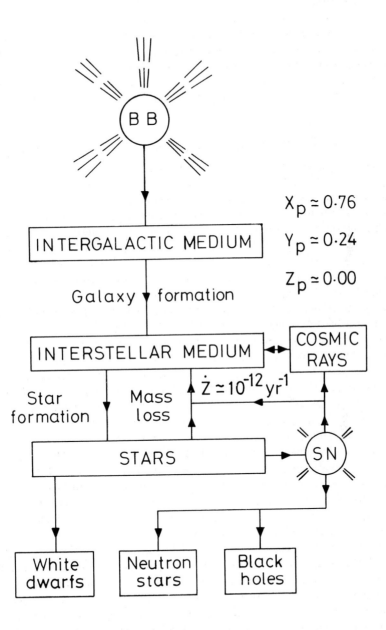

Figure 2.1   A schematic representation of cosmic evolution.

in some red giant stars. Subsequent core–envelope separation may lead to ejection of processed remnants (planetary nebulae, supernova remnants) which ultimately merge with the ISM. Supernovae are also primary sources of cosmic-ray particles, and the interaction of cosmic rays with low-energy interstellar matter leads to further evolution of the elements, notably the production of Li, Be and B, by spallation.

2.1.2   Nucleogenesis

According to standard models for the hot big bang, nuclear reactions occurring at temperatures $\sim 10^9\,$K during the first few minutes of the expansion of the Universe led to the synthesis of substantial quantities of helium. At the commencement of the nucleosynthesis phase of the big bang ($t \sim 20$ seconds), free protons and neutrons are predicted to be present in approximately the number ratio 6:1. The neutrons fused with protons to form deuterons, and further capture of protons, neutrons and deuterons led to the production of ^{3}H, ^{3}He, and ultimately ^{4}He. The primordial helium abundance is fixed by the initial proton-to-neutron ratio, as the free neutrons are captured and processed to ^{4}He on a timescale short compared with their decay time. Heavier elements are not produced in the big bang, except for trace amounts of ^{7}Li with an abundance relative to H of $\sim 10^{-10}$. This conclusion is insensitive to the assumed density during the nucleogenesis phase and thus to whether the Universe is predicted to be open or closed. It arises because there are no stable elements of atomic mass 5 or 8: ^{4}He cannot form a stable product by the capture of a proton or another ^{4}He nucleus. The production of heavier elements such as carbon, requiring the fusion of three ^{4}He nuclei, can occur only under physical conditions which are reached in stellar interiors and not in the big bang: densities several orders of magnitude higher than those pertaining during the nucleosynthesis phase of the big bang are required for such three-body processes to proceed at a significant rate. The rates of all fusion reactions drop rapidly to negligible levels as the density of the Universe falls due to expansion, and after $t \sim 10$ minutes, no further synthesis is likely to occur until the first generations of stars form.

The primordial mass abundance of helium predicted by the big bang ($Y_P \simeq 0.24$) is in good agreement with observations of gaseous nebulae, which suggest (through extrapolation to $Z = 0$) a pre-galactic value of $Y \simeq 0.22$ (Peimbert 1975). This result is one of the corner-stones of the big-bang model for the origin of the Universe. Helium is subsequently produced and destroyed in stars such that its average abundance increases very slowly with time. It is safe to assume that the observed He abundance cannot be explained in terms of stellar nucleosynthesis assuming an initial composition of pure hydrogen.

2.1.3   Stellar Nucleosynthesis

The interaction of atomic nuclei is governed by two opposing forces, the strong nuclear interaction, which is attractive but has very short range, and electrostatic repulsion caused by the fact that they bear positive charge. It is the former which leads to nuclear fusion and which binds nuclei together. In order for two nuclei to become sufficiently close for fusion to occur, they must approach each other at high velocity. The rate at which a given fusion reaction proceeds is critically dependent on both temperature and density. As we consider successive phases of nucleosynthesis in stars, and the production of progressively more massive nuclei, the required temperature increases with the nuclear charge of the 'target' element.

The first nuclear reactions to occur in a newly formed star involve the rare lithium group of elements (Li, Be and B), which are converted to ^{4}He at temperatures $T \sim 10^6$ K. Hydrogen burning, the staple source of energy for stars during their main-sequence lifetime, commences at $T \geq 10^7$ K. This process may be summarized by the equation

$$4\ {}^1\mathrm{H} \rightarrow {}^4\mathrm{He} + 2\mathrm{e}^+ + 2\nu \qquad (2.1)$$

i.e, four protons are converted into an α-particle, two positrons and two neutrinos. The chains of reactions which may lead to this conversion are diverse and include two distinct processes (e.g. Tayler 1975). Of these, the proton–proton (pp) chain requires no raw materials heavier than H and can thus operate in stars condensed from primordial material. The CNO cycle requires the pre-existence of ^{12}C (from helium burning), and can therefore only operate in stars condensed from matter which has been enriched by earlier generations of stars. ^{12}C is consumed and produced by the CNO cycle, and can therefore be regarded as a catalyst:

$$
\begin{aligned}
{}^{12}\mathrm{C} + \mathrm{p} &\rightarrow {}^{13}\mathrm{N} + \gamma \\
{}^{13}\mathrm{N} &\rightarrow {}^{13}\mathrm{C} + \mathrm{e}^+ + \nu \\
{}^{13}\mathrm{C} + \mathrm{p} &\rightarrow {}^{14}\mathrm{N} + \gamma \\
{}^{14}\mathrm{N} + \mathrm{p} &\rightarrow {}^{15}\mathrm{O} + \gamma \\
{}^{15}\mathrm{O} &\rightarrow {}^{15}\mathrm{N} + \mathrm{e}^+ + \nu \\
{}^{15}\mathrm{N} + \mathrm{p} &\rightarrow {}^4\mathrm{He} + {}^{12}\mathrm{C}.
\end{aligned}
\qquad (2.2)
$$

An important aspect of the CNO cycle is that it can lead to significant net production of nitrogen as well. Nitrogen is the fifth most abundant element, yet it is not the end-product of any of the principal burning reaction sequences. However, a hydrogen-burning star composed initially of material enriched in ^{12}C (see reaction (2.3) below) will initially undergo rapid C $\rightarrow$ N conversion by means of the first three reactions in the cycle (2.2) above. Although nitrogen is subsequently consumed, its abundance under equilibrium remains higher than that of carbon during this phase of stellar evolution due to its lower proton capture cross-section.

Helium burning is the first crucial stage in the production of elements heavier than those already available through nucleogenesis. This occurs in the cores of stars in which the central supply of H has been exhausted; the central temperature of the star rises due to contraction caused by the resulting imbalance of gravitational and thermodynamic forces, whilst the envelope expands and cools as the star evolves away from the main sequence. The core typically contains $\sim 10\%$ of the star's mass. When a core temperature of $T \geq 2 \times 10^8$ K is reached, helium burning commences. As the ^{8}Be nucleus is unstable, this requires the quasi-simultaneous fusion of three He nuclei (the triple-α process):

$$3 \; ^4\text{He} \rightarrow \; ^{12}\text{C} + \gamma. \tag{2.3}$$

This may be followed by the capture of a further α-particle to produce oxygen:

$$^{12}\text{C} + \; ^4\text{He} \rightarrow \; ^{16}\text{O} + \gamma. \tag{2.4}$$

Thus, the result of helium burning is a mixture of carbon and oxygen, the relative amounts depending on the temperature at which the process occurs. It is notable that no nitrogen is produced: the major source of nitrogen appears to be the CNO cycle (reactions (2.2) above).

The exhaustion of He fuel results in a further increase in core temperature as the star again attempts to re-establish equilibrium between gravitational and thermodynamic forces. If the star is sufficiently massive, temperatures are reached at which carbon (5×10^8 K) and oxygen (10^9 K) are ignited. Fusion reactions for these elements are capable of producing elements with atomic weights in the range 20–32 by reactions such as:

$$
\begin{aligned}
^{12}\text{C} + \; ^{12}\text{C} &\rightarrow \; ^{24}\text{Mg} + \gamma \\
&\rightarrow \; ^{23}\text{Na} + \text{p} \\
&\rightarrow \; ^{20}\text{Ne} + \; ^4\text{He}
\end{aligned}
\tag{2.5}
$$

and

$$
\begin{aligned}
^{16}\text{O} + \; ^{16}\text{O} &\rightarrow \; ^{32}\text{S} + \gamma \\
&\rightarrow \; ^{31}\text{P} + \text{p} \\
&\rightarrow \; ^{31}\text{S} + \text{n} \\
&\rightarrow \; ^{28}\text{Si} + \; ^4\text{He}.
\end{aligned}
\tag{2.6}
$$

An intermediate stage between reactions (2.5) and (2.6) may involve neon burning, in which α-particles removed from ^{20}Ne nuclei by energetic photons subsequently fuse with ^{20}Ne or ^{24}Mg. The existence of a number of competing reactions is a common feature of nucleosynthesis involving relatively massive nuclei. The protons, neutrons and α-particles released by reactions (2.5) and (2.6) rapidly undergo further processing, and, typically, the main product of carbon and oxygen burning is ^{28}Si, which is a particularly strongly bound nucleus.

Following the production of nuclei in the vicinity of Si in the periodic table, the final phase of the stellar nucleosynthesis chain is the synthesis of elements in the region of iron. However, reactions such as ^{28}Si + ^{28}Si → ^{56}Ni + γ do not proceed directly as the nuclear charges involved are now so high that direct association of nuclei is improbable. At $T > 2 \times 10^9$ K, ambient thermal photons have sufficient energy ($h\nu > 2\times 10^5$ eV) to remove α-particles and protons from heavy nuclei such as Si, and these may then combine with other nuclei to produce heavier products. These destructive and constructive reactions operate in parallel. In the case of Si, this may be represented symbolically by:

$$^{28}\text{Si} + \gamma's \rightarrow 7\ ^4\text{He}$$
$$^{28}\text{Si} + 7\ ^4\text{He} \rightarrow\ ^{56}\text{Ni} + \gamma's \tag{2.7}$$

followed by β-decay of ^{56}Ni to ^{56}Fe.

The fusion reactions which produce elements up to Fe in the periodic table are exothermic: there is a net release of energy, as each successive compound nucleus is more tightly bound than its parent nuclei. Fe has the greatest binding energy per nucleon of all the elements, and so there are no exothermic reactions which can fuse Fe to form still heavier elements. The production of elements heavier than Fe is thought to depend on neutron capture reactions. Free neutrons are by-products of some fusion processes (e.g. the third reaction in (2.6) above), and these may occasionally fuse with heavy nuclei, resulting in a unit increase in atomic mass. As the neutron has no charge, there is no electrostatic potential barrier to overcome. The resultant nucleus is generally unstable to β-decay, leading to a unit increase in atomic number. A specific example is the production of ^{59}Co from ^{58}Fe:

$$^{58}\text{Fe} + \text{n} \rightarrow\ ^{59}\text{Fe}$$
$$^{59}\text{Fe} \rightarrow\ ^{59}\text{Co} + \text{e}^- + \bar{\nu}. \tag{2.8}$$

Neutron capture is normally a slow process as the number density of free neutrons is normally low, but in some circumstances, such as a supernova explosion, neutrons may be produced rapidly. In this situation, the mean free time between n captures may be similar to, or less than, the decay half-life, and a nucleus may undergo several captures before decaying to a stable form. Isotopes produced by slow neutron capture (s-process) tend to have relatively large numbers of protons, whereas rapid (r-process) capture leads to isotopes rich in neutrons.

2.1.4   Enrichment of the Interstellar Medium

The sequential production of heavy elements by the exothermic fusion re-actions (2.1)–(2.7) discussed above proceeds in massive stars ($M > 10\,\mathrm{M_\odot}$) until an iron-rich core is produced. The outer layers of such highly evolved stars are generally still hydrogen-rich, closely resembling the initial composition. Beneath their surfaces are successive shells of enriched material of increasing mean atomic weight surrounding the core. Such a star is destined to become a type II supernova. No further energetically favourable nuclear reactions can occur in the core itself, which rises in temperature as it contracts until the ambient photon field is sufficiently energetic to cause photodestruction of the Fe nuclei to α-particles and neutrons, *absorbing* energy and leading to catastophic implosion of the core to form a neutron star. Gravitational energy thus released ejects the outer layers in a supernova explosion. Ironically, the immediate prelude to the collapse of the core is thus a reversal of the previous cycle of energy-releasing nuclear reactions, for which the debt is paid by gravity. Supernovae are undoubtedly important sources of heavy elements in the interstellar medium, their expanding remnants containing both the ashes of previous burning cycles and the products of r-process evolution in the supernovae themselves. However, it should be noted that there is surprisingly little direct observational evidence for the expected enrichment of supernova remnants (e.g. Trimble 1975) and none for the production of dust in supernovae (Gehrz 1989; Dwek 1989).

The major fraction of recycled material entering the interstellar medium comes from stars of intermediate mass ($M \sim 1$–$10\,\mathrm{M_\odot}$), which are more numerous than high-mass stars, evolve more slowly, and lose mass copiously during the red giant and asymptotic giant branch phases of their evolution. Models suggest that nucleosynthesis in such stars does not generally progress beyond the He-burning phase; they do not become supernovae, but evolve into white dwarfs, often with the ejection of a planetary nebula. Red giant winds and planetary nebulae are major contributors to the total enrichment of the interstellar medium, and are also likely sources of dust, as discussed in detail in Chapter 7 (§7.1).

The products of nucleosynthesis reach the surface of an evolved star only if enriched material is present in the convective zone, or if the outer layers are stripped off. The structure of an evolving star is determined primarily by its age and mass, but is also influenced by a number of other factors, including pulsational instability and, in the case of close binaries, tidal effects and mass exchange. A single star on the asymptotic giant branch has a very compact, degenerate C-rich or O-rich core, surrounded by thin He-burning and H-burning shells, and a fully convective envelope (see, for example, Shu 1982). Temporary instabilities may lead to the episodic establishment of convection in the shell zone, resulting in the dredge-up of

C, the product of He burning, to the surface. Many red giant atmospheres appear to be enriched in this way. If the C abundance is enhanced to the extent that it exceeds that of O, this has a profound effect on the chemistry of the stellar atmosphere and on the composition of solid condensates in the stellar wind.

2.2 THE SOLAR SYSTEM ABUNDANCES—A STANDARD REFERENCE

2.2.1 Significance and Methodology

The element abundances in the Solar System provide a standard reference set which is invaluable in astrophysics, both as a test for models of nucleosynthesis and as a basis for comparison with other regions of the Universe†. The Solar System is the natural choice for this purpose because abundances may be determined more accurately for more elements than is the case for any other location. The Sun's atmosphere is likely to contain a representative cross-section of virtually all the elements present at the time of its formation. Hydrogen burning in the core will have no effect on the abundances available to measurement. Results are thus expected to represent abundances in the solar nebula for all stable elements, and these may, in turn, be representative of abundances in the local interstellar medium at the birth of the Solar System.

Although the solar spectrum provides reliable data for many elements, it is equally important for non-volatile elements to obtain laboratory measurements of abundances in appropriate solids. Sample selection is clearly critical. The Earth's crust has been modified by gravitational fractionation as well as loss of volatiles, although useful data may be obtained over a limited range of atomic mass. Similar comments apply to lunar samples. The most primitive solids available for analysis are meteoritic. C-type meteorites (carbonaceous chondrites) are preferred, as they show less evidence of thermal processing compared with other meteoritic materials and terrestrial igneous rocks. Their spectra resemble those of asteroids at distances > 5 AU from the Sun. They contain minerals with low condensation temperatures (~ 500–1000 K) and high volatile content, and have a granular structure suggestive of formation by an accretion process. Significantly, isotopic abundance anomalies indicate the inclusion of some grains of presolar origin (§7.3.3). It is therefore realistic to regard the C-type meteorites as tangible specimens of solid condensates and accretion products

† The Solar System abundances are often, misleadingly, referred to in the literature as 'cosmic' abundances.

from the early Solar System. The time elapsed since the epoch of condensation is determined rather precisely by radiometric dating techniques to be 4.57 ± 0.03 Gyr (Kirsten 1978), which is the generally accepted value for the age of the Solar System.

2.2.2   Results

The Solar System abundances are thus based on two types of measurement, remote sensing of the solar atmosphere and laboratory studies of selected meteorites. Figure 2.2 shows the correlation of results from the two methods. It is evident that the correspondence is excellent for most elements, supporting the view that the meteoritic solids condensed from gas of essentially solar composition. The most notable exceptions are the noble gases (He, Ne and Ar) and other volatile elements (H, C, N and, to a lesser extent, O); these lie above and to the left of the correlation line in figure 2.2, suggesting underabundance in meteorites. These elements would be expected to remain in the gas unless chemically bonded into condensible compounds. This cannot, of course, occur for the noble gases, explaining their low abundance in meteorites, but the HCNO group of elements are partially condensed in, for example, hydrated silicate and carbonate minerals, organic matter, and solid carbon. The solar abundance is taken to be appropriate for these elements. The only element plotted in figure 2.2 which is significantly anomalous in the opposite sense, i.e. underabundant in the Sun compared with the meteoritic value, is lithium; as discussed in §2.1 above, Li is easily destroyed in stars, and its general rarity in the photospheres of the Sun and other main-sequence stars implies that this process is operating in material which is being transported by convection currents to and from the surface. In this case, the meteoritic value is more likely to represent the true initial abundance.

In order to compile a set of average abundances for the Solar System, it is necessary to calibrate solar and meteoritic results relative to one another. Astrophysical abundances are generally expressed relative to hydrogen, whereas this normalization is obviously inappropriate for meteoritic data, which are better expressed relative to a condensible heavy element such as Si (as in figure 2.2). Calibration is achieved by taking the average of the meteoritic to solar abundance ratios for the most common elements believed to be fully condensed in the former, namely Mg, Al, Si, Ca, Fe and Ni.

Anders and Grevesse (1989) have carried out a critical assessment of the literature concerned with measurements of Solar System abundances. These authors have compiled a new tabulation of the abundances, combining previous results, and this is adopted here as the set of standard abundances. Table 2.1 lists data for the 14 most abundant elements excluding noble gases: atomic number (z), atomic weight (m) in g mol^{-1},

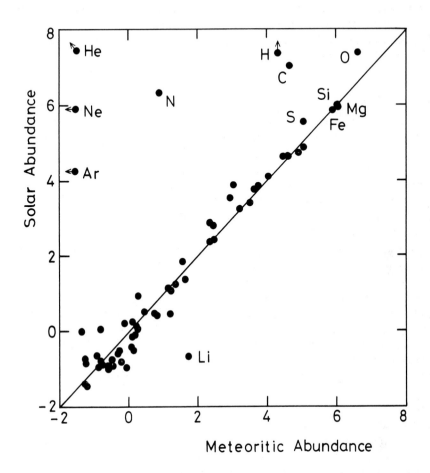

Figure 2.2   The correlation of abundances in the solar atmosphere and those in a typical carbonaceous chondrite. Abundances are by number, logarithmic on the Si = 6 scale. The straight line represents exact agreement.

abundance by number (A_N) and abundance by mass (A_M). All isotopes are summed for individual elements. The abundances are traditionally expressed logarithmically relative to a hydrogen abundance of 10^{12}, i.e., the listed abundances by number and by mass for element X are

$$A_N(X) = 12 + \log \left\{ \frac{N_X}{N_H} \right\}_\odot \qquad (2.9)$$

and

$$A_M(X) = 12 + \log \left\{ \frac{m_X N_X}{m_H N_H} \right\}_\odot \qquad (2.10)$$

Table 2.1   Abundance data for the 14 most abundant elements[†].

Element	z	m	A_N	A_M
H	1	1.01	12.00	12.00
C	6	12.01	8.56	9.64
N	7	14.01	8.05	9.20
O	8	16.00	8.93	10.13
Na	11	22.99	6.31	7.67
Mg	12	24.31	7.59	8.97
Al	13	26.98	6.48	7.91
Si	14	28.09	7.55	9.00
P	15	30.97	5.57	7.06
S	16	32.06	7.27	8.77
Ca	20	40.08	6.34	7.94
Cr	24	52.00	5.68	7.40
Fe	26	55.85	7.51	9.25
Ni	28	58.71	6.25	8.02

† Noble gases are excluded.

where the subscript specifies Solar System values.

The mass fraction of heavy elements may be obtained from the measured abundances by evaluating

$$Z_\odot = 0.71 \sum \left\{ \frac{m_X N_X}{m_H N_H} \right\}_\odot$$
$$\simeq 0.017 \qquad\qquad (2.11)$$

where the factor 0.71 allows for the contributions of the most abundant noble gases (He and Ne) to the total mass. (The contributions of all other elements omitted from table 5.1 are negligible.) The value of $Z_\odot$ in equation (2.11) is effectively an upper limit on the dust to gas ratio for Solar System abundances, subject only to the possibility that enough hydrogen may be condensed to make a significant contribution to the dust mass. For example, a mean hydrogenation factor of 3 (i.e., an average of three H atoms are bonded to each heavy element) would increase $Z_\odot$ to 0.020.

Figure 2.3 shows a plot of A_N against atomic number z for elements in the range $1 \leq z \leq 82$. The general trend is an exponential decline from the very high abundances of the lightest elements, H and He, to the low abundances of the elements with $z > 30$, with a total range of

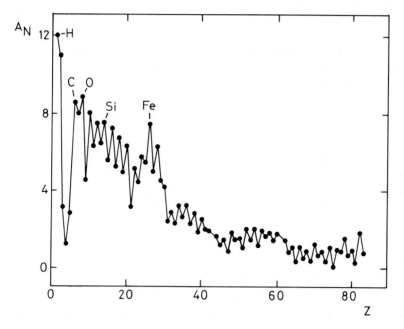

Figure 2.3   Plot of mean numerical abundances (A_N) in the Solar System against atomic number.

~ 12 dex. Structure in the curve supports the view that the heavy elements present in the Solar System are the products of nucleosynthesis within earlier generations of stars (§2.1.3). The trough at z-values 3–5 (lithium group) reflects the intrinsic fragility of these elements. Peaks occur for nuclei composed of integral numbers of α-particles (^{12}C, ^{16}O, ^{20}Ne, ^{24}Mg, ^{28}Si and ^{32}S), and the prominent iron peak centred at $z = 26$ represents the build-up of elements at the end-point of exothermic nucleosynthesis.

2.3   ABUNDANCE TRENDS IN THE GALAXY

It was argued in §2.2.1 above that the abundances measured in the Solar System are likely to be representative of those in the initial cloud (the solar nebula) from which it formed. Similarly, when we consider other stars, our knowledge of stellar structure and evolution suggests that the products of intrinsic internal nucleosynthesis will not generally be apparent in photospheric spectra: the measured abundances reflect initial compositions in most instances, and the exceptions are clearly identifiable and can be considered as distinct groups. This axiom provides a basis for the investigation of systematic abundance trends in the Galaxy.

When comparing solar abundances with those measured in other stars, it is important to distinguish between temporal and spatial effects. There is a natural tendency for heavy-element abundances to increase steadily with time as successive generations of stars evolve and return material to the interstellar medium. Systematic trends with location in the Galaxy are also apparent.

2.3.1   Temporal Variation

The halo population (population II) of the Galaxy contains globular clusters which formed at an early epoch some 15 Gyr ago. The abundances of heavy elements in globular clusters are, as expected, generally much lower than those in the Sun, with $Z_{\rm pop.II}/Z_\odot \sim 0.01$ typically. The presence of detectable heavy-element abundances at any level in the oldest population II stars is significant and implies that a very early phase of enrichment must have occurred in the Galaxy (but see Kraft 1979 for a discussion of the possible significance of intrinsic variations in globular cluster metallicities). At the opposite extreme, the youngest OB stars in the disk (population I) have abundances consistent with those in the Sun ($Z_{\rm pop.I}/Z_\odot \sim 1$). This is a surprising result, as it suggests that little enrichment has occurred in the past 4.6 Gyr since the formation of the Solar System. As a key issue in the study of depletion onto grains is the relevance of solar abundances to the ISM as a whole, we examine this question in some detail.

OB stars have condensed from interstellar clouds within the last 100 Myr, and their atmospheres should thus provide a reliable guide to current abundance levels in the ISM. Support for this hypothesis is provided by investigations of H II regions in the solar neighbourhood: for a range of elements, including C, N, O, Ne, S, and Ar, abundances in H II regions are consistent with those in OB stars to within errors of measurement, and both are approximately solar (e.g. Trimble 1975). Anomalously low metallicities might arise in OB stars if there is a tendency for separation of light and heavy elements during massive star formation due to expulsion of grains by radiation pressure, and this anomaly might be present in the surrounding H II region as well if the more robust grains survive. It is therefore significant that the list of elements showing good agreement with solar values includes volatile species, notably inert gases, as well as those likely to be present in refractory dust.

Twarog (1980) has measured metallicities in a large sample of F-type main-sequence stars in the galactic disk, using the Strömgren (1966, 1987) technique based on intermediate passband photometry. The ages of the stars in the sample are estimated with reference to theoretical isochrones. The logarithmic abundance of element X relative to its solar abundance is given by

$$\left[\frac{X}{H}\right] = \log\left\{\frac{N_X}{N_H}\right\} - \log\left\{\frac{N_X}{N_H}\right\}_{\odot} \qquad (2.12)$$

and we denote the metallicity by $[Fe/H]$, determined from the Strömgren metallicity parameter. Figure 2.4 plots $[Fe/H]$ against age for groups of stars (binned according to age). There is a strong trend of increasing stellar metallicity with decreasing stellar age, which we interpret as a trend of increasing *interstellar* metallicity with time from an early epoch of the Galaxy. Extrapolation of the trend to the present day suggests a modest enhancement over solar abundances; however, the rate of increase may have declined in the past 4 Gyr, which would be consistent with the detection of near-solar abundances in OB stars and their H II regions.

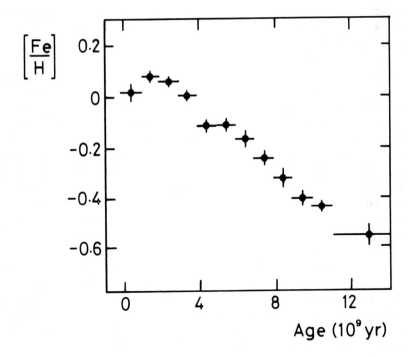

Figure 2.4   Temporal variation in the abundances. Metallicity $[Fe/H]$ is plotted against age for a sample of stars of intermediate spectral type (data from Twarog 1980).

Atoms in the interstellar medium absorb X-rays by the photoelectric effect irrespective of their physical state, and observations thus provide a means of estimating abundances for the gas and dust combined. X-ray data for supernova remnants within 5 kpc of the Sun (Ryter *et al* 1975) indicate abundances for C, N and O which are consistent with solar values to within $\sim 35\%$.

In summary, several independent lines of evidence support the view that the abundances of the elements in the Solar System closely resemble those in the present interstellar medium in our region of the galactic disk. A scenario in which the heavy-element abundances remain almost static on a timescale comparable with the age of the Solar System is difficult to interpret in terms of models for galactic chemical evolution (e.g. Pagel 1987), and it is perhaps more probable that the Sun formed with a metallicity greater than the mean for the ISM at the time of its formation.

2.3.2   Spatial Variation

Spatial abundance trends in galactic disks may be investigated by studying objects of similar age over a range of galactocentric distances, R_G. Figure 2.5 shows plots of oxygen and nitrogen abundances against R_G for our Galaxy, based on observations of H II regions (Shaver *et al* 1983). These show clear systematic trends of increasing abundance with decreasing R_G, with gradients for both elements of approximately -0.08 dex kpc^{-1}. Other heavy elements show similar trends. Extrapolation to $R_G = 0$ suggests a fourfold increase in metallicity at the galactic centre compared with the location of the Sun ($R_G \simeq 7.7$ kpc: Reid 1989). In terms of the evolution of the elements, the nuclear region of the Galaxy appears to have reached greater 'maturity' than the outer arms (Wannier 1989). Observations of external systems such as M33 suggest that this is a common characteristic of spiral galaxies (Pagel and Edmunds 1981).

An increase in metallicity is expected to lead to an increase in dust to gas ratio, as more heavy elements are available to condense into solid particles in stellar outflows, and to attach themselves to existing grain surfaces in the ISM itself. The trend in metallicity towards the galactic centre is in qualitative agreement with the increase in the rate of extinction compared with the solar neighbourhood, discussed in (§1.1.2). Issa, MacLaren and Wolfendale (1990) examine in detail the observed spatial variation of both metallicity and dust to gas in the Milky Way and in nearby external galaxies, and demonstrate a close correlation. This is illustrated in figure 2.6 for galactocentric distances equivalent to that of the Sun.

Systematic changes in heavy-element abundances may lead to corresponding changes in the quality as well as the quantity of dust in the interstellar medium. The relative number densities of carbon-rich and oxygen-rich red giants are sensitive to their initial metallicities: as the natural excess of O over C is enhanced at high metallicity, a greater quantity of C must be dredged up to the surface to produce a carbon star

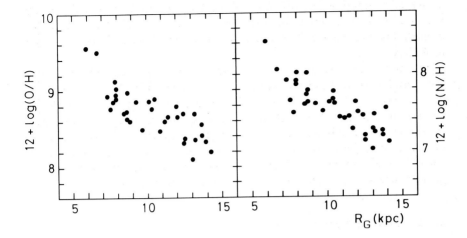

Figure 2.5 Systematic variations with respect to galactocentric distance of abundances in H II regions: oxygen (left) and nitrogen (right). Data are from Shaver *et al* (1983) and references therein.

with $N(C) > N(O)$. Observational estimates of space densities confirm this (Thronson *et al* 1987). Thus, the ejection rates for carbonaceous and oxygen-rich dust are predicted to vary with R_G, the latter dominating near the galactic centre.

2.4   THE OBSERVED DEPLETIONS

2.4.1   Methods

The term depletion refers to the underabundance of gas phase elements with respect to the solar standard as a result of their assumed presence in dust. The fractional depletion of element X is given by

$$\delta(X) = 1 - 10^{D(X)} \tag{2.13}$$

where $D(X)$ is the depletion index given by

$$D(X) = \left[\frac{X}{H} \right] = \log \left\{ \frac{N_X}{N_H} \right\} - \log \left\{ \frac{N_X}{N_H} \right\}_\odot \tag{2.14}$$

(see equation (2.12)). Noting that $D(X)$ is negative for depletion, the fractional depletion is bound by the limits $\delta(X) = 0$ (all atoms in the gas)

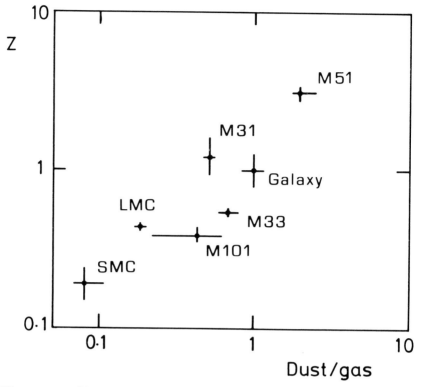

Figure 2.6   Plot of the mass fraction of heavy elements (Z) against the dust to gas ratio for nearby galaxies. Each quantity is evaluated at a galactocentric distance equivalent to that of the Sun and normalized to our Galaxy. (Data from Issa *et al* 1990.)

and $\delta(X) = 1$ (all atoms in the dust). The column densities N_X and N_H (equation (2.14)) are evaluated from analysis of interstellar absorption lines in stellar spectra using the curve of growth technique (e.g. Spitzer 1978, pp 52–55), and normalized to the equivalent solar values (equation (2.9) and table 2.1). Some elements, notably Ca and Na, may be studied in the optical waveband with ground-based telescopes, but most gas phase atomic interstellar absorption lines occur in the satellite ultraviolet.

Problems of spectral-line saturation have proved particularly troublesome in studies of depletion. The column density of the absorber producing an unsaturated absorption line of equivalent width W_ν is

$$N_X = \left(\frac{4\epsilon_0 m_e c}{e^2}\right)\frac{W_\nu}{f} \qquad (2.15)$$

where f is the oscillator strength of the transition. Equation (2.15) describes the linear region of the curve of growth: each absorbing atom along the line of sight sees essentially the full continuum level, and an increase in N_X, the number of absorbers, would give rise to a proportionate increase in W_ν. However, for stronger absorption (optical depth $\tau_\nu \geq 1$ at the line centre), equation (2.15) underestimates the true column density due to saturation, leading to the possibility of systematic errors. Column densities deduced from the curve of growth for saturated lines are, in any case, inherently less accurate (typically by a factor ~ 5) than those deduced from unsaturated lines. As a result, the depletions of some elements are known to considerably greater precision than others.

Carbon is the most problematical of the elements expected to contribute significantly to the grain mass, which is unfortunate, as the depletion of this element is, in principle, an important discriminator between grain models (Whittet 1984a; Duley 1987). As a consequence of the fact that its first ionization potential (11.3 eV) is somewhat less than that of H I (13.6 eV), most of the available gas phase carbon is in C II in H I clouds and in C I (or CO) in H_2 clouds. The abundance of C I is relatively easy to measure, that of C II much more difficult because the permitted resonance lines at 1036 and 1335 Å are generally saturated. Jenkins *et al* (1983) deduce total C abundances towards a number of stars from observations of unsaturated C I lines using ionization equilibrium calculations to estimate the contribution of C II. Observations of an unsaturated semi-forbidden C II line at 2325 Å are available for one line of sight (Hobbs *et al* 1982). In contrast to the situation for carbon, oxygen (13.6 eV) and nitrogen (14.5 eV) have higher first ionization potentials than hydrogen, and only the neutral species need be considered. Most metals have values $\sim$ 5–8 eV and are thus usually singly ionized in interstellar clouds.

Interstellar extinction places practical constraints on the lines of sight in which depletions can be investigated. Results are available from Copernicus and the International Ultraviolet Explorer Satellite for a variety of lightly and moderately reddened stars ($E_{B-V} < 0.5$) obscured predominantly by diffuse clouds and intercloud medium, although some data exist for lines of sight which contain appreciable molecular material. Most pathlengths studied are relatively short ($L < 2\,\mathrm{kpc}$) and systematic trends in metallicity are not normally important. Spitzer (1985) has argued that the environment sampled by a given line of sight can be characterized by the mean hydrogen density

$$\langle n_H \rangle = \frac{N(\mathrm{H\,I}) + 2N(\mathrm{H_2})}{L}. \tag{2.16}$$

As discussed in §2.4.3 below, the observed depletions for a number of elements correlate with $\langle n_H \rangle$. A column of low mean density, $\langle n_H \rangle < 0.2 \times 10^6 \, \mathrm{m}^{-3}$, is unlikely to intercept a cloud containing a cool phase (see §1.2.2). The presence of clouds of various densities and filling factors along a line of sight elevate $\langle n_H \rangle$ to values typically $\sim 1 \times 10^6 \, \mathrm{m}^{-3}$, and exceptionally $> 3 \times 10^6 \, \mathrm{m}^{-3}$.

2.4.2   The Average Depletion Pattern for Diffuse Clouds

It was shown by Field (1974) that the depletions of the elements observed in the interstellar medium show a correlation with condensation temperature, T_C. For a given element, T_C is defined as the temperature at which 50% of the atoms condense into the solid phase in some form, under thermodynamic equilibrium, assuming solar abundances. A plot of D against T_C (figure 2.7), known as the depletion pattern, provides a convenient means of displaying the depletions for the various elements and may have physical significance: for example, a correlation would be expected if grain formation occurs under equilibrium conditions in circumstellar shells around cool stars. The data in figure 2.7 are appropriate to a mean density $\langle n_H \rangle = 3 \times 10^6 \, \mathrm{m}^{-3}$, and density effects (§2.4.3) may thus be ignored here. There is a clear trend towards higher depletion for the more refractory elements, with a strong dependence of D on T_C above $\sim 800 \, \mathrm{K}$. The degree of depletion is almost total for the metallic elements such as Al, Ca, Fe and Ni: in the case of iron, $D(\mathrm{Fe}) \simeq -2$ corresponds to 99% depletion (equation (2.13)). The correlation of $D(X)$ with T_C, although pronounced, shows scatter in excess of observational error: for example, P and Fe have rather similar values of T_C, but their depletions differ by more than a factor of 10.

The observed depletions may be used to estimate the dust density and dust to gas ratio on the assumption of solar abundances for the ISM as a whole. Results are displayed in table 2.2. The total interstellar density of element X (solid plus gas phase) may be related to the hydrogen density:

$$\rho_{\mathrm{ism}}(X) = \left\{ \frac{m_X N_X}{m_H N_H} \right\}_{\odot} \rho_H \tag{2.17}$$

where $\rho_H \simeq 1.8 \times 10^{-21} \, \mathrm{kg \, m}^{-3}$ (§1.2.1) and the quantity in brackets on the RHS is evaluated from the appropriate solar mass abundance (table 2.1). Values of $\rho_{\mathrm{ism}}(X)$ are given in column 2 of table 2.2. The contribution of element X to the dust density is then

$$\rho_d(X) = \delta(X) \, \rho_{\mathrm{ism}}(X) \tag{2.18}$$

(column 4 of table 2.2). Summing the contributions to ρ_d, we obtain a total density of depleted material

$$\rho_d = \sum \rho_d(X) = (21 \pm 8) \times 10^{-24} \, \mathrm{kg \, m}^{-3} \tag{2.19}$$

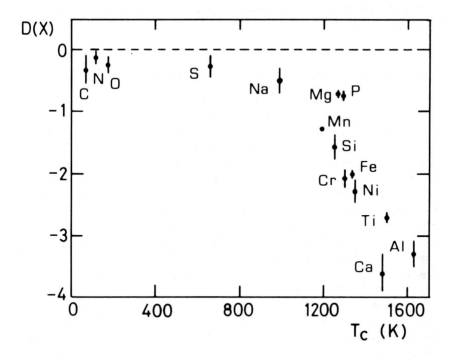

Figure 2.7   The depletion pattern, plotting $D(X)$ at a representative mean density $\langle n_H \rangle \sim 3 \times 10^6 \, m^{-3}$ against condensation temperature T_C (K) for various elements (based on data from Jenkins 1989, Whittet 1984a, and references therein).

and a dust to gas ratio

$$Z_d = 0.71 \frac{\rho_d}{\rho_H} \simeq 0.008 \tag{2.20}$$

where, as before, the factor 0.71 allows for the contributions of the noble gases to the total mass of gas. Comparison with equation (2.11) shows that roughly 50% of the available heavy elements are depleted into dust.

The individual contributions to ρ_d have different degrees of uncertainty. The group of refractory elements (Mg, Al, Si, Ca, Fe and Ni) contribute $\sim 7 \times 10^{-24} \, kg \, m^{-3}$ to ρ_d, a result which is 'robust' in the sense that the fractional depletions of these elements are almost always high ($\delta \geq 0.8$) and relatively insensitive to observational error. The main uncertainty in the evaluation of ρ_d by this method (other than the possibility of systematic departures from solar abundances) arises due to errors in the depletions of O and C. The extreme range of measurements for these two elements give $0.2 < \delta(O) < 0.6$ and $0 < \delta(C) < 0.7$, leading to the estimated error in ρ_d (equation (2.19)).

Table 2.2   Depletions and abundance limitations on grain density for key heavy elements. Densities are in units of 10^{-24} kg m^{-3}.

Element	ρ_{ism}	δ	ρ_{d}
C	7.8	0.5	3.9
N	2.9	0.2	0.6
O	24.3	0.4	9.7
Mg	1.7	0.8	1.4
Al	0.2	1.00	0.2
Si	1.8	0.97	1.7
S	1.1	0.4	0.4
Ca	0.2	1.00	0.2
Fe	3.2	0.98	3.1
Ni	0.2	1.00	0.2
$\sum \rho$	43.4		21.4

2.4.3   Dependence on Cloud Density

The degrees of depletion for many elements show a tendency to correlate with density, most pronounced for the more refractory (and more heavily depleted) elements. As an example, results for titanium are illustrated in figure 2.8: the correlations with column density N_{H} and mean density $\langle n_{\text{H}} \rangle$ are compared. Which of these two measures of density provides the better representation of environment may depend on the particular line of sight: N_{H} may be preferable for those which intercept a single cloud out of the galactic plane, where $\langle n_{\text{H}} \rangle$ (equation (2.16)) may be 'diluted' by inclusion of intercloud gas beyond the cloud towards distant stars; on the other hand, $\langle n_{\text{H}} \rangle$ should be preferable for lines of sight which intercept many clouds in the galactic plane, where high values of N_{H} may arise by accumulation over long path-lengths rather than by inclusion of dense gas. In the cases of refractory elements like Ti, the degree of correlation is invariably higher with $\langle n_{\text{H}} \rangle$ than with N_{H}. In contrast, the depletions of more volatile elements such as O and N show virtually no correlation with $\langle n_{\text{H}} \rangle$ but do show appreciable correlation with N_{H}.

Harris, Gry and Bromage (1984) investigated the correlations of a number of elements with N_{H} and $\langle n_{\text{H}} \rangle$. The correlation coefficient of D with N_{H}, r_N, is typically in the range 0.3 to 0.8, and may possibly be inflated by observational selection (such as insufficient representation of lines of sight with high depletion and relatively low column density in the sample). The correlation coefficient of D with $\langle n_{\text{H}} \rangle$, r_n, shows greater element to

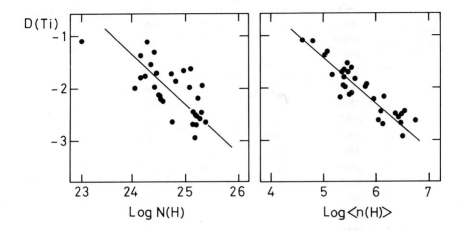

Figure 2.8 Correlation of the depletion index of titanium with column density (left) and mean number density (right). The correlation coefficients are $r_N = 0.70$ and $r_n = 0.91$, respectively. (Data from Harris *et al* 1984.)

element dispersion, varying between 0.1 (for N and O) and 0.9 (for Mg, Ca and Ti). Interestingly, a strong systematic trend occurs in the ratio r_n/r_N with depletion, shown in figure 2.9. The elements plotted clearly separate into groups: high depletion and $r_n/r_N > 1$; low depletion and $r_n/r_N < 1$ (with Cl and P as intermediate cases). It seems probable that this effect is actually dominated by the sensitivity of depletion to the mean density $\langle n_H \rangle$. This correlation extends over two decades of density, and is apparent even at $\langle n_H \rangle < 3 \times 10^5 \, \mathrm{m}^{-3}$ (figure 2.8). No major discontinuity is apparent between depletions in diffuse clouds and those in the intercloud medium.

2.4.4   Implications for Grain Models

The observed depletions provide constraints on models for the composition of interstellar grains and on their origin, growth and destruction. The principal results to be considered are (i) the existence of a correlation with condensation temperature (figure 2.7); (ii) the strong dependence on gas density for the more refractory elements (figures 2.8 and 2.9); and (iii) the inferred distribution of elements in the dust (table 2.2).

The correlations with condensation temperature and with cloud density may be understood if grains condense in stellar atmospheres, are injected into the interstellar medium, and subsequently grow by random accretion of atoms from the gas in interstellar clouds. One can thus envisage two components of depletion, an underlying component due to grain formation in stellar outflows, and a variable component due to environmental factors

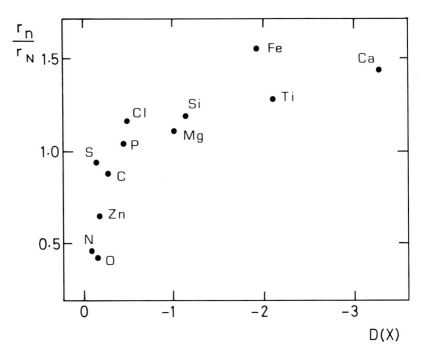

Figure 2.9   Plot of the correlation coefficient ratio r_n/r_N against depletion index D for elements in the study of Harris *et al* (1984).

in the ISM. The correlation between depletion and cloud density has a natural explanation if the underlying depletion is modified by surface accretion and grain destruction by shock sputtering (Barlow and Silk 1977a; Barlow 1978b; Phillips *et al* 1984). For the most refractory elements, high levels of depletion occur at all levels of cloud density. For example, with reference to figure 2.8 (right) and equation (2.14), the gas phase abundance of Ti is $\sim 10\%$ of solar at $\langle n_H \rangle \sim 3 \times 10^4 \, \mathrm{m}^{-3}$ and $\sim 0.1\%$ of solar at $\langle n_H \rangle \sim 5 \times 10^6 \, \mathrm{m}^{-3}$. The significance of the density correlations lies in their implications for grain growth and destruction. The behaviours of rare elements such as Ti are important in constraining models for these processes even though their depletions contribute little to the total grain density. As the underlying depletions of the more abundant heavy elements such as Fe and Si are also high, and as the depletion of O is insensitive to density†, the density correlation has a relatively minor effect on dust to gas ratio over the range of density considered here.

The chemical composition of stellar atmospheres suggests that distinct

---

† In this section, we ignore the growth of ice mantles, which enhances the depletion of oxygen in molecular clouds (§5.3.3).

populations of oxygen-rich and carbon-rich dust are ejected into the inter-stellar medium, although it is not clear to what extent these populations subsequently retain separate identities. The depletion results strongly sug-gest that interstellar dust is predominantly oxygen-rich (see table 2.1). If the carbon depletion were increased to $\delta(C) = 0.70$ (as suggested by some authors and predicted by some grain models), $\delta(O)$ would have to be re-duced to ~ 0.2 for O and C to have *comparable* abundance in the dust.

The optical properties of the dust reviewed in the following chapters allow constraints to be placed on the likely chemical arrangement of the elements in the dust. We note here that the grains responsible for the ex-tinction and polarization at visible wavelengths appear to be predominantly dielectric (non-absorbing) in character. At first sight, this is apparently in-consistent with the result that a substantial fraction of the grain density (approximately one third of ρ_d) is contributed by the metallic elements, as pure metals are strongly absorbing. The ubiquity of oxygen provides a solution to this apparent dilemma: the metals are likely to become chemi-cally bonded into O-rich compounds such as silicates, oxides and sulphates (Jones 1990), which have more dielectric optical properties. These mate-rials are major components of carbonaceous chondrites and interplanetary dust: indeed, if interstellar grains were composed entirely of C1 meteoritic material, the depletions of O, Mg, Si, S and Fe would be very close to those actually observed in the interstellar medium (Whittet 1984a).

Current grain models discussed in §1.4 are consistent with the observed depletions, with the important qualification that the carbon depletion is allowed to be at the higher end of its observed range. Both O-rich and C-rich material are assumed to contribute to the optical extinction, and the former is generally assumed to be predominantly silicate. Approximately 75% of the depleted oxygen is required to oxidize the metals fully, and if all the available Mg, Si and Fe atoms are tied up in particles composed of silicates such as $MSiO_3$ and M_2SiO_4 (where M = Mg or Fe), the density of silicates is predicted to be

$$\rho_d(\text{sil.}) = \rho_d(Mg) + \rho_d(Si) + \rho_d(Fe) + 0.75\rho_d(O)$$
$$\simeq 13 \times 10^{-24} \text{ kg m}^{-3}. \tag{2.21}$$

The balance of the total grain density (equation (2.19)) is presumably car-bonaceous but may have a variety of chemical forms, of which the most widely discussed are graphite, amorphous carbon and organic refractory mantles. If, for example, the depleted carbon is bonded with the depleted nitrogen and with the component of depleted oxygen not tied up in sili-cates, then abundances would allow a total density of organic refractory mantles of

$$\rho_d(\text{org.}) = \rho_d(C) + \rho_d(N) + 0.25\rho_d(O)$$
$$\simeq 7 \times 10^{-24} \text{ kg m}^{-3} \tag{2.22}$$

ignoring the mass contribution of hydrogen.

The abundances and depletions of the elements provide a searching test for the proposal (§1.4) that biota are ubiquitous in the interstellar medium. The biological model for interstellar grains would require a higher degree of C depletion than is generally observed, in common with other grain models, and does not readily account for the high depletions of the metals. A more critical objection concerns the element phosphorus (Duley 1984; Whittet 1984a). Phosphates are an integral part of nucleic acid chains, and the numerical abundance of P in bacteria relative to that of C is typically $[N(P)/N(C)]_B \simeq 0.02$. Assuming solar abundances for the ISM and allowing for the observational result that the depletion of P typically exceeds that of C by a factor ~ 2, the ratio in interstellar dust is $[N(P)/N(C)]_d \simeq 0.002$. Thus, no more than $\sim 10\%$ of the depleted C, or $\sim 2\%$ of the total grain density, can be in bacterial grains. Similar arguments apply to viruses. If biota were widespread in molecular clouds and desiccated in the lower-density clouds and intercloud gas available to depletion analysis, then desorption should lead to gas phase abundances of P significantly *enhanced* over the solar value, contrary to observations.

3

Interstellar Extinction and Scattering

"Further, there is the importance of getting an insight into the true spectrum of the stars, freed from the changes brought about by the medium traversed by light on its way to the observer."

J C Kapteyn (1909)

Extinction occurs whenever electromagnetic radiation is propagated through a medium containing small particles. In general, the transmitted beam is reduced in intensity by two physical processes, absorption and scattering. The energy of an absorbed photon is converted into internal energy of the particle, which is thus heated, whilst a scattered photon is deflected from the line of sight. The spectral dependence of continuum extinction, or extinction curve, is a function of the composition and size distribution of the particles. The polarization of starlight provides evidence that at least one component of the grains responsible for the extinction has anisotropic optical properties, most probably due to elongated shape and alignment by the galactic magnetic field, a topic discussed in detail in Chapter 4. In considering the extinction properties of the dust, spherical grains may be assumed in model calculations without loss of generality. In this chapter, we begin by outlining the theoretical basis for models of extinction and scattering. The relevant observations are are then reviewed, considering, firstly, the average extinction curve for the interstellar medium within a few kiloparsecs of the Sun, and, secondly, departures from the average curve in lines of sight which sample different interstellar environments. The final section discusses attempts to match observations with theory.

3.1   THEORY AND METHODS

3.1.1   Extinction by Spheres

Suppose that spherical dust grains of radius a are distributed uniformly with number density n_d per unit volume along the line of sight to a distant star. The number of grains contained within a cylindrical column of length L and unit cross-sectional area is $N_d = n_d L$. Considering a discrete element of column with length dL, the fractional reduction in intensity of starlight at a given wavelength due to extinction within the element is

$$\frac{dI}{I} = -n_d C_{ext} \, dL \qquad (3.1)$$

where C_{ext} is the extinction cross-section. Integrating equation (3.1) over the entire path-length gives

$$I = I_0 \, e^{-\tau} \qquad (3.2)$$

where I_0 is the initial value of I $(L = 0)$, and

$$\begin{aligned} \tau &= n_d C_{ext} L \\ &= N_d C_{ext} \end{aligned} \qquad (3.3)$$

is the optical depth of extinction due to dust. Expressing the intensity reduction in magnitudes, the total extinction at some wavelength λ is given by

$$\begin{aligned} A_\lambda &= -2.5 \log\left(\frac{I}{I_0}\right) \\ &= 1.086 N_d C_{ext} \end{aligned} \qquad (3.4)$$

using equations (3.2) and (3.3). A_λ is more usually expressed in terms of the extinction efficiency factor Q_{ext}, given by the ratio of extinction cross-section to geometric cross-section:

$$Q_{ext} = \frac{C_{ext}}{\pi a^2}. \qquad (3.5)$$

Hence,

$$A_\lambda = 1.086 N_d \pi a^2 Q_{ext}. \qquad (3.6)$$

If, instead of grains of constant radius a, we have a size distribution such that $n(a)\,da$ is the number of grains per unit volume in the line of sight with radii in the range a to $a + da$, then equation (3.6) is replaced by

$$A_\lambda = 1.086\pi \int a^2 Q_{\text{ext}}(a) n(a)\,da. \tag{3.7}$$

The problem of evaluating the expected spectral dependence of extinction A_λ for a given grain model (assumed composition and size distribution) is essentially that of evaluating Q_{ext}. The extinction efficiency is the sum of corresponding factors for absorption and scattering,

$$Q_{\text{ext}} = Q_{\text{abs}} + Q_{\text{sca}}. \tag{3.8}$$

These efficiencies are functions of two quantities, a dimensionless size parameter,

$$x = \frac{2\pi a}{\lambda} \tag{3.9}$$

and a composition parameter, the complex refractive index of the grain material,

$$m = n - ik. \tag{3.10}$$

Q_{abs} and Q_{sca} may, in principle, be calculated for any assumed grain model, and the resulting values of total extinction compared with observational data. The problem is that of solving Maxwell's equations with appropriate boundary conditions at the grain surface. A solution was first formulated by Mie (1908) and independently by Debye (1909), resulting in what is now known as the Mie theory. A detailed treatment of Mie theory is beyond the scope of this book; excellent modern accounts of both the theory and its applications are available in the literature (van de Hulst 1957; Bohren and Huffman 1983), to which the reader is referred for further discussion.

In order to compute the extinction curve for an assumed grain composition, the real and imaginary parts of the refractive index (equation (3.10)) must be specified. These quantities, n and k, somewhat misleadingly called the 'optical constants', are, in general, functions of wavelength. For dielectric materials ($k = 0$) the refractive index is given empirically by the Cauchy formula

$$m = n \simeq c_1 + c_2 \lambda^{-2} \tag{3.11}$$

where c_1 and c_2 are the Cauchy constants. In general, $c_1 \gg c_2$ and so n is only weakly dependent on λ for dielectrics. Ices and silicates are examples of astrophysically significant solids which behave as good dielectrics ($k \leq 0.05$) over much of the electromagnetic spectrum. For strongly absorbing materials such as metals, k is of the same order as n and both may vary strongly with wavelength.

Figure 1.1 shows sample plots of Q_{ext} and Q_{sca} versus x for constant refractive indices $1.6 - 0.0i$ and $1.6 - 0.05i$. In the former case, the particles are purely dielectric, and $Q_{ext} = Q_{sca}$ ($Q_{abs} = 0$); in the latter, the particles are slightly absorbing and $Q_{ext} > Q_{sca}$. For constant particle radius a, the Q_{ext} plots of are equivalent to extinction curves expressed as A_λ ($\propto Q_{ext}$) versus λ^{-1} ($\propto x$). For $x < 3$ ($\lambda > 2a$), extinction increases steadily with λ in both plots, and a region occurs near $x = 1.7$ where $Q_{ext} \propto x$ ($A_\lambda \propto \lambda^{-1}$) to a good approximation. At larger values of x, resonances occur in the extinction curves for single grain radii which are smoothed out when the contributions of grains with many different radii in a size distribution are summed (equation (3.7)). At very large values of x ($a \gg \lambda$), Q_{ext} is constant, and we have neutral extinction.

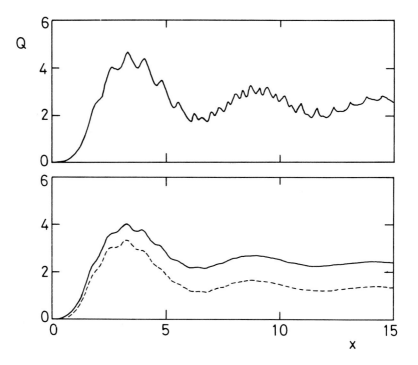

Figure 3.1   Plots of efficiency factors Q_{ext} and Q_{sca} against x for spherical grains. Upper frame: $m = 1.6 - 0.0i$; $Q_{ext} = Q_{sca}$. Lower frame: $m = 1.6 - 0.05i$; solid curve is Q_{ext}, dashed curve is Q_{sca}.

When $x \ll 1$ (i.e., the particles are small compared with the wavelength), useful approximations may be used to give simple expressions for the efficiency factors (see Bohren and Huffman 1983, ch. 5):

$$Q_{sca} \simeq \frac{8}{3} x^4 \left| \frac{m^2 - 1}{m^2 + 2} \right|^2 \tag{3.12}$$

and

$$Q_{abs} \simeq 4x \, \Im \left\{ \frac{m^2 - 1}{m^2 + 2} \right\}. \tag{3.13}$$

For pure dielectric spheres, m is real and nearly constant, as discussed above, and thus $Q_{ext} = Q_{sca} \propto x^4 \propto \lambda^{-4}$ in the small-particle approximation (Rayleigh scattering). More generally, the quantity $(m^2 - 1)/(m^2 + 2)$ is often only weakly dependent on wavelength for materials which are not strongly absorbing, in which case $Q_{abs} \propto \lambda^{-1}$ to a good approximation, whilst $Q_{sca} \propto \lambda^{-4}$, as before. In this case, extinction dominated by absorption in the small-particle limit gives a λ^{-1} dependence, whilst extinction dominated by scattering gives a λ^{-4} dependence.

3.1.2   Albedo, Scattering Function and Asymmetry Parameter

Further quantities of interest, describing the scattering properties of the grains, may be deduced from the Mie theory. The albedo is defined by

$$\gamma = \frac{Q_{sca}}{Q_{ext}} \tag{3.14}$$

and is hence specified as a function of wavelength for a given grain model. As $Q_{ext} \geq Q_{sca} \geq 0$, we have $0 \leq \gamma \leq 1$. For pure dielectics ($k = 0$), $Q_{ext} = Q_{sca}$ and hence $\gamma = 1$. For strong absorbers ($k \sim n$), $Q_{sca} \simeq 0$ and thus $\gamma \simeq 0$. Thus, observational measurements of γ may be used to place constraints on the imaginary (absorptive) component of the refractive index.

The scattering function $S(\theta)$ describes the angular redistribution of light upon scattering by a dust grain. It is defined such that, for light of incident intensity I_0, the intensity of light scattered into unit solid angle about the direction at angle θ to the direction of propagation of the incident beam is $I_0 S(\theta)$ (assuming axial symmetry). The scattering cross-section, defined as $C_{sca} = \pi a^2 Q_{sca}$ by analogy with equation (3.5), is related to $S(\theta)$ by

$$C_{sca} = 2\pi \int_0^\pi S(\theta) \sin \theta \, d\theta. \tag{3.15}$$

The asymmetry parameter is defined as the mean value of $\cos \theta$ weighted with respect to $S(\theta)$:

$$g(\theta) = \langle \cos \theta \rangle$$
$$= \frac{\int_0^\pi S(\theta) \sin \theta \cos \theta \, d\theta}{\int_0^\pi S(\theta) \sin \theta \, d\theta}$$
$$= \frac{2\pi}{C_{\text{sca}}} \int_0^\pi S(\theta) \sin \theta \cos \theta \, d\theta. \qquad (3.16)$$

Calculations for dielectric spheres show that $g(\theta) \simeq 0$ in the small-particle limit, which corresponds to spherically symmetric scattering, whereas $0 < g(\theta) < 1$ for larger particles, indicating forward-directed scattering. As the ratio of grain diameter to wavelength increases from 0.3 to 1.0, a range of particular interest for studies of interstellar dust, the value of $g(\theta)$ increases from 0.15 to 0.75. Hence, the asymmetry parameter is a sensitive function of grain size (Witt 1989).

3.1.3   The Colour-Difference Technique

Consider two stars of identical spectral type and luminosity class but unequal reddening. The apparent magnitude of each star as a function of wavelength is given by

$$m_1(\lambda) = M_1(\lambda) + 5 \log d_1 + A_1(\lambda)$$

and

$$m_2(\lambda) = M_2(\lambda) + 5 \log d_2 + A_2(\lambda) \qquad (3.17)$$

where M represents absolute magnitude, d distance, and A the total extinction (see §1.1.1). The intrinsic spectral energy distribution is expected to be the same for stars of the same spectral classification; thus $M_1(\lambda) = M_2(\lambda)$. If we also assume that $A(\lambda) = A_1(\lambda) \gg A_2(\lambda)$, i.e. the extinction towards star 2 is negligible compared with that towards star 1, then the magnitude difference $\Delta m(\lambda) = m_1(\lambda) - m_2(\lambda)$ reduces to

$$\Delta m(\lambda) = 5 \log \left(\frac{d_1}{d_2} \right) + A(\lambda). \qquad (3.18)$$

The first term on the RHS of equation (3.18) is independent of wavelength and constant for a given pair of stars. Hence, the quantity $\Delta m(\lambda)$ may be used to represent $A(\lambda)$. This technique is termed the 'colour-difference' or 'pair' method.

The interstellar extinction curve has traditionally been studied by the colour-difference method using low-resolution spectrophotometry, broad-band photometry, or some combination of these. When spectrophotometry of resolution $\Delta\lambda < 50\,\text{Å}$ is used, matching of individual stellar spectral lines in the reddened and comparison stars becomes important. Early-type stars (spectral classes O–A0) are generally selected for such investigations as their spectra are relatively simple and thus easy to match, and their intrinsic luminosity and frequent spatial association with dusty regions also render them most suitable for probing the interstellar medium at optical and ultraviolet wavelengths. As an example of the application of the colour-difference method, figure 2.2 plots the observed spectra in the visible (4000–6500 Å) for a matching pair of stars, and the resulting extinction curve. The stars selected are ζ Oph (spectral type O9.5 V, reddening $E_{B-V} = 0.32$) and ν Ori (B0 V, $E_{B-V} = 0.03$). Flux density ($\log F_\lambda$) and extinction, Δm (equation (3.18)) are plotted against wavenumber λ^{-1}. Note the cancellation of stellar spectral lines to produce a smooth extinction curve.

The constant term in equation (3.18) may be eliminated by means of normalization with repect to two standard wavelengths λ_1 and λ_2:

$$
\begin{aligned}
E_{\text{norm}} &= \frac{\Delta m(\lambda) - \Delta m(\lambda_2)}{\Delta m(\lambda_1) - \Delta m(\lambda_2)} \\
&= \frac{A(\lambda) - A(\lambda_2)}{A(\lambda_1) - A(\lambda_2)} \\
&= \frac{E(\lambda - \lambda_2)}{E(\lambda_1 - \lambda_2)}.
\end{aligned}
\tag{3.19}
$$

The normalized extinction E_{norm} should be independent of stellar parameters and determined purely by the extinction properties of the interstellar medium. Normalized curves for different stars may be superposed and compared. Theoretical extinction curves deduced from equation (3.7) for a given grain model may also be normalized in the same way to allow direct comparison between observations and theory. When broadband photometry tied to a standard photometric system is used, it is not essential to observe unreddened comparison stars, as the reddened stars may be compared with tabulated intrinsic colours (e.g. Johnson 1966). Extinction curves are commonly normalized with respect to the B and V passbands in the Johnson system.

Two difficulties associated with the colour-difference method should be noted. Firstly, there is a scarcity of suitable comparison stars for detailed spectrophotometry, as relatively few OB stars are close enough to the Sun or at high enough galactic latitude to have negligible reddening. The problem is most acute for supergiants, and these are often excluded from studies of extinction in the ultraviolet where mismatches in spectral line strengths can be particularly troublesome. Secondly, many early-type stars have infrared excess emission, due to thermal re-radiation from circumstellar dust or free–free emission from ionized gas, and if one attempts to derive the extinction curve for such an object by comparing it with a normal star

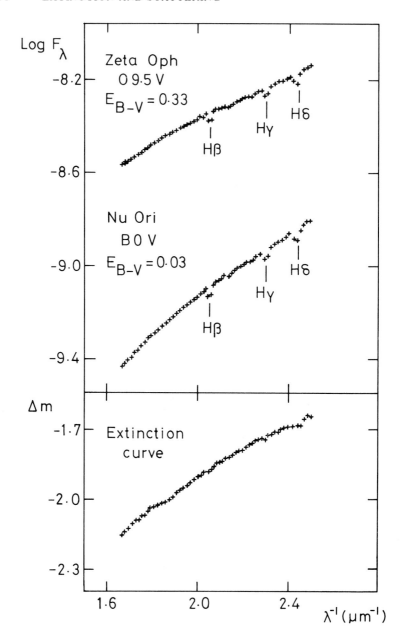

Figure 3.2   An illustration of the colour-difference method. Upper frame: plot of $\log F_\lambda$ (W m^{-2} μm^{-1}) against λ^{-1} (μm^{-1}) for the reddened star ζ Oph and the intrinsically similar unreddened star ν Ori (data from Willstrop 1965). Stellar spectral lines in the Balmer series are labelled. Lower frame: the resulting interstellar extinction curve for the dust towards ζ Oph.

or with normal intrinsic colours, the derived extinction curve will be distorted in the spectral bands at which significant emission occurs. If, for example, there is emission in the K-band at $2.2\,\mu$m, then the colour excess ratio E_{V-K}/E_{B-V} is anomalously large (because K is numerically too small compared with V). This can lead to systematic overestimates of the ratio of total to selective extinction (§3.2.4). Fortunately, shell stars can usually be identified spectroscopically by the presence of optical emission lines, providing a means of discrimination (Whittet and van Breda 1978).

3.2   AVERAGE PROPERTIES

3.2.1   The Mean Extinction Curve

Reliable data on the wavelength dependence of extinction are available in the spectral region from 0.1 to $5\,\mu$m. Studies of large samples of stars have shown that the extinction curve takes the same general form in many lines of sight. Regional variations are apparent, particularly in the blue–ultraviolet region of the spectrum, which we discuss in §3.3 below, but the average extinction curve for many stars provides a valuable benchmark for comparison with curves deduced for individual stars and regions, and a basis for modelling. The best available data, plotted in figure 3.3 and listed in table 3.1, are taken from the compilations of Whittet (1988) for the infrared (1–5 μm) and Savage and Mathis (1979) for the optical and ultraviolet (0.1–1 μm)†. The stars in the sample are reddened predominantly by diffuse clouds within 2–3 kpc of the Sun.

The values of extinction presented in table 3.1 make use of standard normalizations. The relative extinction (replacing the labels λ_1 and λ_2 in equation (3.19) with B and V) is

$$\frac{E_{\lambda-V}}{E_{B-V}} = \frac{A_\lambda - A_V}{E_{B-V}}$$

$$= R_V \left\{ \frac{A_\lambda}{A_V} - 1 \right\}. \tag{3.20}$$

Thus, the *absolute* extinction A_λ/A_V may be deduced from the relative extinction if $R_V = A_V/E_{B-V}$, the ratio of total to selective extinction, is known. The values of A_λ/A_V in table 3.1 are deduced for a value of $R_V = 3.05$ (see §3.2.4 below).

---

† A correction has been applied to remove a spurious bump near $6.3\,\mu\text{m}^{-1}$, which arose due to mismatched stellar C IV lines (Massa *et al* 1983) in part of the data set used by Savage and Mathis.

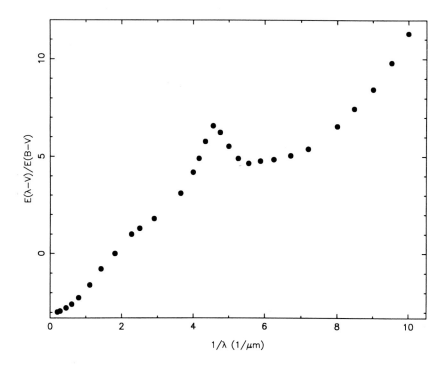

Figure 3.3   The average interstellar extinction curve ($E_{\lambda-V}/E_{B-V}$ versus λ^{-1}) plotted over the entire wavelength range for which data are available, using data from table 3.1.

The mean extinction curve shows a number of distinctive features. It is almost linear in the visible from 1 to $2\,\mu m^{-1}$, with a change in slope in the blue near $2.2\,\mu m^{-1}$. This section of the curve resembles the dependence of Q_{ext} on λ^{-1} for a single grain size (figure 3.1), and for a refractive index $m \simeq 1.6 - 0i$, we deduce from equation (3.9) that grains of radius $a \simeq (2\pi)^{-1} \simeq 0.15\,\mu m$ would reproduce the approximate form of the extinction curve. At shorter wavelengths, this comparison breaks down. The most prominent characteristic of the observed extinction curve is a broad, symmetric peak in the mid-ultraviolet centred at $\sim 4.6\,\mu m^{-1}$. Beyond this, a trough occurs near $\sim 6\,\mu m^{-1}$, followed by a steep rise into the far ultraviolet $(\lambda^{-1} > 6\,\mu m^{-1})$.

3.2.2   Scattering Characteristics

The scattering properties of the grains, as revealed by observations of the diffuse galactic light (DGL), reflection nebulae and cloud haloes, are

Table 3.1 The average extinction curve with B, V normalization.

λ (μm)	λ^{-1} (μm^{-1})	$E_{\lambda-V}/E_{B-V}$	A_λ/A_V
∞	0	-3.05	0
4.8	0.21	-2.98	0.02
3.5	0.29	-2.93	0.04
2.22	0.45	-2.77	0.09
1.65	0.61	-2.58	0.15
1.25	0.80	-2.25	0.26
0.90	1.11	-1.60	0.48
0.70	1.43	-0.78	0.74
0.55	1.82	0.00	1.00
0.44	2.27	1.00	1.33
0.40	2.50	1.30	1.43
0.344	2.91	1.80	1.59
0.274	3.65	3.10	2.02
0.25	4.00	4.19	2.37
0.24	4.17	4.90	2.61
0.23	4.35	5.77	2.89
0.219	4.57	6.57	3.15
0.21	4.76	6.23	3.04
0.20	5.00	5.52	2.81
0.19	5.26	4.90	2.61
0.18	5.56	4.65	2.52
0.17	5.88	4.77	2.56
0.16	6.25	4.85	2.59
0.149	6.71	5.05	2.66
0.139	7.18	5.39	2.77
0.125	8.00	6.55	3.15
0.118	8.50	7.45	3.44
0.111	9.00	8.45	3.77
0.105	9.50	9.80	4.21
0.100	10.00	11.30	4.70

reviewed by Witt (1988, 1989). Of all the phenomena which contribute to our understanding of interstellar grains, the DGL is perhaps the most difficult to observe (because of its intrinsic faintness and the numerous sources of contamination) and to analyse. The spectral dependence of the DGL in the satellite ultraviolet has been investigated in detail by Lillie and

Witt (1976) and Morgan *et al* (1978), and additional optical data have been presented by Toller (1981). Results yield the spectral dependence of the albedo, plotted in figure 3.4. The analysis depends on an idealized plane-parallel model for radiative transfer in the galactic plane, and requires extrapolation of the spectral dependence of the interstellar radiation field from the visual into the ultraviolet. Since the stars which contribute to this radiation field at different wavelengths have different spatial distributions, the geometry of DGL models is itself wavelength dependent. In view of these difficulties, the results must be treated with due caution. However, the relatively high albedo measured in the blue–visible for the DGL ($\gamma \simeq$ 0.6–0.7) is consistent with independent measurements based on studies of reflection nebulae and the surface-brightness profiles of dark clouds (Witt 1989), and this result thus appears to be well established.

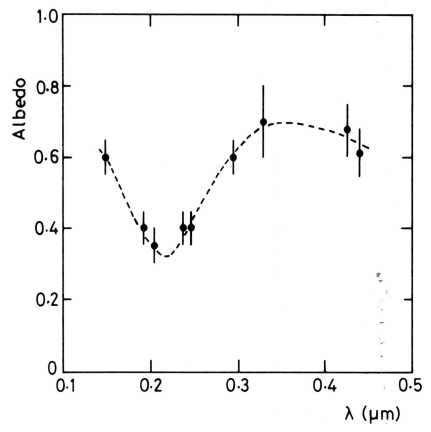

Figure 3.4   The spectral dependence of the albedo from 0.1 to 0.5 μm, deduced from observations of the diffuse galactic light (Lillie and Witt 1976; Toller 1981).

The asymmetry factor $g(\theta)$, derived in the visible region from observations of reflection nebulae and cloud surface-brightness profiles, is typically 0.6–0.8, indicative of predominantly forward-scattering. The value of $g(\theta)$ in the ultraviolet cannot be obtained reliably independently of γ from observations of the DGL, but data for bright reflection nebulae (Witt 1988 and references therein) indicate that $g(\theta)$ is significantly lower (0.2–0.4 for $\lambda < 2000$ Å) than at longer wavelengths, implying a trend towards isotropic scattering.

Several important conclusions may be drawn from these results. Firstly, the albedo is clearly quite high in the blue region of the visible spectrum. If it were not, reflection nebulae would not be seen! This implies that at least one component of the dust has optical properties which are predominantly dielectric. The high values found also for the asymmetry factor indicate that these grains are forward-scattering, and are thus classical ($a \sim 0.1$–$0.3\,\mu$m) in size. In the ultraviolet, γ displays a distinct minimum at $\lambda \sim 0.22\,\mu$m, a wavelength which corresponds to the peak in the extinction curve. This strongly suggests that the peak is a pure absorption feature. At shorter wavelengths, scattering is less forward-biased, implying that much smaller grains are involved. As $g(\theta)$ is determined principally by the ratio of particle size to wavelength, we have a situation where, at progressively shorter wavelengths, the scattering is dominated by grains which decline in size faster than the wavelength itself. This places rather stringent constraints on the grain size distribution and highlights the importance of very small grains for an understanding of the extinction and scattering in the ultraviolet.

3.2.3 Broadband Structure in the Visible

Studies of the extinction curve with spectral resolution better than $0.2\,\mu$m^{-1} have demonstrated the occurrence of significant structure in the visible region (e.g. Whiteoak 1966; van Breda and Whittet 1981). This effect is illustrated in figure 3.5, which plots residuals with respect to a linear fit to the extinction curve from 1.3 to $2.2\,\mu$m^{-1} against wavenumber. The resulting profile shape, characterized by a trough centred near $1.75\,\mu$m^{-1} and/or a peak centred near $2.05\,\mu$m^{-1}, is termed the very broadband structure (VBS).

The origin of the VBS is unknown. It could perhaps be attributable to a symmetrical absorption feature $\sim 0.4\,\mu$m^{-1} wide centred at $2.05\,\mu$m^{-1}, but no satisfactory identification has been proposed. Note that the drop in residuals with increasing wavenumber beyond $2.05\,\mu$m^{-1} is caused, at least in part, by the change in the slope of the extinction curve in the blue–violet region, attributed on the basis of Mie theory calculations (e.g. Whittet, van Breda and Glass 1976) to corresponding changes in the extinction efficiency for classical-sized dielectric grains as the wavelength becomes smaller than

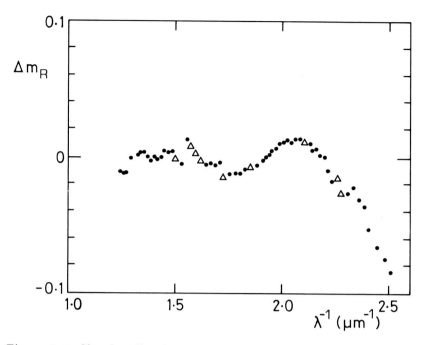

Figure 3.5 Very broadband structure in the visible region of the extinction curve: residuals (Δm_R) with respect to a linear fit to the mean normalized extinction curve for 20 stars are plotted against λ^{-1}. Triangles denote points corrected for diffuse interstellar band absorption (§5.2). Data are from van Breda and Whittet (1981).

typical grain dimensions. Structure in the optical constants of a continuous absorber such as magnetite (Fe_3O_4) may produce VBS in the extinction curve at visible wavelengths (Huffman 1977), but attempts to model the profile are unconvincing (Millar 1982). Perhaps a more plausible alternative is that the VBS arises due to *emission* near $1.75\,\mu m^{-1}$: a possible mechanism is discussed in §6.3.3.

3.2.4   The Infrared Extinction Curve and Evaluation of R_V

In the absence of neutral extinction (see §3.2.5 below) R_V is related formally to the normalized relative extinction by the limit

$$R_V = -\left[\frac{E_{\lambda-V}}{E_{B-V}}\right]_{\lambda\to\infty} \tag{3.21}$$

and may thus be deduced by extrapolation of the observed extinction curve with reference to some model for its behaviour at wavelengths beyond the range for which data are available.

Assuming that the small-particle approximation applies at sufficiently long wavelengths in the infrared, we expect extinction $\propto \lambda^{-4}$ for scattering and $\propto \lambda^{-1}$ for absorption if m is independent of λ (§3.1.1: equations (3.12) and (3.13)). The observed mean curve in the infrared is, in fact, very closely represented by a power law with an index intermediate between these extremes: a non-linear least-squares fit from 0.7 to $5\,\mu$m yields

$$\frac{E_{\lambda-V}}{E_{B-V}} = 1.19\,\lambda^{-1.84} - 3.05 \qquad (3.22)$$

(Martin and Whittet 1990) with formal errors in the fitting constants (amplitude, spectral index and intercept) of $\sim 1\%$. The first term on the RHS of equation (3.22) is A_λ/E_{B-V} (see equation (3.20)), and the second is the intercept $-R_V$. Although we cannot assume that the same power law holds at wavelengths beyond $5\,\mu$m, R_V is not very sensitive to changes in spectral index within the limits for small particles noted above. The value of R_V implied by equation (3.22) is consistent with results deduced by fitting theoretical extinction curves to the data (e.g. Whittet *et al* 1976): values of R_V for individual stars reddened by diffuse clouds are typically in the range $2.9 \le R_V \le 3.3$. This spread probably reflects real variations (Whittet 1977) which are hidden in the average curve for many stars. We adopt

$$R_V = 3.05 \pm 0.15 \qquad (3.23)$$

as the most likely average value for R_V in diffuse clouds.

A useful approximation may be applied to relate R_V to the relative extinction in an infrared passband such as K in a manner consistent with models for the behaviour of the long-wavelength extinction, i.e.

$$R_V \simeq 1.1\,\frac{E_{V-K}}{E_{B-V}}. \qquad (3.24)$$

The value of the constant in equation (3.24) is only weakly model dependent for theoretical curves which fit the observations, and is applicable over a wide range in R_V (Whittet and van Breda 1978). Thus, to estimate R_V for a reddened star of known spectral type, photometry only in the three passbands B, V and K is required.

3.2.5   Neutral Extinction

Neutral (wavelength-independent) extinction is produced by particles which are large compared with the wavelength considered. Any neutral extinction which may occur in the interstellar medium due to 'giant grains' (comparable, say, with interplanetary dust grains which are typically 1–$100\,\mu$m in size) would be undetected by the colour-difference method, yet its presence would affect distance deteminations. Evaluation of R_V by extrapolation of the extinction curve (equation (3.21)) assumes implicitly that the size distribution $n(a) \to 0$ (and hence that $A_\lambda \to 0$) as a increases to sizes large compared with the longest wavelengths at which extinction data are available. This assumption requires justification. Fortunately, R_V may be evaluated independently by a method which includes any contribution by neutral extinction, based on the analysis of open-cluster diameters using the technique pioneered by Trumpler (see Harris 1973). The total visual (neutral plus wavelength-dependent) extinction averaged over a cluster is given by

$$\langle A_V \rangle = \langle V - M_V \rangle - \left\{ 5 \log \left(\frac{D}{\theta} \right) - 5 \right\} \tag{3.25}$$

where θ is the angular diameter of the cluster in radians, and D the linear diameter in parsecs deduced from Trumpler's morphological classification technique. Thus, D/θ is the geometric distance (independent of extinction). $\langle V - M_V \rangle$ is determined by photometry and spectral classification of individual cluster members, from which the mean reddening $\langle E_{B-V} \rangle$ is also deduced. A plot of $\langle A_V \rangle$ against $\langle E_{B-V} \rangle$ for many clusters yields a linear correlation *passing through the origin* to within observational error, and the slope gives $R_V = 3.15 \pm 0.20$, consistent with equation (3.23) above. This demonstrates that neutral extinction is negligible. The interstellar medium does not contain a substantial population of giant grains, a result which is in agreement with abundance considerations (§2.4).

3.2.6   Dust Density and Dust to Gas Ratio

An estimate of the amount of grain material required to produce the observed mean rate of extinction with respect to distance in the galactic plane may be deduced from general principles described by Purcell (1969). The integral of Q_{ext} over all wavelengths can be obtained from the Kramers–Krönig relationship

$$\int_0^\infty Q_{\text{ext}} \, d\lambda = 4\pi^2 a \left\{ \frac{m^2 - 1}{m^2 + 2} \right\} \tag{3.26}$$

for spherical grains of refractive index m. The density of dust in a column of length L is

$$\rho_{\text{d}} = \frac{N_{\text{d}} m_{\text{d}}}{L} \tag{3.27}$$

where

$$m_{\text{d}} = \frac{4}{3} \pi a^3 s \tag{3.28}$$

is the mass of a spherical dust grain composed of material of specific density s. Using equation (3.6) to relate Q_{ext} to A_λ in equation (3.26) and substituting for N_d and m_d in equation (3.27), we have

$$\rho_d \propto s \left\{ \frac{m^2 + 2}{m^2 - 1} \right\} \int_0^\infty \frac{A_\lambda}{L} \, dL. \tag{3.29}$$

From a knowledge of the observed mean extinction curve, ρ_d may be expressed approximately in terms of $\langle A_V/L \rangle$ in mag kpc^{-1} (e.g. Spitzer 1978, p 153):

$$\rho_d \simeq 1.2 \times 10^{-27} s \left\{ \frac{m^2 + 2}{m^2 - 1} \right\} \left\langle \frac{A_V}{L} \right\rangle. \tag{3.30}$$

From observations of reddened stars, $\langle A_V/L \rangle \sim 1.8$ mag kpc^{-1} in the diffuse ISM (§1.1.2), and if we assume that $m = 1.50 - 0i$ and $s \simeq 2500$ kg m^{-3}, appropriate to low-density silicates, then equation (3.30) gives

$$\rho_d \simeq 18 \times 10^{-24} \text{ kg m}^{-3}. \tag{3.31}$$

This result is insensitive to particle shape: non-spherical grains of the same total volume would yield essentially the same result (Greenberg and Hong 1975). However, it is somewhat dependent on the assumed composition; for example, ice grains ($m = 1.33 - 0i$, $s = 1000$ kg m^{-3}) would yield a value $\sim 40\%$ less. It should also be noted that much larger values of A_V/L, and hence of ρ_d, occur locally within individual clouds.

The dust to gas ratio, allowing for the presence of helium in the gas, is

$$Z_d = 0.71 \frac{\rho_d}{\rho_H} \simeq 0.007 \tag{3.32}$$

where $\rho_H \simeq 1.8 \times 10^{-21}$ kg m^{-3} (§1.2.1). This result is consistent with the value $Z_d \simeq 0.008$ deduced independently from the depletions of gas phase elements with respect to solar abundances (§2.4.2, equation (2.20)). Thus, about 40% by mass of the available heavy elements (equation (2.11)) are needed in grains to account for the observed rate of extinction. A similar calculation for reddened stars in the Large Magellanic Cloud (Koornneef 1982) yields $Z_d \sim 0.002$, a significant discrepancy which is qualitatively consistent with the low metallicity of that galaxy compared with the Milky Way (see figure 2.6).

3.3 SPATIAL VARIATIONS IN THE EXTINCTION CURVE

3.3.1 The Blue–Ultraviolet

Regional variations in the optical properties of interstellar dust in response to environmental influences were first discussed by Baade and

Minkowski (1937), who found that the extinction curves for stars in the Orion Nebula (M42) differ from the mean curve for more 'normal' regions in a manner consistent with the selective removal of small particles from the size distribution. Such an effect can be produced by a number of physical processes (§7.2), including grain growth by coagulation, size-dependent destruction, and selective acceleration of small grains by radiation pressure in stellar winds. Star to star variations are most conspicuous at ultraviolet wavelengths: for example, Witt *et al* (1984) find significant departures from the mean curve in $\sim 70\%$ of cases studied. Anomalous extinction is most frequently observed in lines of sight which sample individual clouds associated with current or recent star formation, whereas extinction curves integrated over a number of clouds naturally tend to resemble the mean curve more closely.

Variations in the morphological appearance of the extinction curve at $\lambda^{-1} > 2.0\,\mu\mathrm{m}^{-1}$ may be characterized by three effects: changes in the slope of the blue to near-ultraviolet (NUV) curve from 2 to $3\,\mu\mathrm{m}^{-1}$; changes in the strength and width of the peak at $4.6\,\mu\mathrm{m}^{-1}$ (the $\lambda2175$ feature), and changes in the slope of the far-ultraviolet (FUV) rise ($\lambda^{-1} > 6\,\mu\mathrm{m}^{-1}$). Variations in the blue–NUV slope have been discussed by Whittet *et al* (1976) and references therein, and probably arise due to regional fluctuations in the size distribution of the grains responsible for the visual and infrared extinction. These changes also lead to variability in R_V (§3.3.2 below), which is generally enhanced to values ~ 4 and exceptionally ~ 5 in dense clouds associated with recent star formation, compared with the usual value of ~ 3.1 (equation (3.23)).

Star to star variations in the $\lambda2175$ feature and the FUV rise are illustrated in figure 3.6. This sample includes lines of sight which probe a variety of environments, including dense clouds (HD 147701 and HD 147889 in the ρ Oph region) and H II regions (Herschel 36 in M8; HD 37022 = θ^1 Ori C in M42) as well as those obscured by more typical diffuse clouds. A variety of morphologies are apparent. The nature and origin of the $\lambda2175$ feature is discussed in detail in Chapter 5 (§5.1). We note here that its strength shows variations which are apparently independent of changes in the FUV extinction (compare, for example, the curves for HD 204827, HD 37367 and HD 37022 in figure 3.6).

Fitzpatrick and Massa (1986, 1988) have shown that the ultraviolet extinction curve is well-represented mathematically by the formula

$$\frac{E_{\lambda-V}}{E_{B-V}} = c_1 + c_2\lambda^{-1} + c_3D + c_4F \qquad (3.33)$$

where c_1, c_2, c_3 and c_4 are constants for a given line of sight, D is the Drude function representing the profile of the $\lambda2175$ feature (§5.1), and F is a polynomial function representing the FUV extinction rise. The latter

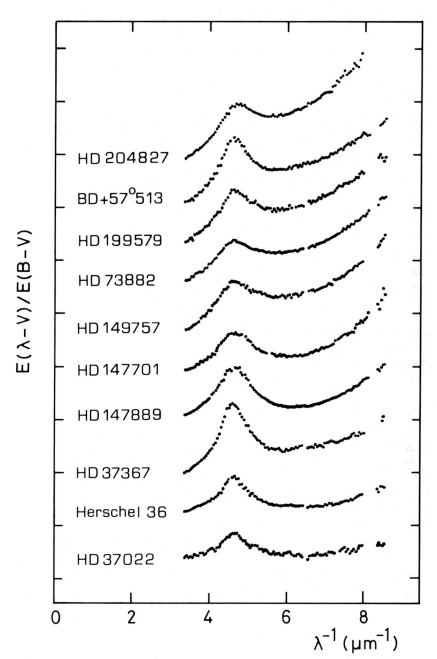

Figure 3.6   Comparison of the ultraviolet extinction curves of 10 stars observed with IUE (Fitzpatrick and Massa 1986, 1988). The vertical axis is $E_{\lambda-V}/E_{B-V}$ (one division = 4 magnitudes), individual curves being displaced vertically for display.

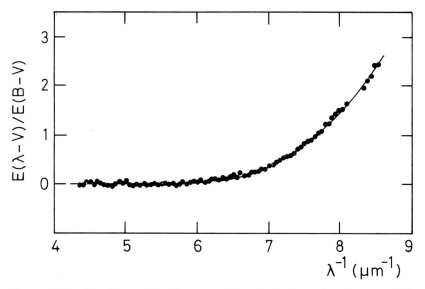

Figure 3.7   The shape of the far-ultraviolet rise in the extinction curve (Fitzpatrick and Massa 1988). Observational data for 18 stars are averaged and the residuals (points) are plotted after extraction of a linear background and a Drude profile representing the $\lambda 2175$ absorption. The smooth curve is a fit based on the polynomial function in equation (3.34).

is expressed in terms of $y = \lambda^{-1}$ as

$$F(y) = 0.539(y - 5.9)^2 + 0.0564(y - 5.9)^3 \qquad (3.34)$$

for $y > 5.9\,\mu\text{m}^{-1}$, with $F = 0$ for $y \leq 5.9\,\mu\text{m}^{-1}$. A fit based on this functional form to the average residual curve for 18 stars, with the linear background $(c_1 + c_2\lambda^{-1})$ and the $\lambda 2175$ feature $(c_3 D)$ removed, is shown in figure 3.7. The quantity c_4 characterizes the amplitude of the FUV rise. The *shape* of the FUV component of the curve (equation (3.34)) is the same for all stars in the sample, regardless of environmental factors or the morphology of the extinction curve at longer wavelengths. This suggests that the FUV extinction rise is not an artifact of the size distribution but a distinct optical property of some physical component of the dust. One possibility, discussed by Fitzpatrick and Massa (1988), is that the FUV extinction is the long-wavelength tail of a resonance absorption in the extreme ultraviolet, produced by a mechanism analogous to that proposed for $\lambda 2175$ (§5.1.2).

Detailed studies of interstellar extinction in external galaxies are currently restricted by observational constraints to the two Magellanic Clouds (Fitzpatrick 1989). Results are displayed in figure 3.8 (LMC) and figure 3.9 (SMC) and compared with the mean curve for the Galaxy. In the case of the LMC, fits using the same functional form derived empirically for extinction

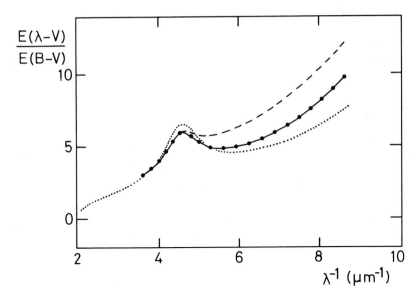

Figure 3.8   Ultraviolet extinction curves for the Large Magellanic Cloud (LMC), based on the analytical fitting procedure of Fitzpatrick (1986). The average curve for stars widely distributed in the LMC (curve with solid circles) is compared with that for the 30 Dor region of the LMC (dashed curve) and the Milky Way (dotted curve).

curves in the Galaxy (equation (3.33)) can be applied. Two distinct curves are shown in figure 3.8. That for stars in the vicinity of the 30 Doradus complex is anomalous in that λ2175 is weak and the FUV rise is particularly rapid, compared with the Galaxy. Stars more widely distributed in the LMC show more normal extinction. Indeed, it is remarkable that the general LMC curve is so *similar* to that of the Milky Way, resembling, particularly, curves for individual stars with diffuse-cloud reddening (e.g. HD 204827 in figure 3.6). Clearly, the ingredients of the ISM which lead to λ2175 absorption and the FUV rise are also present in the LMC. The situation in the SMC is less certain as there are fewer luminous stars with significant reddening. The curve in figure 3.9 is an average for just three stars. If this is truly representative of the SMC, comparison with the galactic curve is striking. The λ2175 feature is apparently absent in the SMC, and the FUV extinction remarkably high, rising more steeply than in any known line of sight in the Galaxy.

It is likely that the comparison between extinction curves for the Galaxy and the Magellanic Clouds contains important clues on the nature of the grains responsible for the various features of the curve. Attention has been drawn (e.g. Nandy 1984) to the observed trend of decreasing metallicity and dust to gas ratio accompanying the trend of increasingly anomalous

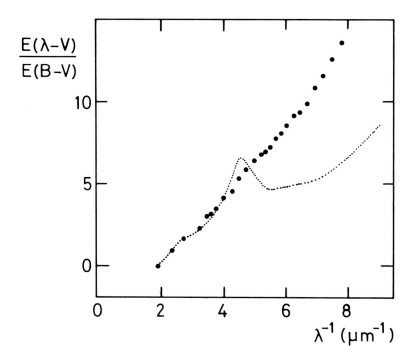

Figure 3.9   The ultraviolet extinction curve for the Small Magellanic Cloud (SMC, points), based on data from Prévot *et al* (1984) for three stars. The average for the Milky Way is also shown (dotted curve).

extinction in the sequence Galaxy–LMC–SMC. The interpretation of these results is not yet clear.

3.3.2   The Red–Infrared

A number of investigations have demonstrated that the extinction law in the spectral range 0.7–$5.0\,\mu$m is constant to within observational error (e.g. Martin and Whittet 1990; Cardelli *et al* 1989; Koornneef 1982). In contrast to the situation at shorter wavelengths, no significant differences are generally apparent between different regions in the solar neighbourhood, or between the Milky Way and the Magellanic Clouds. Apparent variations in earlier work actually arose due to the change in slope of the extinction law in the blue–visible region, affecting the differential extinction between B and V, and were thus an artifact of the customary choice of normalization to unit E_{B-V} (Clayton and Mathis 1988). This effect becomes apparent when normalization is carried out with respect to passbands with $\lambda > 0.8\,\mu$m. Figure 3.10 compares the extinction curve for the ρ Oph dark cloud with the average curve, both normalized (equation (3.19)) with $\lambda_1 = 1.25\,\mu$m

(J) and $\lambda_2 = 2.2\,\mu$m (K). A common power law is fitted to the data in the infrared (0.9–5 μm), which takes the form (analogous to equation (3.22)):

$$\frac{E_{\lambda-K}}{E_{J-K}} = 2.33\,\lambda^{-1.84} - 0.54. \tag{3.35}$$

At shorter wavelengths ($\lambda^{-1} > 1.2\,\mu$m^{-1} in figure 3.10) the curves diverge. It is well known that the ρ Oph cloud has an elevated value of the ratio of total to selective extinction in B, V normalization ($R_V \simeq 4.3$ compared with the diffuse cloud value of 3.05; e.g. Whittet $et\ al$ 1976); however, the equivalent ratio $R_K = A_K/E_{J-K}$ in J, K normalization has a common value of 0.54 ± 0.03 for both ρ Oph and the general ISM. This convergence of extinction laws in the infrared appears to be universal and independent of R_V (Martin and Whittet 1990).

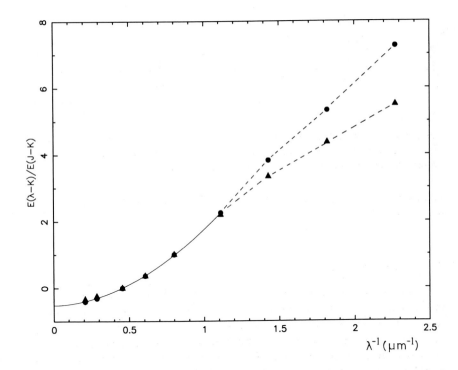

Figure 3.10   The visible–infrared extinction curve in the form $E_{\lambda-K}/E_{J-K}$ versus λ^{-1} (μm^{-1}), comparing the average interstellar medium (circles) with the ρ Oph dark cloud (triangles). The two data sets follow a common power law in the infrared (continuous curve) given by equation (3.35). Data from Martin and Whittet (1990) and references therein.

Investigation of the spectral dependence of extinction in dense molecular clouds with high visual extinction is feasible only at infrared wavelengths.

Whittet (1988) compared values of the ratio E_{J-H}/E_{H-K} in a number of dense and diffuse clouds with that towards the galactic centre. Results confirm the general uniformity of the galactic extinction curve in the infrared, and show that it extends to dense regions where mantle growth is expected to occur. This apparent immunity to environmental factors has practical applications, as a simple mathematical form (equation (3.35)) may be used to deduce the intrinsic spectral energy distributions of highly obscured infrared sources from observed fluxes.

Visual extinctions may be estimated from infrared colour excesses assuming a form for the extinction law from $0.55\,\mu$m to the spectral region of convergence. Whittet (in preparation) has deduced an empirical formula

$$A_V = r E_{J-K} \qquad (3.36)$$

where

$$r = \frac{2.332}{0.778 - 1.164\,R_V^{-1}}. \qquad (3.37)$$

For the average diffuse-cloud extinction curve, $r \simeq 5.9$, and for the average ρ Oph extinction curve, $r \simeq 4.6$. Equations (3.36) and (3.37) provide a useful method of evaluating A_V towards objects too heavily obscured to be observable at visual wavelengths.

3.4   MODELS FOR INTERSTELLAR EXTINCTION

The nature of the grains responsible for the observed extinction must be investigated by a trial and error process, in which model calculations are carried out for candidate materials with laboratory-measured optical properties and an assumed size distribution, and the results compared with the observations. In addition to reproducing the observed form of the extinction curve and its spatial variability, a viable grain model must be consistent with abundance constraints (§2.4.4) and the observed albedo and asymmetry factor (§3.2.2). In principle, the visible–infrared segment of the extinction curve may be explained by a single material with dielectric optical properties and grain sizes ranging up to values comparable with the wavelength of visible light: historically, these requirements were met by the Oort and van de Hulst ice model (§1.4). Observations in the ultraviolet led to the development of multicomponent grain models. These typically include: (i) 'large' grains which contribute to the visible–infrared extinction; (ii) 'small' grains which absorb at 2175 Å; and (iii) an additional population of small grains which account for the FUV extinction. Graphite is capable of explaining component (ii), subject to constraints on the distribution of particle sizes and shapes (§5.1), and is an essential ingredient of models proposed by a number of authors. Another common feature of

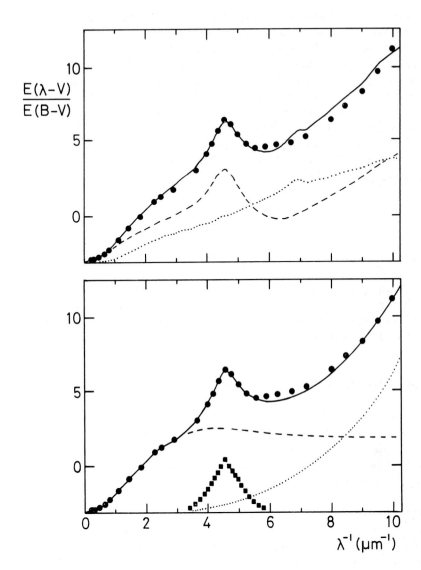

Figure 3.11   Model fits to the observed extinction curve. The mean curve (table 3.1) is plotted as solid circles, as in figure 3.3. Upper frame: the 'MRN' two-component model in the version of Draine and Lee (1984), in which graphite (dashed curve) and silicate (dotted curve) contribute to the total extinction (continuous curve). Lower frame: a more general three-component model (e.g. Greenberg 1973), in which 'classical' dielectric grains produce wavelength-dependent visual extinction and neutral ultraviolet extinction (dashed curve) and very small grains contribute the far-ultraviolet rise (dotted curve). The λ2175 feature is produced by an independent absorbing component (squares).

current models is the assumption that silicates contribute some fraction of the visible extinction (and polarization). Silicates are optically dielectric in the visible, consistent with observations of the albedo.

Figure 3.11 illustrates two attempts to fit the average extinction curve for the Galaxy. A two-component model, based on uncoated refractory particles following a power-law size distribution $n(a) \propto a^{-3.5}$, was formulated by Mathis et al (1977; MRN) and refined by Draine and Lee (1984). In this model (figure 3.11, upper frame), graphite contributes to the extinction at all wavelengths and requires $\sim 70\%$ depletion of carbon into this form of dust, compared with a minimum of $\sim 20\%$ required in small particles to explain the $\lambda 2175$ feature (§5.1.2). The absence of correlated variations in the amplitudes of $\lambda 2175$ and the FUV extinction (§3.3.1) suggests that a third component is required. In the lower frame of figure 3.11, a bimodal size distribution is assumed. The optical–IR extinction is fitted by a 'classical' component ($a \sim 0.2\,\mu$m) which is dielectric but otherwise of unspecified composition; minor fluctuations in the size distribution of this component (possibly involving the accretion of smaller grains onto larger ones) account for the changes in R_V and the slope of the NUV extinction. The $\lambda 2175$ absorption and FUV extinction rise are provided by distinct populations of very small grains ($a < 0.02\,\mu$m), which respond differently to environmental influences and thus account for morphological variations in the ultraviolet extinction curves.

The grain models currently under discussion (Duley et al 1989a; Greenberg 1989; Mathis and Whiffen 1989) are all capable of reproducing the essential features of the extinction curve over the entire spectral range available. As discussed in §1.4, the ability of a grain model to fit the extinction curve is recognized as a necessary, but not a sufficient, condition for general acceptance.

4

Interstellar Polarization and Grain Alignment

"...needle-like grains tend to spin end-over-end, like a well-kicked American football."

C Heiles (1987)

The interstellar medium is responsible for the partial plane polarization of starlight. The interstellar origin of this phenomenon is not in doubt, as, in general, only reddened stars are affected and there is a positive correlation between polarization and reddening. The accepted model for interstellar polarization is linear dichroism (directional extinction) resulting from the presence of non-spherical grains which are aligned by the galactic magnetic field. If the direction of alignment changes along the line of sight, the interstellar medium also exhibits linear birefringence, producing a component which is circularly polarized. Studies of interstellar linear and circular polarization are important because they provide information on grain properties (size, refractive index) and also on the galactic magnetic field. In this chapter, we begin by extending the discussion of extinction by small particles (Chapter 3) to include the non-spherical case. Observational results and their implications are discussed in detail in §4.2 and §4.3, and the problem of the alignment mechanism is considered in the final section. We are concerned here only with continuum polarization in the transmitted radiation from reddened stars: polarization associated with discrete spectral features and with continuum emission is discussed in Chapters 5 and 6.

4.1 EXTINCTION BY NON-SPHERICAL PARTICLES

For a non-spherical particle, the efficiency factors defined in §3.1.1 depend on the orientation of the particle with respect to both the direction of the E-vector and the direction of propagation of the radiation. Real interstellar grains may assume a wide variety of shapes, and it is necessary to generalize in order to render model calculations practicable. Elongated, axially symmetric particles are usually assumed, either prolate spheroids or 'infinite' cylinders (length $\gg$ radius). Consider the case where such a particle is orientated with its long axis, the axis of symmetry, perpendicular to the direction of propagation of the incident radiation: we may define $Q_{\parallel}$ and $Q_{\perp}$ as the values of Q_{ext} when the long axis is parallel and perpendicular to the E-vector, respectively. More generally, when the long axis of the grain is not parallel to the wavefront, $Q_{\parallel}$ and $Q_{\perp}$ are defined analogously in terms of orientation of the E-vector with respect to the plane formed by the long axis and the direction of propagation. The anisotropy in the physical shape of the particle produces a corresponding anisotropy in the extinction produced. For a specific grain model (size, shape and refractive index), $Q_{\parallel}$ and $Q_{\perp}$ may be evaluated as a function of wavelength by means of Mie theory computations (e.g. Greenberg 1968). This is illustrated in figure 4.1, which plots $Q_{\parallel}$ and $Q_{\perp}$ against $x = 2\pi a/\lambda$ for infinite cylinders of radius a and refractive index $1.33 - 0.05i$. Note that $Q_{\parallel} \geq Q_{\perp}$ for all values of x. The equivalent calculation of Q_{ext} for spheres (§3.1.1) would show a dependence on x qualitatively similar to that of the mean value $(Q_{\parallel} + Q_{\perp})/2$.

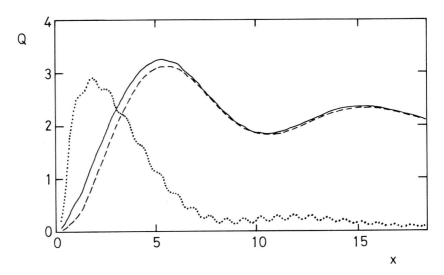

Figure 4.1   Extinction efficiency factors $Q_{\parallel}$ (continuous curve) and $Q_{\perp}$ (dashed curve) plotted against $x = 2\pi a/\lambda$, using Mie calculations for infinite cylinders of radius a and refractive index $1.33 - 0.05i$. The dotted curve is the difference $(Q_{\parallel} - Q_{\perp})$ scaled up by a factor of 8.

Considering parallel and perpendicular cases in turn, the extinction produced by a medium containing identical, perfectly aligned particles of column density N_d is

$$A_{\parallel} = 1.086 N_d \sigma Q_{\parallel}$$
$$A_{\perp} = 1.086 N_d \sigma Q_{\perp} \tag{4.1}$$

by analogy with equation (3.6), where σ is the cross-sectional area of a particle in the plane of the wavefront. A measure of the degree of linear polarization in magnitude units is then given by the quantity

$$p = A_{\parallel} - A_{\perp}$$
$$= 1.086 N_d \sigma (Q_{\parallel} - Q_{\perp}), \tag{4.2}$$

and thus $p \propto \Delta Q$ for a given grain size, where $\Delta Q = Q_{\parallel} - Q_{\perp}$. An example of the dependence of ΔQ on x is shown in figure 4.1. Considering cylinders of constant radius a, we note that polarization becomes vanishingly small at wavelengths long compared with a ($x \to 0$), as expected from the behaviour of extinction; however, in contrast to extinction, the polarization also becomes very small at wavelengths short compared with a ($x > 8$ in this example). A peak in ΔQ appears at an intermediate value of x, and, comparing results for dielectric cylinders of differing refractive index n, this is found to occur when $x(n-1) \sim 1$.

Figure 4.1 also illustrates that polarization is, in general, small compared with extinction ($Q_{\parallel} \sim Q_{\perp} \gg \Delta Q$). At a given wavelength, the ratio of polarization to mean extinction is given by

$$\frac{p}{A} = 2 \left\{ \frac{Q_{\parallel} - Q_{\perp}}{Q_{\parallel} + Q_{\perp}} \right\} \tag{4.3}$$

where A is $(A_{\parallel} + A_{\perp})/2$. The quantity p/A is often referred to as the alignment efficiency, although it depends on the nature of the grains and the viewing geometry as well as on the efficiency with which they are aligned. The most efficient polarizing medium conceivable would contain infinite cylinders with diameters comparable to the wavelength considered, perfectly aligned such that their long axes are parallel to one another and perpendicular to the line of sight. Mie calculations for such a model place a theoretical upper limit on the polarization efficiency; at visual wavelengths, particles with dielectric optical properties give

$$\frac{p_V}{A_V} \leq 0.3. \tag{4.4}$$

The corresponding observational result is deduced in the following section.

4.2   VISUAL POLARIMETRY OF REDDENED STARS

In its simplest form, an astronomical polarimeter may be considered as a photoelectric photometer with a broadband filter and an analyser in the light path†. In this section, we consider observations obtained with a single colour filter such as the Johnson V. When a reddened star is observed, intensity maxima and minima are recorded in orthogonal directions as the analyser is rotated, and the degree of polarization, P, expressed as a percentage, is defined by

$$P = 100 \left\{ \frac{I_{max} - I_{min}}{I_{max} + I_{min}} \right\}. \qquad (4.5)$$

An alternative definition is the polarization in magnitudes, denoted by the lower-case symbol

$$p = 2.5 \log \frac{I_{max}}{I_{min}}. \qquad (4.6)$$

This latter quantity is equivalent to the polarization defined in terms of model-dependent parameters in equation (4.2). We may easily show that P is proportional to p to a close approximation if the polarization is sufficiently small: from equation (4.6),

$$\frac{I_{max}}{I_{min}} \simeq 1 + \left\{ \frac{\ln 10}{2.5} \right\} p,$$

neglecting p^2 and higher powers; with $I_{max} \simeq I_{min}$ in equation (4.5),

$$P \simeq 50 \left\{ \frac{I_{max}}{I_{min}} - 1 \right\}$$
$$\simeq 46.05p. \qquad (4.7)$$

This approximation is generally valid for interstellar polarization. In addition to the degree of polarization, the position angle of the measurement is determined by the orientation of the analyser for maximum intensity, relative to some reference direction. It is convenient to select a reference with respect to the orientation of the Galaxy: the galactic position angle (θ_G) is defined formally as the angle between the plane of vibration of the $\boldsymbol{E}$-vector projected on to the celestial sphere and the great circle from the star to the North Galactic Pole.

The degree of polarization in the visual waveband shows a distinct, but by no means perfect, correlation with reddening, as shown in figure 4.2.

---

† For a discussion of modern instrumentation, see (e.g.) Hough *et al* (1991).

The scatter is much greater than can be accounted for by observational errors alone, and shows that the polarization efficiency of the interstellar medium is intrinsically non-uniform. The points in figure 4.2 are bounded by the straight line

$$\frac{P_V}{E_{B-V}} = 9.0\% \text{ mag}^{-1} \tag{4.8}$$

representing optimum alignment. This may be expressed as an observational upper limit for the ratio of polarization to extinction

$$\frac{p_V}{A_V} \leq 0.064 \tag{4.9}$$

using equations (4.7) and (3.23). This result is consistent with the theoretical upper limit in equation (4.4). If real interstellar grains resembled infinite cylinders, the alignment efficiency would not need to be high to explain the observed polarization; more realistically, irregular or moderately elongated particles may suffice if alignment of the longest axes is fairly efficient.

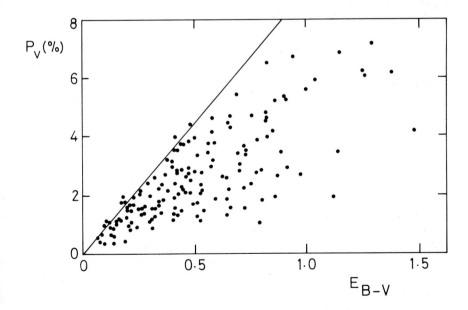

Figure 4.2   A plot of visual polarization (P_V) against reddening (E_{B-V}) for a typical sample of field stars in the Milky Way. The straight line represents optimum alignment efficiency (equation (4.8)). Data are from Serkowski *et al* (1975) and references therein.

The distribution of polarization vectors with respect to galactic coordinates has been investigated by a number of authors (Axon and Ellis 1976 and references therein), of which the most extensive data sets are those published by Mathewson and Ford (1970) and Klare *et al* (1972). Figure 4.3 illustrates results for two distinct distance régimes. As expected from our discussion of the distribution of reddening material (§1.1.2), the stars with strongest interstellar polarizations are to be found either behind relatively nearby dark clouds, including some at intermediate galactic latitude, or at greater distances within a few degrees of the galactic plane. The upper frame of figure 4.3 shows the effect of dust towards stars within 400 pc of the Sun: considerable structure is evident, such as a loop extending northwards from the galactic plane at $\ell \sim 30°$. The most highly polarized stars tend to lie North of the plane towards the galactic centre ($\ell = 0°$) and South of the plane towards the anticentre ($\ell = 180°$) consistent with an origin in dust associated with Gould's Belt (see §1.1.2 and §4.3.4). The lower frame in figure 4.3 plots data for much more distant stars (2–4 kpc) reddened by the cumulative effect of many clouds. Here, the polarization vectors show a tendency to align parallel to the galactic plane in certain longitude zones (e.g. near $\ell \simeq 120°$ and $\ell \simeq 300°$), and to be randomly orientated in others ($\ell \simeq 15°$, $75°$, and $265°$). This effect appears to be related to galactic structure.

It may be deduced from inspection of figure 4.3 that if an average of the position angle θ_G were taken for different zones of longitude, its standard deviation would vary systematically with ℓ. However, in order to investigate the variation in alignment with galactic coordinates statistically, it is preferable to adopt quantities which depend on both P and θ_G, namely, the Stokes parameters Q and U. For partially plane-polarized light, the four Stokes parameters are given by (e.g. Hall and Serkowski 1963):

$$I = I_{max} + I_{min}, \tag{4.10}$$

$$Q = PI \cos 2(\theta_G - 90), \tag{4.11}$$

$$U = PI \sin 2(\theta_G - 90), \tag{4.12}$$

$$V = 0. \tag{4.13}$$

V/I denotes the circular polarization, which does, in reality, have a small but finite value for many lines of sight in the interstellar medium (§4.3.2). Q and U may also be expressed in magnitude units,

$$q = p \cos 2(\theta_G - 90), \tag{4.14}$$

$$u = p \sin 2(\theta_G - 90), \tag{4.15}$$

and evaluated from measured values of p and θ_G. The behaviour of the alignment with respect to galactic structure may be described conveniently

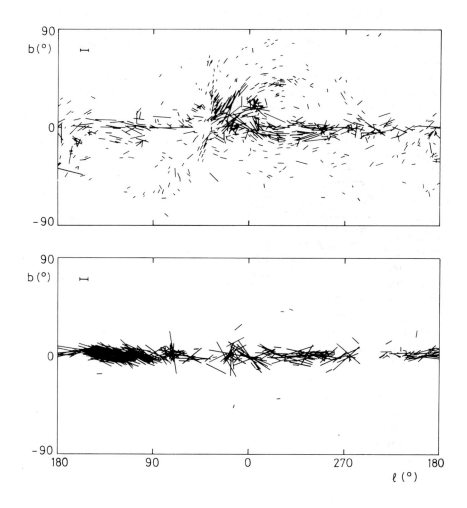

Figure 4.3   Distribution of polarization vectors in galactic coordinates (ℓ, b), based on data from the catalogue of Axon and Ellis (1976). Upper frame: nearby stars $(d < 400\,\mathrm{pc})$, illustrating structure in the magnetic field associated with local clouds. Lower frame: stars in the distance interval $2 < d < 4\,\mathrm{kpc}$, illustrating the effect of alignment averaged over many clouds. The length of the bar at top left in each frame indicates 1% polarization. Only stars with $P_V > 0.2\%$ are plotted.

in terms of the dependence of $\langle q \rangle$, the mean value of q (equation (4.14)) for stars in a given region of the Milky Way, on galactic longitude ℓ. For alignment with the $\boldsymbol{E}$-vector predominantly parallel to the galactic plane $(\theta_G \simeq 90°)$, we expect $\langle q \rangle$ to be positive and similar in magnitude to the mean value of p, whereas for random orientation, $\langle q \rangle \sim 0$, and $\langle q \rangle$ would be negative for net alignment perpendicular to the galactic plane. Figure 4.4 plots $\langle q \rangle$ against ℓ for various longitude zones along the Milky Way. A systematic variation is apparent in $\langle q \rangle$ between zero and positive values, the sense of which is loosely represented by a sine wave with minima at $\ell \sim 50°$ and $230°$, directions which are roughly parallel to the Cygnus–Orion spiral arm. This dependence relates to the net direction of the magnetic field responsible for alignment. As discussed in §4.4, the grains tend to align such that their longest axes are perpendicular (and hence the **E**-vectors parallel) to the magnetic field, and we have $\langle q \rangle \geq 0$ for a net field direction parallel to the galactic plane. A peak in $\langle q \rangle$ occurs for directions which cross field lines, whereas $\langle q \rangle \simeq 0$ for directions along field lines, suggesting a field orientation parallel to the local spiral arm in the solar neighbourhood of the Milky Way. These results are consistent with data for external galaxies (see Sofue *et al* 1986 for a review). A good example of similar Hubble type to the Milky Way is the edge-on spiral NGC 4565, in which the **E**-vectors are observed to be aligned parallel to the long axis of the optical dust lane, and thus the average magnetic field direction is likely to be parallel to the disk of the galaxy (Jones 1989a).

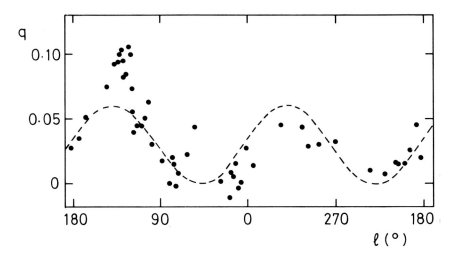

Figure 4.4   Plot of Stokes parameter q (equation (4.14)) against galactic longitude ℓ. Each point represents an average of approximately 25 stars, and only stars with $b < 3°$ and $d > 600\,\mathrm{pc}$ are included. A sine wave with minima at $\ell = 50°$ and $\ell = 230°$ is shown (adapted from Hall and Serkowski 1963).

Although the macroscopic properties of interstellar polarization in the Milky Way within a few kiloparsecs of the Sun are described adequately in terms of a unidirectional magnetic field, this apparent uniformity breaks down when path-lengths comparable with the separations of individual clouds are considered. Regional variations occur in the alignment efficiency (P/A). These may be due to changes in magnetic field strength or in grain properties in different environments: examples of physical processes which tend to reduce alignment efficiency in dense clouds include mantle growth (due to accretion from the gas) and increased grain–grain collision rates. However, apparent variations in P/A also occur due to geometrical effects. To illustrate this, consider the idealized situation where initially unpolarized radiation from a distant star is partially plane-polarized by transmission through a single cloud with uniform grain alignment; the beam then encounters a second cloud of similar optical depth and uniform alignment in a different direction. The emergent light is in general elliptically polarized, the interstellar medium behaving as an inefficient waveplate. This effectively introduces a weak component of circular polarization and causes depolarization of the linear component. In the extreme case where the alignment axes of the two clouds are orthogonal, high extinction can be produced with no net polarization. Real interstellar clouds frequently exhibit complex magnetic field structure (Vrba *et al* 1976); thus, changes in alignment geometry along a line of sight can arise because of twisted magnetic field lines within a single cloud, resulting in a reduction in the apparent efficiency of alignment. In view of these diverse effects, it is not unexpected that plots of linear polarization against reddening for heterogeneous groups of stars, as in figure 4.2, show scatter considerably in excess of observational errors. The straight line of equation (4.8) represents optimum alignment efficiency for a uniform magnetic field perpendicular to the line of sight.

Studies of interstellar polarization at visible wavelengths are limited by sensitivity considerations to lines of sight towards stars with modest degrees of extinction ($A_V < 10$ mag). In order to investigate the behaviour of the magnetic field deep within molecular clouds and at large distances within the galactic plane, measurements at infrared wavelengths are needed. Extensive data are available in the K ($2.2\,\mu$m) passband for a variety of galactic sources (Jones 1989b and references therein). At this wavelength the extinction is a factor of 11 less than in the visual (table 3.1). Figure 4.5 plots P_K against optical depth τ_K (where $\tau_K = A_K/1.086 \simeq 0.084A_V$). Approximately 90 sources are included, covering a total range in optical depth of more than a factor of 100. The sample includes moderately reddened stars with $0.05 < \tau_K < 1.0$, which are also represented in the visual sample (figure 4.2), sources associated with the galactic centre ($\tau_K \sim 3$), and young stellar objects embedded in molecular clouds ($1 < \tau_K < 11$). The solid curve in figure 4.5 is derived from equation (4.8), transformed to P_K versus τ_K, and, as in figure 4.2, this provides an upper bound to the distribution of points representing optimum alignment. In view of the diversity of environments sampled, the level of correlation in figure 4.5 is remarkable, suggesting that the alignment mechanism is generally effective to a comparable degree in both dense and diffuse clouds.

The interstellar magnetic field may be described as consisting of a uniform component and a random component (e.g. Heiles 1987 and references therein). The uniform component represents the general galactic magnetic field, upon which the random component is superposed. The latter is attributed to local structures in the magnetic field of scale size $\sim 100\,\mathrm{pc}$, possibly associated with old supernova remnants. The general trend of P_K with τ_K is consistent with a model based on this hypothesis (Jones 1989b; dashed curve in figure 4.5). In this model, the optical path is divided into sequential segments $\Delta\tau_K = 0.1$; in each segment, the magnetic field has a uniform component (constant position angle for all segments) and a random component (allowed to take any position angle). Vector addition of the two components results in a net field strength and direction. The model illustrated in figure 4.5 assumes that the grains have alignment efficiency given by $Q_\perp/Q_\parallel = 0.9$ and that the magnitude of the random component of the magnetic field is 45% of the total for each segment. At low optical depth (few segments), the random component strongly affects the net polarization. As many segments are accumulated, the random component eventually averages out to a small net effect, and the polarization is dominated by the uniform component. The model implies that the typical optical depth interval over which the galactic magnetic field changes geometry is $\tau_K \simeq 0.05$–0.10 (equivalent to $A_V \simeq 0.5$–1.1), as increments for $\Delta\tau_K$ outside this range produce significantly poorer fits. In the general ISM, this extinction corresponds to path-lengths of a few hundred parsecs, in agreement with other estimates of the scale length for variations in the galactic magnetic field (Heiles 1987).

4.3   THE SPECTRAL DEPENDENCE OF POLARIZATION

4.3.1   Linear Polarization and the Serkowski Law

When the degree of polarization is measured through a number of passbands, systematic variations with wavelength are apparent. The spectral dependence of linear polarization (the polarization curve, usually plotted as P_λ versus λ^{-1}) has been well-studied at visible and near-infrared wavelengths, displaying a broad, asymmetric peak in the visible for most stars. The wavelength of maximum polarization, λ_{max}, varies from star to star, and is typically in the range 0.4 to 0.8 μm with a mean value of 0.55 μm. If P_λ and λ are normalized with respect to P_{max} and λ_{max} (where P_{max} is the

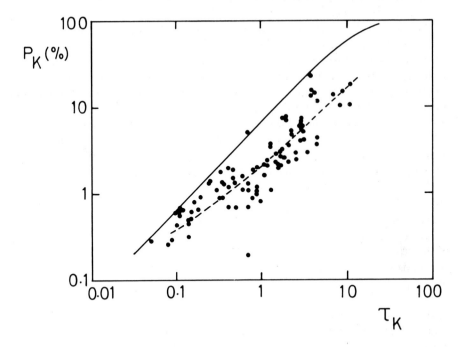

Figure 4.5   Plot of P_K against τ_K, using data from Jones (1989b). The solid curve represents optimum alignment (the equivalent form of equation (4.8)) and the dashed curve represents the model discussed in the text.

value of P_λ at $\lambda = \lambda_{max}$), an empirical formula may be used to describe the polarization curve in a general form:

$$\frac{P_\lambda}{P_{max}} = \exp\left\{-K \ \ln^2\left(\frac{\lambda_{max}}{\lambda}\right)\right\} \tag{4.16}$$

(Serkowski 1973; Coyne, Gehrels and Serkowski 1974; Serkowski, Mathewson and Ford 1975). The parameter K, which determines the width of the peak in the curve (figure 4.6), was originally taken to be constant with a value of $K = 1.15$: we refer to equation (4.16) with this value of K as the Serkowski law. This form generally provides an adequate description of the observed polarization in the visible region, but Codina-Landaberry and Magalhães (1976) noted that the goodness of fit may be further improved by allowing K to vary. Extension of the wavelength coverage to $2.2\,\mu$m in the infrared showed discrepancies at wavelengths greater than $1\,\mu$m (Dyck and Jones 1978) which led to a refinement of the empirical law (Wilking et al 1980, 1982; Whittet et al 1992). With K treated as a free parameter, least-squares fits of equation (4.16) to data for stars with a range of λ_{max}

show that K and $\lambda_{\max}$ are linearly related:

$$K = c_1\lambda_{\max} + c_2 \qquad (4.17)$$

where the current best values for the slope and intercept are $c_1 = 1.66 \pm 0.09$ and $c_2 = 0.01 \pm 0.05$ (Whittet *et al* 1992). We refer to equation (4.16) with K given by equation (4.17) as the Wilking law. The dependence of K on $\lambda_{\max}$ implies a systematic decrease in the width of the polarization curve with increasing $\lambda_{\max}$, as illustrated in figure 4.6, comparing data and empirical fits for two stars with contrasting $\lambda_{\max}$ values. The Wilking law provides a consistent representation of the systematic changes occurring within individual regions with rather different environments (Whittet *et al* 1992).

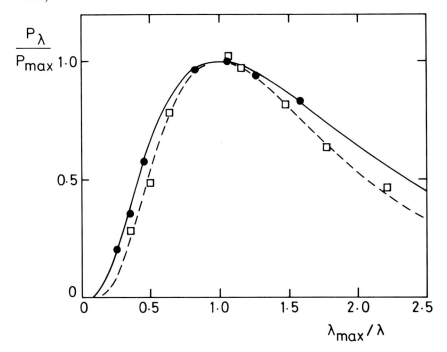

Figure 4.6   Normalized interstellar linear polarization curves for two stars with contrasting values of $\lambda_{\max}$. Observational data (points) are compared with empirical fits based on the Wilking law for HD 42087 ($\lambda_{\max} = 0.57\,\mu$m, filled circles and continuous curve), and HD 147889, ($\lambda_{\max} = 0.80\,\mu$m, open squares and dashed curve). Data are from Wilking *et al* (1980) and references therein.

Whilst the mathematical representation of the polarization law provided by equations (4.16) and (4.17) has practical applications (allowing, for example, accurate determination of $P_{\max}$ and $\lambda_{\max}$ by least-squares fits to broadband polarimetry), it is not clear that it has physical significance. As discussed below, λ_{max} is related to the average size of the polarizing grains, and the trend of K with λ_{max} might be interpreted in terms of systematic changes in sphericity or a narrowing of the size distribution of the grains as the size increases in response to environmental influences. However, the form of the correlation could also depend on purely subjective factors such as the number and choice of filter wavelengths in the data set, and the accuracy of the measurements (Clarke and Al-Roubaie 1983; Clarke 1984). Predictably, the spectral ranges over which the Serkowski and Wilking laws represent good fits to the data correspond to the spectral ranges of the data available at the time of their formulation (0.35–1.0 μm and 0.35–2.2 μm, repectively), an effect illustrated in figure 4.7 for the reddened supergiant Cyg OB2 no.12. It should be noted that this star is exceptional in having an unusually low value of λ_{max} which tends to emphasize the difference between the Serkowski and Wilking fits.

Recent observations at longer infrared wavelengths ($> 3 \mu$m) than those available to Wilking et al show significant departures from levels of polarization predicted by an extrapolation of the Serkowski/Wilking formulae (Martin 1989; Nagata 1990; Martin and Whittet 1990). As discussed in §3.2.3, the infrared $extinction$ curve is well-described by a power law, and, as polarization is differential extinction, a similar dependence for polarization might be expected if the same grain population is responsible. Observations in the 2–5 μm region tend to confirm this. Assuming a form

$$P_\lambda = P_1 \lambda^{-\alpha} \qquad (4.18)$$

for $\lambda > 2 \mu$m, where P_1 is constant for a given line of sight, Martin and Whittet (1990) find evidence for a common power law of index $\alpha = 1.8 \pm 0.2$, consistent with data for both diffuse and dense clouds: power-law fits to infrared data for Cyg OB2 no.12 and the BN object in Orion (representative of the two environments) are illustrated in figures 4.7 and 4.8, respectively. The latter shows peaks associated with ice and silicate absorption features at 3.0 μm (0.33 μm^{-1}) and 9.7 μm (0.10 μm^{-1}), discussed in §5.3; the power law closely matches the underlying continuum polarization.

The interstellar polarization curve may be modelled with reasonable precision by means of Mie theory computations for dielectric cylinders or spheroids aligned by the galactic magnetic field (e.g. Mathis 1979, 1986; Hong and Greenberg 1980). The general form of the polarization curve is consistent with dielectric optical properties for the aligned grains. It is difficult to reproduce the observed spectral dependence (e.g. figure 4.6), with its smooth peak in the visible and steady decline in the infrared, using a material in which the refractive index varies strongly with wavelength (see Martin 1978), a result which discriminates against metals and other strong absorbers such as graphite and magnetite. The value of the albedo deduced from its scattering properties (§3.2.2) leads us to conclude that at least one

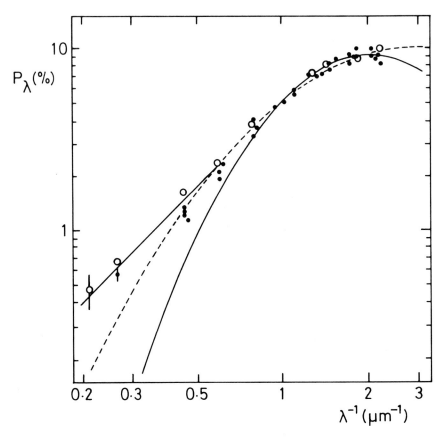

Figure 4.7   The polarization curve for Cygnus OB2 no.12, compared with empirical fits. Observational data are from Nagata (1990) and references therein (small filled circles) and Martin *et al* (1992, open circles). The parameters of the fits are: Serkowski law (continuous curve, McMillan and Tapia 1977) $P_{max} = 9.2\%$, $\lambda_{max} = 0.49\,\mu m$, $K = 1.15$; Wilking law (dashed curve, Wilking *et al* 1980) $P_{max} = 10.2\%$, $\lambda_{max} = 0.33\,\mu m$, $K = 0.57$; power law (straight line, equation (4.18)) $P_1 = 5.06\%$, $\alpha = 1.6$.

component of the dust is dielectric at visible wavelengths; the dielectric nature of the grains responsible for polarization is confirmed by results for circular polarization (§4.3.2 below), and by the presence of polarization enhancement in the 9.7 μm feature (§5.3.2) identified with silicates (which are optically dielectric). On this basis, strongly absorbing materials are excluded from models for the polarization: if they contribute to the visible extinction, it must be assumed that they are tied up in particles which do not produce net polarization, due either to sphericity or failure to align.

A specific attempt to fit the observed linear polarization is illustrated

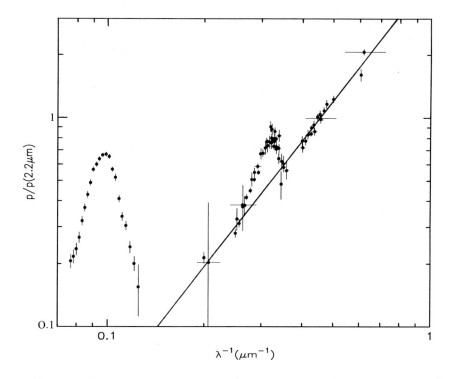

Figure 4.8   The interstellar polarization curve for the Becklin–Neugebauer (BN) infrared source in the Orion Nebula (Martin and Whittet 1990 and references therein). The peaks at 0.33 and 0.10 μm^{-1} are associated with ice and silicate absorption features. A power law of index $\alpha = 2$ (line) is fitted to the continuum polarization.

in figure 4.9. This is based on the MRN grain model (Mathis *et al* 1977), assuming that polarization is produced only by silicates with a power-law size distribution $n(a) \propto a^{-3.5}$ (§3.4). Mathis (1979) showed that a good fit is obtained for infinite cylinders, provided that only those with radii above a certain threshold value are assumed to be aligned. Mathis (1986) developed this model by proposing that the probability of alignment increases with the volume of the grain (due to the presence of ferromagnetic inclusions), providing a physical basis for the assumption that only the larger grains tend to be aligned. In figure 4.9, the observed mean polarization is represented by the empirical curve in the Wilking form, and this is compared with the calculated values at specific wavelengths for prolate spheroids (Mathis 1986). It should be noted that this fit in only valid in the range shown (from the near ultraviolet to 2.2 μm in the infrared). As discussed above, the Wilking curve is no longer a fair representation of the observed polarization at longer wavelengths (e.g. figure 4.7); no model has yet been proposed which reproduces the $\sim \lambda^{-1.8}$ dependence of polarization observed at $\lambda > 2\,\mu$m, and this will be an important test for future models.

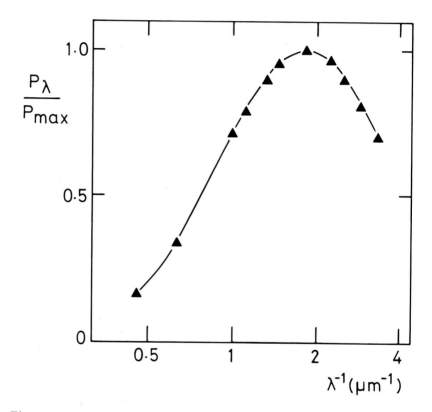

Figure 4.9   A model for interstellar linear polarization in the spectral range 0.3 to 2.2 μm. The curve is the empirical Wilking law fit to observational data for $\lambda_{max} = 0.55\,\mu$m. The triangles are calculated values of the polarization produced by aligned prolate silicate spheroids (Mathis 1986). All data are normalized to unity at peak polarization.

Mie theory computations also demonstrate that the wavelength of maximum polarization is related to grain dimensions. Dielectric cylinders of radius a and refractive index n polarize most efficiently when the quantity $2\pi a(n-1)/\lambda$ is close to unity (see §4.1), and thus

$$\lambda_{max} \simeq 2\pi a(n-1). \qquad (4.19)$$

As noted above, the range of grain sizes which contribute to the polarization may be restricted by the alignment mechanism. If a in equation (4.19) is taken to represent the mean 'characteristic' size, $\langle a \rangle$, of the grain population which produces polarization, grains much smaller than $\langle a \rangle$ are unlikely to be aligned in the Mathis model, and grains much larger than $\langle a \rangle$ are rare (due to the form of the size distribution). In the diffuse ISM, $\lambda_{max} \simeq 0.55\,\mu m$, and hence $\langle a \rangle \simeq 0.15\,\mu m$ for $n = 1.6$ (typical of silicates). Increases in λ_{max} in dark clouds (§4.3.4) are usually interpreted simply in terms of grain growth (e.g. Whittet and Blades 1980), but if the growth process involves the accretion of mantles from the gas, changes in the refractive index are also likely to occur.

4.3.2   Circular Polarization

The circular polarization produced by interstellar birefringence is weak ($|V/I| < 10^{-3}$), and consequently difficult to measure. Complementary investigations of the spectral dependences of linear and circular polarization in the same lines of sight have great potential for the elucidation of grain properties. Results for circular polarization have been published for a number of stars (Avery $et\ al$ 1975, Martin and Angel 1976, and references therein), but few have wide spectral coverage. For those in which the data are sufficiently extensive, V/I is found to vary strongly with wavelength, exhibiting opposite handedness in the blue and red spectral regions. A typical example is shown in figure 4.10, which compares observations of linear and circular polarization for o Sco. V/I changes sign at a distinct wavelength λ_c (the cross-over wavelength), which is close to the peak wavelength for linear polarization. The value of λ_c is determined reliably in only six lines of sight (Martin and Angel 1976; McMillan and Tapia 1977), and is found to be identical to λ_{max} to within observational error:

$$\lambda_{max}/\lambda_c = 1.00 \pm 0.05. \qquad (4.20)$$

This result provides a significant constraint on the composition of the polarizing grains, as Martin (1974) has shown that the ratio λ_{max}/λ_c is sensitive to k, the imaginary component of the grain refractive index (equation (3.10)). Theory predicts a value of unity ($\lambda_{max} = \lambda_c$) for $k = 0$, as illustrated by the model in figure 4.10. As the absorption increases ($k > 0$), λ_c is predicted to increase relative to λ_{max}: the precise relation depends on the real as well as the imaginary part of the refractive index, but, typically, λ_{max}/λ_c is reduced to ~ 0.9 for $k = 0.05$ and to ~ 0.7 for $k = 0.3$ (see Aannestad and Greenberg 1983). Thus, the existing observations suggest $k < 0.03$, i.e. the aligned grains are good dielectrics.

Although the evidence for dielectric grains appears to be strong, a notable exception has been found in the case of the reddened supergiant HD 183143, which has a λ_{max} value of $0.56\,\mu m$, and no evidence for a cross-over in V/I in the wavelength range 0.35–$0.8\,\mu m$ (Michalsky and Schuster

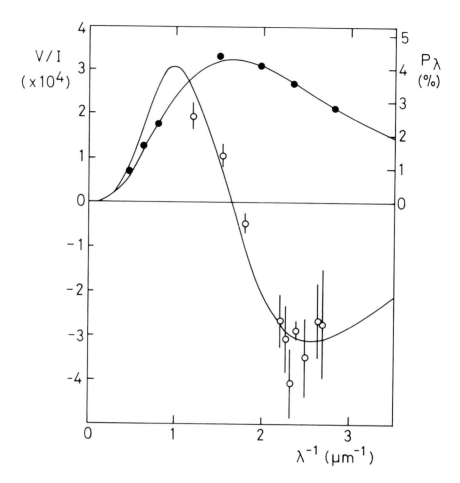

Figure 4.10   Linear and circular polarization curves, comparing observational data for *o* Scorpii (Wilking *et al* 1980 and references therein) with a theoretical model (curves) for grains with $m = 1.50 - 0i$ (Martin 1974). Circular polarization (left-hand scale): open circles with error bars; linear polarization (right-hand scale): solid circles.

1979). It is unlikely that this anomaly is caused by circular polarization intrinsic to the star, and a peculiar grain composition may be implied.

Measurements of circular polarization towards a much larger sample of stars are required to exploit the potential of this technique for studying environmental variations in grain properties. Observations are needed for stars exhibiting a wide spread in λ_{max} of linear polarization: this should allow an assessment of the extent to which changes in optical properties of the grains, normally attributed to biasing of the grain size distribution, are influenced by changes in composition.

4.3.3   The Relation between Polarization and Extinction Curves

Variations in the ratio of total-to-selective extinction, R_V, occur predominantly as a result of changes in the optical properties of the grains in the blue–visible region of the spectrum (see §3.3). These variations are likely to be caused by changes in the size distribution of the grains in response to environmental influences, and Mie calculations indicate that increases in R_V may accompany a reduction in the relative number of smaller grains. It follows that correlated changes in the polarization and extinction curves, characterized by the behaviour of the parameters λ_{max} and R_V, are expected if the polarizing grains contribute significantly to visual extinction. Such a correlation is, indeed, observed (figure 4.11). Making use of the colour excess ratio E_{V-K}/E_{B-V} (equation (3.24)) to represent R_V, Serkowski *et al* (1975) found an approximately linear dependence on λ_{max}, but with scatter in excess of observational errors. Whittet and van Breda (1978) and Clayton and Mathis (1988) have tightened the correlation by means of improved infrared photometry and a selection procedure which discriminates against shell stars. As discussed in §3.1.3, erroneous values of R_V will result for stars with fluxes contaminated by circumstellar infrared emission. Intrinsic polarization may also be produced by scattering in the circumstellar shell, although, in general, λ_{max} is probably a more reliable grain size parameter than R_V. Omitting stars known to have shell characteristics such as optical emission lines, a linear fit to the data set of Whittet and van Breda (1978) gives

$$R_V = (5.6 \pm 0.3)\lambda_{max} \qquad (4.21)$$

(with λ_{max} in μm). Thus, $\lambda_{max} = 0.55\,\mu$m corresponds to $R_V = 3.05$, the 'normal' average values of these parameters applicable to the ISM outside molecular clouds (represented by the cross in figure 4.11). The other points in figure 4.11 are based on data for individual stars obscured by denser clouds: R_V and λ_{max} tend to rise to higher values in such regions, where the mean size of the dust grains is apparently larger by up to $\sim 70\%$ (see equation (4.19)). The fact that λ_{max} and R_V respond in a similar way to environmental influences strongly supports the view that the grains responsible for the polarization are also responsible for at least part of the visual extinction. McMillan (1978) has calculated the theoretical relationship between R_V and λ_{max} for simple grain models, and, as expected from other lines of evidence discussed above, dielectric grains give the best fit to the observations (see figure 4.11).

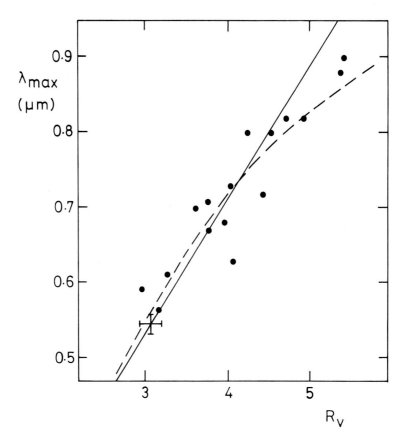

Figure 4.11   Correlation of $\lambda_{\max}$ with R_V. The cross is the mean value for 60 galactic field stars with diffuse-cloud extinction, and the circles represent lines of sight to individual stars in dense clouds (data from Vrba *et al* 1981; Whittet and Blades 1980 and references therein). The continuous line is the linear relation (equation (4.21)) deduced by Whittet and van Breda (1978), and the dashed curve represents a model for the growth of dielectric grains (McMillan 1978).

4.3.4   Regional Variations in the Galaxy

It follows from the above discussion that multi-waveband polarimetry yielding $\lambda_{\max}$ values provides a potentially powerful method of studying spatial variations in the grain size distribution. In this section, we examine the evidence for both macroscopic and cloud-to-cloud variations.

The possibility of systematic, macroscopic changes in the optical properties of the polarizing dust has been investigated by Whittet (1977, 1979). As noted in Chapter 1, variations in the extinction curve could be a serious source of error in distance measurements (via the parameter R_V), with implications for galactic structure, as originally discussed by Johnson (1968 and references therein). Gross variations on the scale proposed by Johnson can now be ruled out, but a significant second-order effect is apparent in data for both R_V and $\lambda_{\max}$. The latter is illustrated in figure 4.12, which plots $\lambda_{\max}$ against galactic longitude, each point representing the mean for a region of the Milky Way. There is evident variation with a 360° period, which may be described by the empirical law

$$\lambda_{\max} = 0.545 + 0.045\sin(\ell + 175°).\qquad(4.22)$$

The equivalent variation in R_V may be deduced from equation (4.21): the range is 2.8–3.3, which will have a small effect on photometric distance determinations for reddened stars (comparable with errors from other sources).

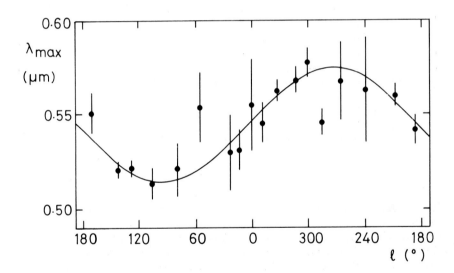

Figure 4.12   Plot of mean $\lambda_{\max}$ against galactic longitude for 18 regions of the Milky Way (Whittet 1977). The sinusoidal curve in equation (4.22) is fitted to the data.

Systematic variations in R_V have been proposed which are related to the alignment of the grains (Wilson 1960; Greenberg and Meltzer 1960; Rogers and Martin 1979). However, the variation in $\lambda_{\max}$ (figure 4.12) is clearly different from that in q (figure 4.4), which has a 180° rather than a 360° period, ruling out the alignment mechanism as the source of the variations. Omission of stars more distant than 1 kpc increases the amplitude of the variation in $\lambda_{\max}$ (Whittet 1979), suggesting that the phenomenon is peculiar to the solar neighbourhood. Its origin may be understood by considering the distribution of individual high-λ_{max} stars with respect to galactic coordinates (figure 4.13). The greatest concentrations of stars with high λ_{max} are to be found towards Orion ($\ell = 208°$, $b = -18°$) and Scorpius–Ophiuchus ($\ell = 353°$, $b = 20°$). These are members of Gould's Belt, the local sub-system of early-type stars. The direction of the centre of the system is shown by a cross in figure 4.13: this coincides precisely with the direction of peak λ_{max} value ($\ell = 270°$) in the galactic plane (figure 4.12). Let us assume that the local ISM contains two populations of grains, 'normal' interstellar grains with $\lambda_{max} \simeq 0.5\,\mu m$, distributed in the galactic plane, and large grains with $\lambda_{max} > 0.6\,\mu m$, distributed in a disk tilted at 18°. The proportion of grains from each population in the column to a distant star in the galactic plane changes with direction. Because the Sun lies towards the edge of Gould's Belt, roughly along the line of intersection of the two planes, the resulting variation in λ_{max} with ℓ (figure 4.12) has a 360° period. As Gould's Belt contains regions of recent star formation, it is probable that the attendant grain population has been processed more recently through molecular clouds than the average interstellar population.

The behaviour of λ_{max} has been investigated in several regions of current or recent star formation. Table 4.1 presents a statistical summary of available data for dark clouds (R CrA, ρ Oph and Chamaeleon I), H II regions (M17 and the Orion Nebula) and open clusters/associations (NGC 7380, Cyg OB2 and the α Per cluster), compared with the diffuse interstellar media in the Galaxy and the Large Magellanic Cloud. Corrections have been applied, where necessary, for foreground polarization. The regions are listed in order of descending mean λ_{max}. There is evidence that this represents an evolutionary sequence; for example, Vrba *et al* (1981) consider the α Per cluster (age $\sim 20\,\mathrm{Myr}$) to be an older analogue of the R CrA cloud (age $< 1\,\mathrm{Myr}$). In general, λ_{max} is highest in the dark clouds, intermediate in the H II regions, and close to the interstellar value in the young clusters. The standard deviation of λ_{max} tends to increase with the mean, indicating a greater spread in λ_{max} (from normal values up to about $1\,\mu m$) in the less evolved regions. There is also a tendency for λ_{max} to increases with A_V in some dark clouds, supporting the view that grain growth is most efficient in the densest, most shielded regions. The effect of subsequent star formation is to return the mean grain size to its normal interstellar value, as indicated by the similarity in table 4.1 between the ISM value and the mean values for the three clusters.

4.3.5   Other Galaxies

Perhaps the most remarkable statistic in table 4.1 is the close agreement in the mean value of λ_{max} comparing the ISM in the Milky Way with that

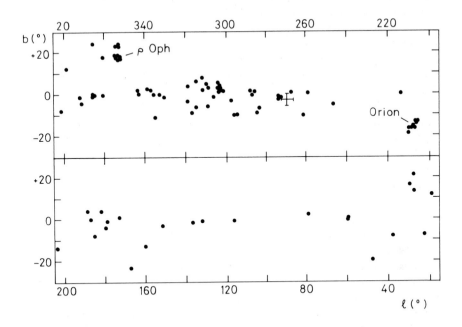

Figure 4.13   The distribution of stars with high λ_{max} ($\geq 0.58\,\mu$m) in galactic coordinates. The cross marks the direction of the centre of Gould's Belt (Whittet 1979).

in the Large Magellanic Cloud: Clayton *et al* (1983) conclude from a detailed study of the polarization curves towards several LMC stars that the aligned grains in the two galaxies have rather similar optical properties. Data for other galaxies are sparse, but, where available, generally support the view that the form of the spectral dependence of polarization is not radically different from that in the Galaxy. Magalhães *et al* (1989) find $\lambda_{max} = 0.48 \pm 0.07\,\mu$m for three stars in the Small Magellanic Cloud, and Martin and Shawl (1982) deduce $\lambda_{max} = 0.45 \pm 0.05\,\mu$m for the integrated light from the globular cluster S78 in M31 (the Andromeda Galaxy). The most precise evaluation of the polarization law in an external galaxy to date is that of Hough *et al* (1987) towards a supernova (1986G) in NGC 5128 (Centaurus A), illustrated in figure 4.14. The supernova occurred within the dust lane of the galaxy, with an estimated reddening $E_{B-V} \simeq 1.6$; the observed polarization from 0.36 to 1.65 μm is consistent with the standard galactic law with $\lambda_{max} = 0.43 \pm 0.01\,\mu$m. This value of λ_{max} can be explained if the polarizing grains are $\sim 20\%$ smaller on average than those in the Milky Way if the refractive index is the same (equation (4.19)). The alignment efficiency ($P_V/E_{B-V} \simeq 3\%$ mag^{-1}) is lower than the maximum

Table 4.1   Statistical summary of λ_{max} data for 8 regions of current or recent star formation, compared with the diffuse interstellar medium in the Galaxy (ISM) and the Large Magellanic Cloud (LMC). Mean and standard deviation are given in μm; n is the number of stars in each sample.

Region	$\langle\lambda_{max}\rangle$	σ	n	Reference[†]
R CrA cloud	0.75	0.09	43	1
ρ Oph cloud	0.65	0.09	15	2, 3, 4
M17	0.63	0.13	11	5
Orion nebula	0.61	0.08	19	6
Chamaeleon I cloud	0.60	0.05	18	7
α Per cluster	0.54	0.07	55	8
NGC 7380	0.51	0.05	10	9
Cygnus OB2	0.51	0.07	16	10
ISM	0.54	0.06	180	1, 2
LMC	0.55	0.10	19	11

[†] References: 1. Vrba *et al* (1981); 2. Coyne *et al* (1974); 3. Serkowski *et al* (1975); 4. Wilking *et al* (1980); 5. Schulz *et al* (1981); 6. Breger *et al* (1981); 7. Whittet *et al* (1986, 1992); 8. Coyne *et al* (1979); 9. McMillan (1976); 10. McMillan and Tapia (1977); 11. Clayton *et al* (1983).

value (equation (4.8)) but consistent with the observed range for stars in the Milky Way (figure 4.2). Within the limits of the available data, the ratios P/E_{B-V} and R_V/λ_{max} in external galaxies are both similar to the solar-neighbourhood values.

4.4   GRAIN ALIGNMENT

The correlations which have been demonstrated between the degree of polarization and reddening or extinction (figures 4.2 and 4.5), and between the environmentally sensitive parameters λ_{max} and R_V (figure 4.11), provide a firm basis for believing that the observed polarization is produced by interstellar grains which are also responsible for some fraction of the visible extinction. For this to occur, two conditions must be satisfied: (i) individual particles must be optically anisotropic, to give the required directional dependence of extinction efficiency; and (ii) net alignment of the axes of anisotropy must be achieved, in order to produce net polarization of a

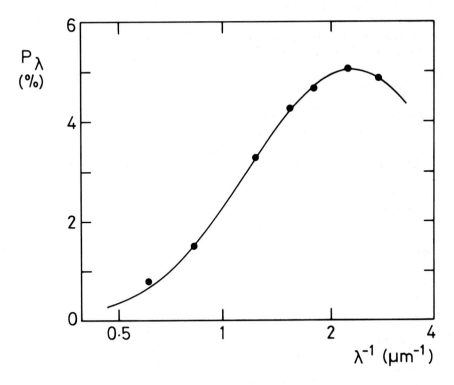

Figure 4.14   The interstellar polarization curve for the supernova 1986G in NGC 5128. The observations (points) are fitted by the Serkowski curve (equation (4.16)) with $K = 1.15$, $\lambda_{max} = 0.43\,\mu m$, and $P_{max} = 5.16\%$.

beam which propagates through a medium containing many particles. It is assumed that the optical anisotropy results from the shape of the particle; the effect could also be produced by birefringence of the grain material, but it is unclear how the optic axes of such particles can be aligned unless they are also elongated in shape.

The results discussed in §4.2, notably the dependence of q on ℓ (figure 4.4), clearly imply that the alignment mechanism is related to some macroscopic property of the Galaxy. Viable models for alignment depend in general on an interaction between the spin of a grain and the galactic magnetic field. A dust grain immersed in a gaseous medium at some temperature T_g is subject to random collisions with the atoms and molecules of the gas, and these impart rotational energy to the grain. For a grain of arbitrary shape in collisional equilibrium with the gas, the energies associated with spin about the three principal axes of inertia are equal: the angular momentum is therefore greatest along the axis of maximum inertia, and so elongated grains tend to spin 'end-over-end'. The problem of alignment is thus reduced to that of aligning the angular momenta of the grains with respect to the galactic magnetic field. Given that the average magnetic field direction is parallel to the galactic plane, the observed behaviour of the polarization vectors is explained if the grains are orientated with their angular momenta predominantly parallel (and hence their long axes predominantly perpendicular) to the magnetic field.

The theory of magnetic alignment was first formulated by Davis and Greenstein (1951), and more recently extended by Jones and Spitzer (1967), Purcell (1979), and other authors (see Johnson 1982 for a review). In this section, we begin by reviewing the basic principles of magnetic alignment, and show that the classical Davis–Greenstein (DG) mechanism requires a magnetic flux density significantly in excess of that observed by other methods. Developments of classical theory which lead to alignment at more realistic field strengths are then discussed.

4.4.1 Davis–Greenstein Alignment

The DG mechanism achieves alignment by paramagnetic relaxation of thermally rotating grains. Assuming a Maxwellian distribution of velocities in the gas, an initially stationary grain is set spinning by random impulsive torques produced by collisions, and its angular velocity increases until it becomes limited by rotational friction with the gas itself. If the collisions are elastic, the rotational kinetic energy of the grain about a principal axis with moment of inertia I is

$$\frac{1}{2}I\langle\omega^2\rangle = \frac{3}{2}kT_{\mathrm{g}}. \tag{4.23}$$

A typical value of the mean angular velocity may be estimated by taking an average moment of inertia $I = \frac{2}{5}m_{\mathrm{d}}a^2$ appropriate to a spherical grain of radius a and mass m_{d}; equation (4.23) then gives

$$\omega_{\mathrm{rms}} \simeq \left(\frac{2kT_{\mathrm{g}}}{a^5 s}\right)^{\frac{1}{2}} \tag{4.24}$$

where s is the density of the grain material. Assuming values of $a \sim 0.2\,\mu\mathrm{m}$, $s \sim 2000\,\mathrm{kg\,m}^{-3}$ and $T_{\mathrm{g}} \sim 80\,\mathrm{K}$ (for diffuse clouds), then $\omega_{\mathrm{rms}} \sim 6 \times 10^4\,\mathrm{rad\,s}^{-1}$.

The presence of an external magnetic field of flux density $\boldsymbol{B}$ causes the induction of an internal field in the solid particle, the strength of which depends on the magnetic susceptibility of the grain material. In a static situation, the internal and external fields would be parallel. However, for a spinning grain, it is impossible for the internal field to adjust itself instantaneously to the direction of the external field, and so there is always a slight misalignment. This results in a dissipative torque about an axis perpendicular to B, tending to bring the angular momentum of the grain into alignment with B. As discussed by Martin (1971), the magnetic field direction should actually become the axis of symmetry for precession of the spinning grain. Alignment is opposed by gas–grain collisions, which tend to restore random orientation; thus, to achieve significant net alignment, the timescale for paramagnetic relaxation (t_r) must not exceed that for collisions (t_c), i.e.

$$\delta = \frac{t_c}{t_r} \geq 1 \qquad (4.25)$$

where t_c is defined as the time taken for the grain to collide with a mass of gas equal to its own mass. We may show that t_c depends on the gas temperature T_g and t_r depends on the dust temperature T_d. In the case of t_c, we have

$$t_c = \frac{4as}{3nm\langle v \rangle} \qquad (4.26)$$

where

$$\langle v \rangle = \left(\frac{8kT_g}{\pi m} \right)^{\frac{1}{2}} \qquad (4.27)$$

and m and n are the mass and number density of the atoms in the gas (see §7.2). Spherical grains of radius a are again assumed for simplicity. In the case of t_r, we have

$$t_r = \frac{\mu_0 a^2 s\omega}{\chi'' B^2 \sin\theta} \qquad (4.28)$$

(see, e.g., Spitzer 1978, pp 187–189), where θ is the angle between B and ω, and χ'' is the imaginary part of the magnetic susceptibility ($\chi = \chi' + i\chi''$). For paramagnetic materials, this is related to the angular velocity and temperature of the grain by

$$\chi'' = \frac{4\pi\omega k_m}{T_d} \qquad (4.29)$$

where k_m is a constant for a given grain material.

It is important to realize that T_g and T_d are, in general, different. Let us assume that the gas and dust are in independent thermal equilibrium with the interstellar radiation field, and that gas–grain collisions are unimportant as a source of grain *heating*, which is likely to be the case in diffuse clouds. Equilibrium temperatures $T_d \sim 20\,\mathrm{K}$ are predicted for the dust (§6.1.1) which are considerably smaller than those typical of the gas ($T_g \sim 80\,\mathrm{K}$, table 1.1). The rotational temperature of a grain about a given axis is (Jones and Spitzer 1967)

$$T_R = \frac{T_g + \delta T_d}{1 + \delta}. \qquad (4.30)$$

The rotation temperature is thus a function of the direction of the principal axis of rotation (θ) with respect to the magnetic field: $T_R \rightarrow T_g$ when $\theta \rightarrow 0$ (equation (4.28)), and, in general, $T_d < T_R < T_g$. It is this effect which leads to an asymmetric distribution of angular momenta in the grain population.

From equations (4.25), (4.26) and (4.28), we have

$$\delta = \frac{4\chi''B^2 \sin\theta}{3\mu_0 amn\omega \langle v \rangle}. \tag{4.31}$$

The requirement $\delta \geq 1$ leads to a lower limit on the magnetic flux density for DG alignment:

$$B \geq \left\{ \frac{3\mu_0 naT_d(2.5kT_g m)^{\frac{1}{2}}}{16\pi k_m} \right\}^{\frac{1}{2}} \tag{4.32}$$

(using equations (4.27) and (4.29)), where $\sin\theta$ has been set to unity. For paramagnetic solids in the limit of low rotational frequency, the constant k_m (equation (4.29)) is insensitive to the precise composition of the grain and is typically $\sim 2.5 \times 10^{-12}$ (Spitzer 1978). For grains of dimensions $a \sim 0.2\,\mu$m and temperature $T_d \sim 20$ K immersed in an H I cloud of density $n_H \sim 3 \times 10^7$ m^{-3} and temperature $T_g \sim 80$ K, we obtain $B \geq 3 \times 10^{-9}$ T†.

For specific grain models, the polarization efficiency p/A may be deduced as a function of B, as illustrated in figure 4.15. Results are plotted for dielectric cuboidal grains of equal volume for two values of the axial ratio (b/a) in the long-wavelength limit (Purcell and Spitzer 1971; Aannestad and Purcell 1973). Calculations for spheroids with the same dimensions would yield closely similar results. Observations at $2.2\,\mu$m (the longest wavelength at which a large data set is available: see figure 4.5) provide a useful comparison. Typically, $p_K/A_K \simeq 0.04$, with a maximum value of ~ 0.12, requiring B to be in the range $(1-6) \times 10^{-9}$ T for $b/a = 5$, with higher values for lower b/a. Thus, on the basis of both estimates, we may take $B \sim 3 \times 10^{-9}$ T as an appropriate lower limit on magnetic field density required for DG alignment.

The average value of the interstellar magnetic field in the solar neighbourhood of the Galaxy may be determined by observations of Faraday rotation and Zeeman splitting (see Heiles 1987 for a review). A value of $B \simeq 4 \times 10^{-10}$ T is obtained. Thus, the minimum required flux density exceeds the measured flux density by a factor of almost 10. A further objection to classical DG alignment is provided by the fact that high levels of polarization are observed in dense molecular clouds. In such regions,

---

† In the astronomical literature, magnetic fields are almost invariably expressed in cgs units (1 Gauss $= 10^{-4}$ T).

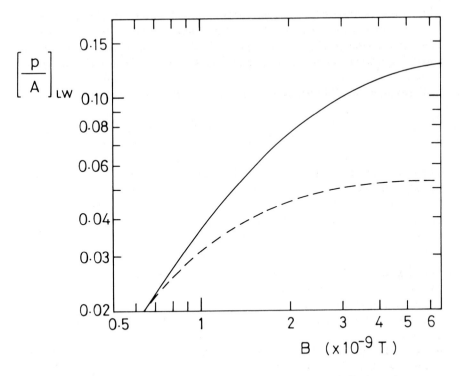

Figure 4.15   Plot of alignment p/A in the long-wavelength limit against magnetic flux density B. Dielectric cuboidal grains of cross-sectional area $a \times a$, length b, and equal volume $a^2 b = 10^{-21}$ m^{-3} are assumed. Plots are shown for a refractive index of 1.5 and axial ratios $b/a = 5$ (continuous curve) and $b/a = 2$ (dashed curve). (Adapted from Aannestad and Purcell 1973.)

gas–grain collisions become important and have the effect of tending to equalize gas and dust temperatures; if collisional thermal equilibrium is reached, $T_R = T_g = T_d$ for all orientations (equation (4.30)) and the mechanism becomes inoperative.

Although the DG mechanism fails quantitatively to explain the required degree of grain alignment in the ambient interstellar magnetic field, it is very successful in accounting for the *geometric* properties of the observed polarization. This strongly suggests that some process analogous to the DG mechanism is operating. The alternative models discussed below may be regarded as developments of DG alignment, in which either the spin velocity or the magnetic susceptibility of the grain material is enhanced

4.4.2   Suprathermal Spin

Real interstellar grains are unlikely to have smooth, uniform surfaces, and collisions between gas and grains are unlikely to be elastic. In particular, it is probable that a hydrogen atom colliding with a grain will stick, and subsequently migrate across the surface until it combines with another to form H_2 with the release of binding energy 4.5 eV. Ejection of the molecule from the surface simultaneously imparts angular momentum to the grain. As discussed by Hollenbach and Salpeter (1971), molecule formation is likely to occur preferentially at active sites (defects or impurity centres) on the grain surface. A migrating H atom, which would be held only by van der Waals forces elsewhere on the grain, becomes trapped at an active centre until recombination with another migrating atom occurs. The number of active centres on a grain may be quite small and their distribution over the surface will determine the spin properties of the particle (Purcell 1975, 1979). The systematic contributions to angular momentum arising from a series of recombination events at a limited number of active sites will lead to angular velocities well in excess of those predicted by random elastic collisions, and may, with a favourable geometry, lead to values as high as $\omega \sim 10^9 \, \mathrm{rad\,s^{-1}}$. Equation (4.29) is no longer valid at such high rotational velocities (Spitzer 1978), and χ'' approaches the static susceptibility (the value of χ' at $\omega = 0$), which is typically $\sim 5 \times 10^{-3}$ for paramagnetic materials at $T_d \sim 20$ K. Purcell (1979) has discussed in detail the effect of suprathermal rotation on grain alignment, and concludes that alignment is achieved on a timescale short compared with that for collisional damping.

Suprathermal rotation appears to provide a plausible route to grain alignment in diffuse H I clouds. However, it appears less viable as a means of explaining the very high levels of polarization observed in dense molecular clouds, where much of the hydrogen is already in molecular form and the grains become coated with mantles dominated by molecules such as H_2O which do not readily desorb. Resurfacing of the grains in molecular clouds will affect the number and distribution of active centres, and it is unclear how high suprathermal rotational velocities can be maintained in such regions.

4.4.3   Superparamagnetic Alignment

If a paramagnetic grain contains clusters of ferromagnetic atoms or molecules (e.g. metallic Fe, Fe_3O_4, or other oxides or sulphides of iron, with ~ 100 Fe atoms per cluster), the value of χ'' may be enhanced by a factor $\sim 10^6$ over that typical of paramagnetic materials (Jones and Spitzer 1967), an effect termed superparamagnetism. It is evident from the discussion in §4.4.1 above that this could lead to very efficient grain alignment in the ambient interstellar magnetic field ($t_r \ll t_c$). Mathis (1986) has developed a model in which a silicate grain is aligned if, and only if, it contains at least one such cluster or 'inclusion'. This model can explain the observed alignment efficiency (p/A) on the assumption that carbonaceous grains which contribute to the extinction do not align, either because they do not contain Fe-rich inclusions or because they are not elongated in shape. The predicted wavelength dependence of polarization depends on only one free parameter, the average size of a grain containing an inclusion. As previously discussed, this closely matches the observations (figure 4.9). The model is 'robust' in the sense that the alignment process is not marginal: it is sufficiently efficient to be insensitive to environmental influences such as mantle growth in dense clouds. It is possible that, in dense regions, weak suprathermal spin-up associated with surface defects in grain mantles combines with superparamagnetism of the cores to align the grains. The assumption that silicate grains do, indeed, contain ferromagnetic inclusions appears to be a reasonable one: indirect support is provided by studies of the composition of meteorites and interplanetary dust, and by the high abundance and depletion of iron in the ISM. It is, perhaps, harder to explain why distinct populations of grains evidently exist which do not share these properties.

4.4.4   Non-magnetic Alignment

Magnetic alignment along the lines proposed by DG is capable of explaining the macroscopic properties of the observed interstellar polarization, subject to the assumption that the efficiency of the process is enhanced in some way, as discussed in the previous sections. Other mechanisms have been proposed, but these are less successful and have not gained acceptance as viable models for the general alignment in the Galaxy (see Aannestad and Purcell 1973 for a critical review). However, at least two of these proposals deserve comment as possible contributors to alignment in specific situations, notably in the environs of luminous stars. We conclude this chapter with a brief discussion of these possibilities.

Gold (1952) pointed out that streaming of dust with respect to gas can lead to alignment of the dust grains. Gas atoms colliding with a grain contribute angular momentum predominantly perpendicular to the streaming velocity v, resulting in net alignment of the long axes predominantly parallel to v. The process in most efficient when the magnitude of v exceeds the thermal speed of the gas atoms. Examples of processes which may cause streaming include cloud–cloud collisions and differential acceleration of gas and dust by radiation pressure. As interstellar grains acquire charge due to the photoelectric effect, their kinematics nevertheless tend to be controlled by the magnetic field, such that significant relative velocities are generally reached only in directions parallel to B. In this case, streaming would tend to produce polarization with position angle orthogonal to that predicted by the DG mechanism, which is incompatible with the observed distribution of polarization vectors in the Milky Way. Aitken *et al* (1986) discuss alignment by streaming in the central region of the Galaxy, but conclude that such a model cannot account for either the degree or the orientation of the observed polarization. In weakly ionized clouds, however, charged grains tend to drift through the gas in a direction *normal* to B due to ambipolar diffusion, at speeds which may be sufficient to produce appreciable alignment (Roberge and Hanany, in preparation). This possibility warrants further investigation.

Harwit (1970) drew attention to the fact that a grain absorbing unpolarized starlight acquires angular momentum anisotropically due to the intrinsic angular momentum of the photons. This effect would tend to align prolate grains with their axes transverse to the direction of propagation of the light, and would be most efficient for the smallest grains. However, under typical interstellar conditions, the alignment mechanism is overwhelmed by the randomizing effect of thermal emission of low-energy photons from the grains (Martin 1972), and momentum transfer is not therefore feasible for general interstellar alignment. It may, however, be significant in the vicinity of luminous cool stars which emit intensely in the infrared. As discussed by Aitken *et al* (1985), the disorienting effect of thermal emission is then less critical as the difference between the colour temperature of the ambient radiation field and the grain temperature is much lower than it is in a typical region of the interstellar medium. Aitken *et al* (1985) show that the polarization of IRc2 in the Orion Nebula may be explained in this way.

5

Spectral Absorption Features

Cosmic dust grains provide an interesting and unusual application of solid-state spectroscopy. Interstellar absorption features in the spectra of reddened stars and infrared sources are attributed to solid particles on the basis of position, width, shape and continuity of profile. The absorption lines due to gas phase atomic or molecular species in the interstellar medium are, in general, extremely sharp (full-width half maxima equivalent to $\sim 1\,\mathrm{km\,s^{-1}}$ or less), reflecting the conditions of low temperature and pressure in which they are formed. Solid-state spectral features are intrinsically broad and continuous, and cannot be resolved into discrete lines, in contrast to the vibration–rotation bands of many gas phase molecules (e.g. Banwell 1983). Of the observed absorption features attributed to interstellar dust, the $\lambda 2175$ feature† in the mid-ultraviolet is the strongest and, in frequency units, the broadest, forming a prominent peak in the interstellar extinction curve (Chapter 3). The other features discussed here are the optical 'diffuse interstellar bands', and the infrared absorption features. All of these absorptions may be regarded as fine structure in the extinction curve. However, it should be noted that identification with some component of the dust is not certain in the case of the optical diffuse bands (§5.2), where an origin in gas phase interstellar molecules is also tenable.

In principle, the dust-related spectral features provide a direct means of identifying the chemical composition of interstellar grains. This potential has yet to be fully realized: the identity of many of the observed features remains controversial. Observationally, sufficient spectral resolution and signal-to-noise are needed to allow structure and blending of line profiles to be investigated, and variations in the observed strength and profile shape in contrasting interstellar environments may be used to test proposed identifications. The interpretation of astronomical spectra in terms of grain composition is not straightforward, depending critically

---

† For convenience, visible and ultraviolet spectral features are identified in this chapter by the symbol λ followed by the wavelength to four significant figures in Ångstroms (10^{-10} m).

on the availability of suitable laboratory data for comparison. Wavelength coincidences are often insufficient to clinch an identification, and detailed modelling of the observed spectrum is generally required. Input to the models must be provided by laboratory measurements of optical constants for candidate materials, made under appropriate physical conditions. At ultraviolet and optical wavelengths, the source generally consists of a star of known spectral type seen through a foreground absorbing cloud, as discussed in Chapter 3. In the infrared, more complex situations frequently arise: the nature of the underlying source may be in doubt, as large visual extinctions are usually required to produce significant optical depths in the absorption features, and normal spectral classification techniques cannot therefore be applied. In addition, the presence of circumstellar dust may result in an emergent spectrum which is a combination of continuum emission, band emission and corresponding band absorption, originating from different regions in the line of sight; a further complication is the possibility that a dust-embedded object may modify the composition of the dust in its vicinity. Infrared spectroscopy of dust is a rapidly developing field, with recent advances in observational capability parallelled by developments in laboratory astrophysics and mathematical modelling techniques. As the data-base on infrared sources increases, it becomes easier to identify suitable candidates for investigation, in which the interstellar component may be extracted reliably (see Whittet 1987). Progress in our understanding of the features at shorter wavelengths is less marked, despite a wealth of observational material and the relative ease with which stellar and interstellar effects may be deconvolved. In particular, the identity of the optical diffuse bands is a fundamental, long-standing problem in astronomical spectroscopy (see Bromage 1987 for an amusing historical review).

This chapter is divided into three sections discussing the principle classes of observed absorption feature attributed to dust: the ultraviolet ($\lambda 2175$) feature (§5.1), the optical diffuse bands (§5.2), and the infrared features (§5.3).

5.1   THE 2175 Å FEATURE

"...a dramatic piece of spectroscopic evidence which should have much to tell us about at least a part of the interstellar grain population."

B T Draine (1989a)

5.1.1   Observations

The presence of a broad absorption feature in the mid-ultraviolet spectra of reddened stars was first suggested by rocket observations made in the 1960s (Stecher 1965) and confirmed by data from several satellites, including OAO-2 (Bless and Savage 1972; Savage 1975), TD-1 (Nandy *et al* 1975, 1976), ANS (Wu *et al* 1980; Meyer and Savage 1981), and IUE (Seab *et al* 1981; Witt *et al* 1984; Fitzpatrick and Massa 1986). These investigations have established the observational parameters of the feature and their variability from star to star, which we review in this section.

The most striking aspects of the λ2175 feature are its ubiquity, strength, stability of central wavelength, and symmetry of profile. It is almost invariably detectable in the spectra of stars with visual reddening $E_{B-V} >$ 0.05 mag, given adequate signal-to-noise. The extinction curves of some typical examples are illustrated in figure 3.6. The central wavelength of the bump, λ_0, has a mean value

$$\langle \lambda_0 \rangle = 2175 \pm 10 \text{ Å}, \qquad (5.1)$$

the error representing a 2σ dispersion of only 0.46%. Comparing data for individual stars, λ_0 is constant to within observational error (Fitzpatrick and Massa 1986) with very few known exceptions (Cardelli and Savage 1988). In the most deviant cases the feature is shifted to shorter wavelength by $\sim 2.5\%$.

The strength of the feature is well correlated with visual reddening. For example, Meyer and Savage (1981) deduce a linear correlation coefficient of 0.98 between the height of the bump and E_{B-V} for 1367 stars observed by ANS. This correlation confirms the interstellar nature of the feature but does not, in itself, prove that the absorbing agent necessarily resides in solid particles. Indeed, it is unlikely (§5.1.2) that λ2175 absorption and visual reddening are produced by the same species. Dust and gas are well mixed in the ISM, as indicated, for example, by the correlation between E_{B-V} and hydrogen column density N_H resulting from the fact that both increase with distance; thus any unsaturated interstellar spectral line or band is likely to show a reasonably strong positive correlation with E_{B-V}.

A mathematical representation of the λ2175 profile shape is useful and may provide physical insight. To a good approximation, the feature is Lorentzian (Savage 1975; Seaton 1979), but Fitzpatrick and Massa (1986) show that an even better fit to the observations is obtained with a model based on the Drude theory of metals (Bohren and Huffman 1983). In this representation, the extinction in the region of the bump is given in terms of $y = \lambda^{-1}$ by

$$\frac{E_{\lambda - V}}{E_{B-V}} = c_1 + c_2 y + c_3 D(y) \qquad (5.2)$$

(omitting the FUV term in equation (3.33), which is zero for $y \leq 5.9\,\mu\text{m}^{-1}$). The first and second terms on the RHS of equation (5.2) denote a linear background, and the third term is the contribution of the feature to the observed extinction, where

$$D(y) = \frac{y^2}{(y^2 - y_0^2)^2 + \gamma^2 y^2} \tag{5.3}$$

is the Drude profile. At a given wavenumber y, $D(y)$ is completely specified by the position, $y_0 = \lambda_0^{-1}$, and width, γ (FWHM), of the feature in wavenumber units.

Traditionally, the strength of the feature has been expressed in terms of its peak intensity in magnitudes, E_{bump}, relative to an assumed continuum level, such as the linear background. Putting $y = y_0$ in equation (5.3), we deduce, with reference to equation (5.2), that

$$\frac{E_{bump}}{E_{B-V}} = \frac{c_3}{\gamma^2} \tag{5.4}$$

with respect to the linear background. However, a more appropriate measure of strength is the quantity

$$A = \int_0^\infty \frac{c_3 y^2}{(y^2 - y_0^2)^2 + \gamma^2 y^2} \, dy = \frac{\pi c_3}{2\gamma}. \tag{5.5}$$

(Fitzpatrick and Massa 1986), which is analogous to the concept of equivalent width, measured in wavenumber units: A represents the area under the $\lambda 2175$ profile in the normalized extinction curve, and is thus a measure of strength per unit E_{B-V}.

The parameters λ_0, γ and A have been evaluated for many lines of sight by least-squares fits of equation (5.2) to observational extinction curves in the range $3 < \lambda^{-1} < 6 \, \mu m^{-1}$. Table 5.1 compares results for a selection of individual stars with mean values for the Milky Way and the Large Magellanic Cloud.

Perhaps the most remarkable observational property of the $\lambda 2175$ profile is the occurrence of variations in width which are unaccompanied by changes in position. Figure 5.1 compares the profiles with largest and smallest γ-values in the group of 45 stars studied by Fitzpatrick and Massa (1986). This property is hard to reconcile with any solid-state model for the feature (§5.1.2). The variations are not related to environment in an obvious way, as the *mean* profile widths for stars obscured by diffuse and dense clouds are essentially identical (Massa, Savage and Fitzpatrick 1983).

Although the strength of the $\lambda 2175$ feature is generally well correlated with reddening, significant variations, characterized by changes in the parameter A, occur from star to star and from region to region (see table 5.1). These may give insight into the nature of the carrier. Stars which show large departures from average $\lambda 2175$ strength are generally associated with dense clouds or nebulosity. Two extreme examples are HD 29647 (in the

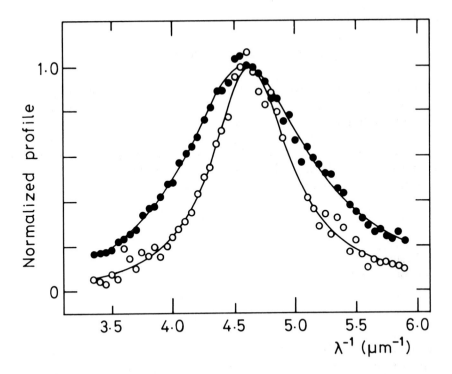

Figure 5.1   Normalized profiles of the λ2175 feature towards two stars, illustrating variation in width: data for ζ Oph (filled circles) and HD 93028 (open circles) are fitted with the Drude function. (Adapted from Fitzpatrick and Massa 1986.)

Taurus dark cloud near TMC-1) and HD 62542 (close to ionization fronts in the Gum Nebula). Comparison with the average ISM profile is shown in figure 5.2. In both cases, the feature is significantly broadened, with an apparent extension to shorter wavelength. It is of interest to compare bump strength with the level of FUV extinction. Towards HD 62542, the FUV extinction is very high, with a much steeper slope than in the average ISM extinction curve; a more modest enhancement is apparent towards HD 29647. Other cases have been reported which show similar, but less dramatic, anomalous behaviour (e.g. Witt, Bohlin and Stecher 1981; Walker *et al* 1980). In contrast, the extinction curves of stars in the ρ Oph dark cloud show nearly normal λ2175 strength (table 5.1), and shallow FUV extinction (e.g. Seab *et al* 1981). Stars such as θ^1 Ori and Herschel 36, embedded in H II regions, show another characteristic anomaly, with comparative weakness in both λ2175 and FUV extinction (Bohlin and Savage 1981; Hecht *et al* 1982). These examples illustrate general variations in the ultraviolet extinction curve (Meyer and Savage 1981; Greenberg and

Table 5.1   Representative values of $\lambda2175$ parameters in units of μm^{-1}.

Star/Region	λ_0^{-1}	γ	A
HD 29647	4.70	1.62	3.35
θ^1 Ori C	4.63	0.84	2.43
HD 37061	4.57	1.00	2.69
HD 37367	4.60	0.91	7.04
HD 38087	4.56	1.00	6.68
HD 62542	4.74	1.29	3.11
HD 93028	4.63	0.79	2.62
HD 93222	4.58	0.81	3.33
ρ Oph	4.60	0.99	5.57
HD 147889	4.63	1.16	7.14
ζ Oph	4.58	1.25	5.71
Herschel 36	4.62	0.88	3.51
HD 197512	4.58	0.96	6.83
HD 204827	4.63	1.12	4.98
Mean ISM (45 stars)	4.598	0.992	5.17
General LMC (13 stars)	4.608	0.994	4.03
30 Dor LMC (12 stars)	4.606	0.894	2.62

Chlewicki 1983; Witt *et al* 1984), variations which have been interpreted as evidence for independent grain populations responsible for the $\lambda2175$ absorption and the FUV extinction, with environmental factors governing their relative contributions: for example, Witt *et al* (1981) argue that $\lambda2175$ strength is dependent on cloud density, whilst FUV extinction is dependent on the local radiation field from OB stars. However, such scenarios are questioned by Fitzpatrick and Massa (1988), who report a correlation between the *width* of the $\lambda2175$ bump and the amplitude of FUV extinction rise (c_4 in equation (3.33)). They suggest that that the latter is produced by another absorption feature, situated at $\lambda < 1000\,\text{Å}$ in the extreme ultraviolet (EUV) beyond the spectral limit of available observations, of which only the long-wavelength wing is seen. If an EUV feature were to be produced by a second resonance in the same carrier (see §5.1.2 below), then a correlated increase in the width of both features would be expected to enhance the FUV extinction.

Observations of stars in the Magellanic Clouds have been used to study the $\lambda2175$ feature in the interstellar media of these galaxies. Both the

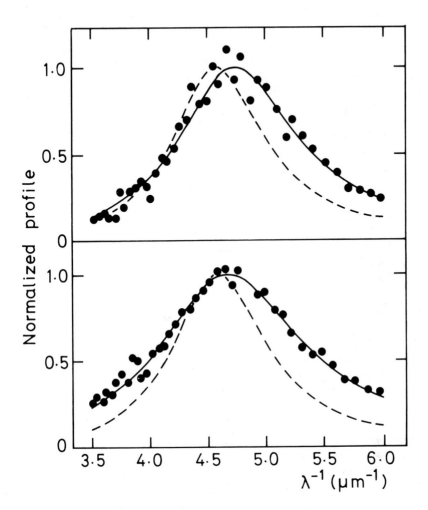

Figure 5.2 Profiles for two stars with extremely unusual λ2175 features: HD 62542 (upper) and HD 29647 (lower), compared with the average profile for the ISM. (Based on data from Cardelli and Savage 1988.)

LMC and SMC extinction curves show departures from the average for the Galaxy (§3.3.1), and in the case of the SMC, λ2175 is extremely weak or absent. These variations have been associated tentatively with the known systematic differences in metallicity (e.g. Nandy 1984; Prévot *et al* 1984). Two independent groups of stars in the LMC are represented in table 5.1: those in the 30 Dor region, and those widely distributed outside this zone, which we call the general LMC (Fitzpatrick 1986). It is apparent that in the general LMC, λ2175 is remarkably similar to the corresponding feature observed in the solar-neighbourhood ISM (see figure 3.8): the position and width are essentially identical to within observational error, whilst the strength is $\sim 20\%$ less. In the 30 Dor region, the position and width are again similar but the strength is lower by $\sim 50\%$. For both LMC samples, all three parameters are within the range of values observed towards individual stars in the Milky Way. The carrier is clearly a characteristic ingredient of dust in both galaxies. It is tempting to associate the weakness in the 30 Dor nebula with that observed in galactic H II regions (comparing results for 30 Dor, θ^1 Ori C and Herschel 36 in table 5.1), and to conclude that environmental influences are more important than metallicity effects (which would also apply to the general LMC, not just 30 Dor). The reason for the absence of the feature in the SMC has yet to be discovered, and this could provide an important clue to the nature of the carrier.

The wavelength dependence of polarization across a dust-related absorption feature is of general interest as a diagnostic of the elongation and alignment of the carrier: excess polarization at the wavelength of the feature is expected for significant alignment of non-spherical grains. In the case of $\lambda 2175$, preliminary results suggest that there is no significant excess above the continuum predicted by an extrapolation of the Serkowski law (§4.3) to ultraviolet wavelengths (Gehrels 1974). If confirmed, this would imply that the carrier particles are spherical or unaligned†. It is also relevant to note that the strength, central wavelength and profile shape of the $\lambda 2175$ spectral feature are independent of the wavelength of maximum linear polarization, which also suggests that the carrier does not reside in grains responsible for extinction and polarization in the visible.

Observations of the grain albedo in the ultraviolet are discussed in Chapter 3 (§3.2.2). Results indicate the presence of a broad minimum in the region of the $\lambda 2175$ feature, as expected for pure absorption. However, studies of two reflection nebulae suggest the presence of scattering in the long-wavelength wing of the feature at $\lambda \simeq 2300\,\text{Å}$ (Witt, Bohlin and Stecher 1986). This would imply that the particles containing the carrier have radii $a \geq 0.01\,\mu m$, as smaller particles are ineffective scatterers at these wavelengths. Further observations of this effect are needed.

Attempts to detect other diffuse absorption features or fine structure in the interstellar extinction curve at ultraviolet wavelengths have been either unsuccessful or inconclusive. Savage (1975) examined OAO-2 data in the 1800–3600 Å spectral region for six stars, and found no evidence for structure greater in amplitude than 5% of the depth of the $\lambda 2175$ feature. Seab and Snow (1985) extended the search to 58 stars observed by IUE (1150–3200 Å) with similarly negative results. The extinction curve has not been

---

† Recent data from the Wisconsin Ultraviolet Photopolarimeter Experiment (Clayton *et al* 1992) are intriguing, with $\lambda 2175$ excess apparently detected in some lines of sight and not others.

studied intensively at wavelengths below 1150 Å, lying beyond the range of sensitivity for satellite spectrometers most commonly used in such investigations, but Copernicus and Voyagers 1 and 2 provided useful data down to the Lyman limit at 912 Å (York *et al* 1973; Longo *et al* 1989), beyond which absorption by interstellar hydrogen gas precludes measurement. Again, no features are readily apparent in the available data, although it has been suggested that the FUV rise may be the wing of a hidden EUV feature, as discussed above. A number of other detections have been claimed but not confirmed. Discrepancies in stellar line strengths between reddened and comparison stars can be particularly troublesome in the FUV: a weak, broad hump near $6.3\,\mu m^{-1}$ in extinction curves derived from TD-1 data is caused by this effect in the $\lambda1550$ C IV lines (§3.2). Carnochan (1989) has argued that a broad, shallow absorption feature centred near 1700 Å $(5.9\,\mu m^{-1})$ is present in TD-1 data, but this is not apparent in IUE extinction curves (Fitzpatrick and Massa 1988). Finally, the apparent absorption at ~ 2700 Å, reported in IUE data and put forward as evidence for interstellar proteins, proved to be of instrumental origin (McLachlan and Nandy 1984; Savage and Sitko 1984). It is concluded that the $\lambda2175$ bump is the only interstellar dust-related spectral feature detectable with confidence to current sensitivity limits in the satellite ultraviolet.

5.1.2   Implications for the Identity of the Carrier

Primary requirements of a $\lambda2175$ absorber as candidate for the interstellar feature are that it is cosmically abundant, sufficiently robust to survive in a variety of interstellar environments, and capable of matching closely the observed profile position, width and shape, without producing significant absorptions at other wavelengths in the ultraviolet.

Let us first consider what constraints may be placed on the abundance of the carrier. The equivalent width of $\lambda2175$ is related to the strength parameter A (equation (5.5)) by

$$W_\nu = cAE_{B-V}. \qquad (5.6)$$

Using $A \simeq 5.2 \times 10^6\,\mathrm{m}^{-1}$ (table 5.1) and $N_\mathrm{H}/E_{B-V} \simeq 5.8 \times 10^{25}\,\mathrm{m}^{-2}$ (§1.2.1), the abundance relative to hydrogen of the carrier (X) is

$$\frac{N_\mathrm{X}}{N_\mathrm{H}} \simeq \frac{10^{-5}}{f} \qquad (5.7)$$

where f is the oscillator strength per absorber associated with the feature (equation (2.15)). The strongest permitted transitions typically have $f \leq 1$, and so the abundance of the carrier of $\lambda2175$ must be $N_\mathrm{X}/N_\mathrm{H} \geq 10^{-5}$ (see Draine 1989a). With reference to the standard element abundances (§2.2),

X must therefore contain one or more from the set {C, N, O, Ne, Mg, Si, S, Fe}. This set can immediately be reduced from eight elements to four: Ne is a noble gas and can therefore be ruled out; S (abundance 1.8×10^{-5}) is only weakly depleted in the interstellar medium and can also be excluded; finally, N and O are rejected as they are electron acceptors. We may conclude that the carrier must contain one or more elements from the set {C, Mg, Si, Fe}.

Graphite is the most widely discussed material amongst candidates for the $\lambda 2175$ feature, and this identification has gained a measure of acceptance. The average profile of the observed feature is well matched by theoretical models involving small graphite grains, as illustrated in figure 5.3 and discussed below. As a form of solid carbon, graphite easily satisfies abundance constraints: the oscillator strength per carbon atom is $f \sim 0.16$ in the small-particle limit (Draine 1989a), and equation (5.7) gives $N_C/N_H \sim 6.3 \times 10^{-5}$, which is $\sim 20\%$ of the solar abundance. Graphite is an optically anisotropic, uniaxial crystal. Once formed, it is sufficiently refractory to survive for long periods in the diffuse ISM. Absorption arises from excitation of electrons with respect to the positive ion background of the solid, to produce resonance peaks in the optical constants, a phenomenon referred to in the bulk material as plasma oscillations. Excitation of π electrons in graphite produces absorption in the mid-ultraviolet. Excitation of σ electrons produces a feature in the EUV, centred near 800 Å, the presence of which could have implications for the shape of the observed extinction from 912 to 1500 Å, as discussed in §5.1.1 above. It should be noted that hydrogenation suppresses absorption by localizing the electrons (Hecht 1986) and the carrier grains must therefore be assumed to have low hydrogen content.

There are two specific problems with the graphite identification. Firstly, the question of its origin is raised by observations of circumstellar dust, which indicate that the solid particles ejected into the ISM by carbon stars are predominantly non-graphitic, a topic reviewed in Chapter 7. A plausible mechanism for the production of graphite in the interstellar environment must therefore be formulated in order to strengthen the case for its inclusion in grain models. Secondly, the observational properties of the $\lambda 2175$ feature place tight, and possibly unrealistic, constraints on the nature of the particles (their size, shape, and the presence or absence of surface coatings). In the remainder of this section, we discuss the feasibility of graphite as the carrier of $\lambda 2175$, and examine some alternatives.

The symmetry of profile, constancy of central wavelength λ_0, and absence of any trend in λ_0 with width strongly suggest that the $\lambda 2175$ absorbers are in the small-particle limit, i.e. they have radii $a \ll \lambda$, which, in practical terms, means $a \leq 0.01 \, \mu\text{m}$ for the graphite feature. At larger radii, λ_0 increases systematically with particle size as a result of scattering which contributes to extinction predominantly on the long-wavelength side

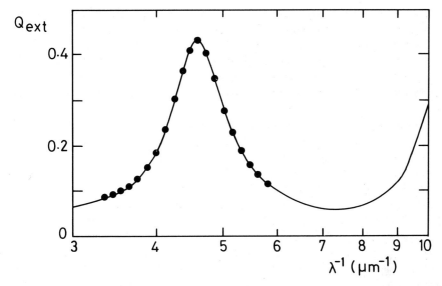

Figure 5.3   Plot of Q_{ext} against λ^{-1} for randomly orientated graphite spheroids with $a = 0.003\,\mu m$ and axial ratio $b/a = 1.6$ (Draine 1989a, curve), compared with the mean observed $\lambda 2175$ profile (data points rebinned from Fitzpatrick and Massa 1986, and normalized to the peak of the feature).

of the absorption peak. This is demonstrated using classical Mie theory calculations for spherical particles in figure 5.4. Graphite spheres in the small-particle limit produce a feature centred at $\lambda_0 \simeq 2080$ Å, significantly below the observed mean value (equation (5.1)). In order to match the observations, a particle size distribution which is sharply peaked at a grain radius of $0.018 \pm 0.002\,\mu m$ is required for spheres, and this must be immune to environmental changes such as mantle growth, in view of the low dispersion in λ_0.

Particle shape also affects the position and profile of the ultraviolet absorption feature in graphite. Spherical graphite grains, considered in the previous discussion, are frequently assumed in model calculations (e.g. figure 5.4). They have the property that the resultant feature is unpolarized, consistent with current observational constraints. However, the assumption of sphericity is highly artificial for grains composed of an anisotropic crystalline material. Non-spherical grains are physically more reasonable as graphite particles minimize their free energy when they are flattened, and they are also capable of matching the observations, subject to restrictions on their size and shape (Savage 1975; Draine 1989a). Spheroidal particles have the notable advantage that an acceptable fit can be obtained *within the small-particle limit*, in contrast to the situation for spheres. The computed profile shown in figure 5.3 assumes spheroidal particles with shape

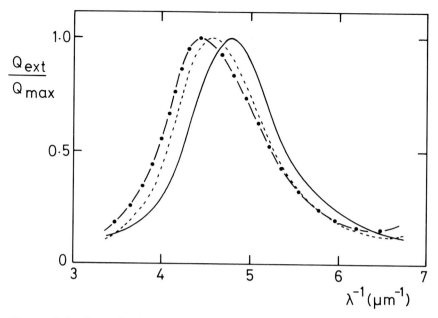

Figure 5.4   Normalized extinction efficiencies for graphite spheres with radii of 0.01 μm (continuous curve), 0.02 μm (dashed curve), and 0.025 μm (dash-dot curve).

given by an axial ratio $b/a = 1.6$ and size $a = 0.003\,\mu$m, where $2a$ and $2b$ are the particle lengths parallel and perpendicular to the axis of symmetry (Draine 1989a). The spheroids are assumed to be orientated randomly in space, producing no net polarization. The fit to the mean observed profile is clearly excellent. However, the goodness of fit depends on both b/a and the orientation of the crystal axis relative to the geometrical axis (see Savage 1975), and so the properties of the particles are highly constrained. As with other models, the observed variation in the width of the profile is hard to explain; it is not clear how graphite grains can account for the narrowest features unless their optical constants differ significantly from those of the bulk material.

One possibility which warrants further investigation is that the width of the feature may vary with the purity of the graphite. Hecht (1986) and Sorrell (1990) suggest that the carrier is formed by graphitization of amorphous carbon. Perrin and Sivan (1990) model the λ2175 feature by means of calculations for graphite with amorphous carbon inclusions; their results indicate that changes in width of $\sim 10\%$ as a function of impurity concentration are accompanied by changes in position of only $\sim 0.5\%$ in the small-particle limit. These calculations were carried out for spheres, and it would be useful to extend them to spheroids.

In view of the crucial rôle played by Mie theory calculations in the identification of $\lambda 2175$, it is appropriate to review the reliability of the modelling process. There are two distinct issues, the propriety of the technique, and the accuracy of the laboratory data on which it depends. The reliability of Mie theory calculations using bulk optical constants must decrease with decreasing particle size: it is intuitively obvious that if a solid is repeatedly subdivided, its optical properties must ultimately deviate from those of the bulk material as molecular orbital theory takes over from solid-state band theory. The small, sub-classical grains invoked to explain $\lambda 2175$ lie in an intermediate zone, where the proximity of surfaces may significantly modify the collective behaviour of electrons in solids (see Bohren and Huffman 1983, p 335). Errors of measurement for the bulk optical constants are a more readily quantifiable source of uncertainty. In the case of graphite, data are required in two planes, parallel and perpendicular to the crystal axis, and in both cases there is substantial variation between data sets published by different authors (see Huffman 1977, 1989, and Draine and Lee 1984, for further discussion and references). Extinction curves measured directly for laboratory-manufactured smokes provide a potentially valuable comparison for those generated from Mie theory, but difficulties are encountered in determining and controlling the size distribution and degree of crystallinity of the samples. Day and Huffman (1973) confirm the presence of an absorption feature near 2200 Å in graphite smoke, but note discrepancies in the shape of the feature which could be due to clumping of particles in the sample. Stephens (1980) presents extinction measurements for amorphous carbon grains which are in reasonable agreement with the equivalent Mie calculations (but which do not match the interstellar feature). In general, differences between the directly measured and calculated extinctions are as likely to be caused by saturation effects, uncertainties in the parameters of the laboratory smokes, or errors in the bulk optical constants used in the calculations, as by a breakdown in Mie theory.

Other carbon-based materials have been suggested as carriers of the $\lambda 2175$ feature. Sakata et al (1977, 1983) demonstrate the presence of absorption near 2200 Å in the spectra of carbonaceous extracts from the Murchison meteorite and synthetic quenched carbonaceous composites (QCCs) containing graphite and hydrocarbons. The spectra of polycyclic aromatic hydrocarbon (PAH) and buckminsterfullerine (C_{60}) molecules have also been discussed (e.g. Léger et al 1989; Krätschmer et al 1990 and references therein). A common characteristic of QCCs and PAH clusters is that they produce absorptions near 2200 Å which are too broad compared with the observed feature. In the case of PAHs, a more crucial objection arises because absorption corresponding to the bump is generally accompanied by a variety of features in the 2400–4000 Å wavelength interval, none of which have been observed in the ISM. For example, coronene has absorption between 2700 and 3450 Å with oscillator strength per C atom

$f \sim 0.06$, which is $\sim 40\%$ of the strength of the feature near 2000 Å in this molecule (Draine 1989a). But no feature is seen near 3000 Å in the observed extinction curve to a limit 5% of the strength of $\lambda 2175$. By analogy with equation (5.7), we deduce that no more than $\sim 2.5\%$ of the available carbon can be in coronene. The calculation may be repeated for many other PAHs with similar results. It is safe to conclude that those PAHs for which ultraviolet laboratory data are available *cannot* contribute to the $\lambda 2175$ absorption, and that their abundances are highly constrained by the lack of any observed near-ultraviolet features. Recent results reported by Krätschmer *et al* (1990) indicate a similar objection to C_{60}.

A number of oxygen-rich materials have also been proposed as the carrier of $\lambda 2175$, including silicates, quartz and magnesium oxide. These materials are dielectrics, and any solid of this nature will absorb continuously in the ultraviolet at sufficiently high photon energies, due to excitation of electrons to the conduction band. The rapid onset of absorption can coincide with a rapid decrease in scattering, so the net effect on the extinction curve may be to simulate a broad peak near the absorption edge. Huffman and Stapp (1971) noted that this occurs near $\lambda 2175$ in enstatite ($MgSiO_3$) spheres of radius $a \simeq 0.06\,\mu$m. However, the feature position and shape are extremely sensitive to particle radius (much more so than in the case of graphite), requiring artificial fine-tuning of the size distribution. This proposal is therefore highly implausible. The most promising O-rich materials proposed for the $\lambda 2175$ feature produce a fit within the small-particle limit, avoiding the need to adjust the size distribution artificially to some preferred value.

The optical constants of bulk MgO show an absorption edge at wavelengths below about 1800 Å, and a resonance peak near 1620 Å (Huffman 1977). The possibility of an identification for $\lambda 2175$ with MgO arises in very small particles due to resonance transitions in low-coordination surface O^{2-} ions (MacLean, Duley and Millar 1982); MacLean *et al* note that irradiated quartz (SiO_2) also produces an absorption feature near $\lambda 2175$, and propose a mixture of these materials. A problem with this hypothesis is that both substances are expected to produce absorption structure at wavelengths shortward of the $\lambda 2175$ peak. Savage (1975) pointed out that irradiated quartz absorbs strongly at ~ 1200 Å, and no corresponding feature is apparent in the observed extinction. The laboratory data of Maclean *et al* confirm the presence of a feature at ~ 1600 Å in small particles (near the position of the resonance in the bulk material): the absence of this feature in the interstellar extinction curve implies that no more than $\sim 3\%$ of the Mg abundance can be in MgO (Massa *et al* 1983).

Steele and Duley (1987) have inferred, on the basis of photoexcitation spectra, the presence of an absorption feature near 2175 Å associated with OH^- ions at low-coordination sites on the surfaces of small hydrogenated magnesium silicate particles. A model is proposed which reproduces the observed λ2175 feature within the small-particle limit, requiring ∼ 30% of the solar abundance of Mg and Si. As no *direct* laboratory measurements of the absorption are yet available to support the identification, this proposal cannot be subjected to detailed critical analysis.

To summarize, the identification of λ2175 with graphite or partly graphitized carbon remains the most likely of the various alternatives. If the feature is purely absorptive in character, then the particles responsible must be small (probably with radii $a \leq 0.005\,\mu m$), and the position of the feature is then independent of size but does depend on shape: the mean profile may be fitted by small spheroids with an axial ratio of 1.6, and some 20% of the solar abundance of carbon is required to be in such particles. Of the alternatives, perhaps the most promising is OH^- absorption on small silicate grains, a proposal which warrants further laboratory investigation. Critical observational tests may be provided by observations designed to detect other predicted spectral signatures. In the case of graphite, the EUV feature near 800 Å and its possible contribution to the FUV extinction have already been mentioned. An infrared feature at $11.5\,\mu m$ in graphite has been discussed by Draine (1984), but this is too weak to be observed in absorption in the interstellar medium. The OH^- hypothesis predicts a broad, weak infrared feature near $2.9\,\mu m$, due to the O–H stretching resonance (Steele and Duley 1987), but this region of the spectrum is difficult to observe to the necessary precision with ground-based telescopes because of strong telluric absorption.

5.2   THE OPTICAL DIFFUSE BANDS

"The chemical identification of these lines has not yet been made."

P W Merrill (1936)

5.2.1   The Diffuse Band Spectrum

First observations of the diffuse interstellar bands (DIBs) date from the pioneering years of stellar spectroscopy. The bands at λ5780 and λ5797 were noted in 1897 as the only dark lines in the spectrum of a Wolf–Rayet star (see Heger 1922), and λ4430 was apparently observed in the spectrum of a reddened supergiant by Annie Cannon (and misidentified as Hγ) during the compilation of of the Henry Draper (HD) catalogue some time before 1920 (see Code 1958). The λ5780 and λ5797 features were seen in spectroscopic binaries by Heger (1922), who noted that they did not share the periodic Doppler motion of the stellar lines. However, the interstellar nature of the diffuse band spectrum was not discussed until the work of Merrill (1934, 1936), who noted the features at $\lambda5780$, 5797, 6284 and 6614 as being of probable interstellar origin.

Since this early work, the list of DIBs has grown steadily as the spectra of reddened stars have been studied in more detail. A comprehensive survey of the diffuse band spectrum in the wavelength range 4400–6850 Å was carried out by Herbig (1975), who identified some 40 features of certain or very probable interstellar origin. More recently, a number of additional features have been reported in the range 6500–8900 Å (Sanner, Snell and Vanden Bout 1978; Herbig and Soderblom 1982; Herbig 1988). As with extinction curve studies, OB stars are generally used for studies of the DIBs, as stellar lines are less troublesome than in cooler stars, but DIBs are present in stars of all spectral types given adequate reddening. However, the suggestion of Seddon (1969) that some DIBs are prominent in the spectra of type I supernovae does not stand up to scrutiny (Herbig 1975) and is not discussed further here.

Table 5.2 presents a catalogue of the principal diffuse bands. The central rest-wavelength and mean error are given in columns 1 and 2: note that the features are identified by their wavelengths to the nearest Ångstrom in the text (e.g. $\lambda6284$ for the feature at 6283.91 Å), with the exception of $\lambda4430$ (centred at 4428 Å), for which the 'classical' wavelength is retained. Columns 3 and 4 of table 5.2 list equivalent widths (W_λ) and line-widths (FWHM) for each DIB in the spectrum of the star HD 183143, a supergiant with high reddening ($E_{B-V} = 1.28$), which was selected by Herbig as a standard reference star for diffuse bands. It should be noted that the line of sight to this star contains discrete clouds with radial velocities differing by $\sim 15\,\mathrm{km\,s^{-1}}$ (Herbig and Soderblom 1982), and the line widths should not therefore be treated as absolute values. Table 5.2 includes only those features with $W_\lambda \geq 0.05$ Å in HD 183143; a number of weaker features of probable interstellar origin are discussed by Herbig (1975, 1988) and Sanner *et al* (1978). The short-wavelength cut-off at 4400 Å appears to be real: no diffuse bands have been detected between the $\lambda4430$ and $\lambda2175$ features.

Some of the bands listed in table 5.2 are illustrated in figure 5.5. Both sharp features such as $\lambda5780$, $\lambda5797$ and $\lambda6203$ (FWHM ~ 2 Å), and broad, shallow ones such as $\lambda5778$ and $\lambda6177$ (FWHM ~ 20 Å) are included. The contrast between $\lambda6177$ and $\lambda6196$ is particularly striking. A generic distinction between diffuse 'lines' and diffuse 'bands' on the basis of width has been suggested by Wu (1972), with $\lambda5780$ and $\lambda5778$ as examples of each type. This distinction is somewhat artificial in that there is a fairly continuous progression in width from the narrowest to the broadest features known. However, there is a notable tendency for diffuse bands of disparate width to occur in pairs, with the sharper feature placed on the long-wavelength edge of the broader one. Herbig (1975) recognized seven such pairs, including $\lambda5797$, $\lambda5778$ and $\lambda6203$, $\lambda6177$ (figure 5.5). It is tempting to associate them with the bandheads and R branches of unresolved vibration–rotation

Table 5.2 The principal optical diffuse interstellar bands. The central wavelength (λ_0), its standard deviation (σ), the equivalent width (W_λ), and full-width half maximum are given in Å for the star HD 183143.

λ_0	σ	W_λ	FWHM
4428	0.5	3.4	20
4501.8	0.2	0.25	3.0
4726	0.5	0.20	5.0
4754.9	0.5	0.17	5.6
4763.0	0.3	0.29	5.3
4779.7	0.2	0.065	1.8
4882	2	1.27	17
5362	1	0.15	4.4
5404.3	0.2	0.07	1.0
5420	1	0.18	11
5449	2	0.56	14
5487.31	0.13	0.30	4.4
5535	5	0.53	23
5705.12	0.08	0.29	3.5
5778.3	1	0.95	17
5780.41	0.01	0.88	2.6
5797.03	0.02	0.39	1.3
5844.1	0.2	0.14	4.5
5849.79	0.03	0.10	1.0
6010.9	0.2	0.19	4.2
6042	2	0.31	14
6113.0	0.2	0.072	0.85
6177.1	1	1.85	30
6195.96	0.01	0.094	0.70
6203.06	0.03	0.43	2.3
6269.77	0.02	0.39	1.4
6283.91	0.02	2.0	3.8
6314	2	0.8	19
6353.5	0.3	0.06	3.1
6376.08	0.04	0.091	1.5
6379.30	0.02	0.16	0.86
6613.58	0.01	0.35	2.1
6660.71	0.06	0.13	2
6699.4	0.2	0.045	1.3
6993.0	0.3	0.16	1.0
7224.01	0.01	0.37	1.1
7562.2	1.0	0.22	4
8620.7	0.3	0.45	5

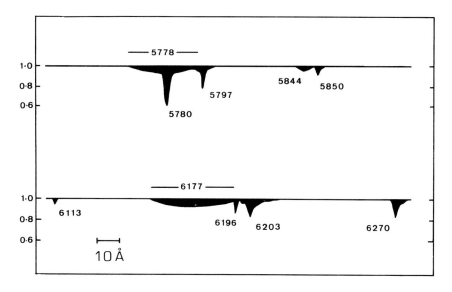

Figure 5.5 A schematic representation of diffuse bands in the yellow–red region of the spectrum, based on intensity traces for the reddened star HD 183143 (Herbig 1975). Interstellar absorptions are shown in the wavelength range 5730–5900 Å (above) and 6110–6280 Å (below). Photospheric and telluric features in the spectra are eliminated with reference to corresponding data for a comparison star (β Orionis) of similar spectral type and low reddening.

bands in gas phase molecules, but this is probably illusory as there is no asymmetry in the sharp feature, and the overall structure does not resemble the PQR profile of typical molecules (e.g., Banwell 1983). In any case, the majority of the DIBs are not associated with pairs.

Herbig (1975) has searched for regularities in the wavenumbers of the diffuse bands such as might be expected if, for example, they arise in transitions from a single lower state to a sequence of simply spaced higher levels. Allowance was made in the analysis for the possibility of missed lines. Although numerous runs were detected, none were convincing when continuity in width or strength along the sequence was demanded. Herbig concluded that if a single carrier is responsible for all of the features, a polyatomic molecule of forbidding complexity is required unless absorption from levels at least 0.1 eV above the ground state is possible.

Attempts to detect enhancement of interstellar polarization at the wavelengths of three diffuse bands ($\lambda4430$, $\lambda5780$ and $\lambda6284$) have been reported by Martin and Angel (1974, 1975) and Fahlman and Walker (1975). In each case, significant negative results were obtained. This implies that the carriers of these DIBs do not reside in the grains responsible for the optical polarization, a result which appears to exclude the aligned component of the large classical grains, and, specifically, the large silicates, as the $9.7\,\mu m$ silicate feature does show polarization enhancement (see §5.3 below).

5.2.2   Correlations

Statistical analyses of diffuse band strengths with respect to reddening and other interstellar parameters provide a useful method of investigating their origin. However, problems arise when data from diverse sources are combined, as systematic errors frequently exist between the different data sets. These may occur in the instrumentation (e.g. resolution differences) or in the analysis (e.g. calibration errors or systematic differences in assumed continuum levels). Quality control is essential in assembling data from the literature. Snow, York and Welty (1977) compiled a catalogue of diffuse band measurements for the extensively observed diffuse bands at $\lambda 4430$, $\lambda 5780$, $\lambda 5797$ and $\lambda 6284$ in approximately 1200 stars, and attempted to eliminate systematic errors by reducing the data to a common measurement system (that of Herbig). The catalogue has recently been revised and updated by Somerville (1989). Another major source of error is blending of stellar, interstellar and telluric features: for example, the $\lambda 6284$ DIB discussed by Snow, York and Welty overlaps telluric O_2 absorption. Blends with weak stellar lines are particularly troublesome, as the contamination may not be apparent unless data of high spectral resolution and signal-to-noise are available.

The interstellar nature of the diffuse bands is established on the basis of a general correlation with reddening. This is illustrated in figure 5.6, which plots the central depth of $\lambda 4430$ against E_{B-V}. Similar plots (but with smaller sample sizes) are obtained for other features. The correlations are not significantly better or worse when infrared colours such as E_{V-K} are used instead of E_{B-V} (Sneden et al 1978). As previously discussed, the existence of such correlations for interstellar features does not inevitably limit potential carriers to a component of the solid particles, as dust and gas are generally well mixed in the interstellar medium. Stars with anomalous band strengths with respect to reddening warrant the greatest scrutiny; Snow et al tabulate those deviating from the least-squares fits by more than 2σ, thus providing a useful target list for observing programmes. These stars, which constitute $\sim 6\%$ of the entire catalogue, include distant supergiants, stars in dark clouds, and a number of Be stars.

The question of whether the various DIBs are produced by a single absorber or many is addressed by the degree of intercorrelation between the strengths of different bands. Herbig (1975) carried out a detailed study of the 17 diffuse bands which have sufficient data available to give meaningful results, using a coherent data set to reduce the effect of systematic errors. He found correlation coefficients (corrected for observational error)

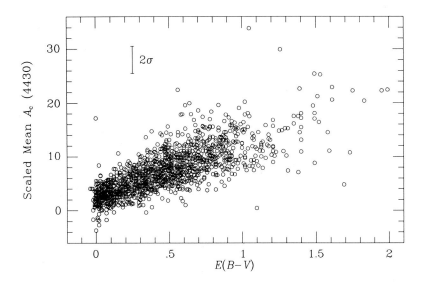

Figure 5.6   The correlation of central depth (A_c) of the $\lambda4430$ feature with reddening (E_{B-V}) for a compilation of data from the literature, merged to a common measurement system. (Diagram courtesy of W B Somerville.)

to be very close to unity in almost every case; the lowest value obtained from the 136 combinations is $r = 0.64$ (between $\lambda6284$ and $\lambda6376$), and the correlation of the broadest line in the sample ($\lambda4430$) with the narrowest ($\lambda6196$) gives $r = 0.98$! In general, the diffuse bands correlate marginally more closely with each other than with reddening. These results led Herbig to conclude that the diffuse bands are essentially a single spectrum. However, a number of cases have been reported in which specific correlations for pairs of bands appear to break down in certain lines of sight (Bromage and Nandy 1973; Baines and Whittet 1983). Krelowski and Walker (1987) suggest, on the basis of a study of the high-latitude star ζ Per, that there are three distinct families of DIBs. This possibility requires more detailed investigation in other lines of sight.

A peculiarity of diffuse band versus reddening correlations reported by a number of observers is the presence of statistically significant non-zero intercepts in unconstrained least-squares fits, usually in the sense that weak residual band absorption is implied for unreddened stars. This phenomenon has been termed the ρ Leonis effect, after a well-known example of a star with low reddening and apparent anomalously strong diffuse bands. Smith, Snow and York (1977) were led to suggest that high-velocity, low-mass clouds, which may be present in the lines of sight to distant stars with low extinction, are more efficient producers of diffuse bands than massive clouds producing high extinction. However, Blades and Somerville (1977, 1981) showed that the anomaly can result from the presence of weak, unresolved stellar lines in the diffuse band profile. This is particularly serious for broad features such as $\lambda 4430$, and in the case of ρ Leonis the anomaly disappears when the contribution of stellar lines is removed, using data of sufficient spectral resolution.

Environmental variations in the diffuse band production efficiency (measured with respect to reddening by the ratio W_λ / E_{B-V}) are potentially of great importance in identifying the carrier. Evidence for a systematic dependence on cloud density has been presented by a number of authors. Somerville (1988) has examined this question in detail, and has shown that previous analyses were influenced by failure to correct for non-zero intercepts in the W_λ versus E_{B-V} relation, which led to anomalously high apparent production efficiencies at low reddening. A general trend with density appears to be ruled out, as does a dependence on iron depletion (which does correlate with cloud density). Federman et al (1984) suggest a correlation with the column density of H_2, but again the effect is less convincing when corrections for non-zero intercepts are applied (Somerville 1988). Moreover, Meyer (1983) has demonstrated that $\lambda 5780$ is present in lines of sight with extremely low $N(H_2)$. At higher reddenings, systematic errors in W_λ / E_{B-V} are less important; there is reliable evidence for anomalies in a number of individual regions, notably those associated with current or recent star formation (Baines and Whittet 1983, and references therein). These may be due to local effects such as the presence or absence of mantles rather than any systematic trend with cloud density.

Correlations of DIB strengths with parameters of the ultraviolet extinction curve have been studied by a number of investigators (e.g. Nandy and Thompson 1975; Dorschner et al 1977; Danks 1980; Wu, York and Snow 1981; Witt, Bohlin and Stecher 1983; Seab and Snow 1984; Benvenuti and Porceddu 1989). The bands are generally well correlated with the $\lambda 2175$ feature, which is expected as both correlate with reddening. Of more interest is the possibility that individual stars which deviate significantly from the normal band strength versus E_{B-V} correlation may show similar deviations for $\lambda 2175$. Witt et al (1983) find a marginally significant positive correlation between the strengths of $\lambda 4430$ and $\lambda 2175$, and a marginally significant negative correlation between $\lambda 4430$ and the FUV extinction rise, each quantity being normalized to $E_{B-V} = 1$. Krelowski and Walker (1987) argue that $\lambda 2175$ belongs to a family of diffuse bands which also includes $\lambda 5797$, $\lambda 5850$, $\lambda 6376$, $\lambda 6379$ and $\lambda 6614$, but Benvenuti and Porceddu (1989) conclude that the $\lambda 2175$ and DIB carriers, although coexisting in the ISM, have an independent identity and history.

5.2.3 Profiles

High-resolution studies of diffuse band profiles are valuable as potential discriminators between proposed absorption mechanisms. If the carrier resides in dust, then profile shapes provide diagnostic information on grain sizes, as previously discussed for the $\lambda 2175$ feature, whereas a gas phase molecular carrier may produce resolvable fine-structure lines. For broad features such as $\lambda 4430$, reliable profile measurements are difficult because of contamination from stellar lines and the problem of establishing an accurate continuum level over a large wavelength interval. Recent investigations have thus naturally tended to concentrate on the sharper diffuse features.

The $\lambda 5780$ feature is particularly suitable for profile investigation, as it is the strongest of the sharper diffuse bands, and is not blended with any strong stellar or telluric features. However, it must be borne in mind that underlying absorption occurs throughout the $\lambda 5780$ region, produced by the broader, shallower interstellar band centred at $\lambda 5778$ (see figure 5.5). High-resolution observations of $\lambda 5780$ have been carried out by a number of investigators (Wu 1972; Savage 1976; Danks and Lambert 1976; Snell and Vanden Bout 1981; Westerlund and Krelowski 1988). Representative profiles are shown in figure 5.7, ordered in sequence of increasing reddening. The mean spectral resolution of the data is $0.07\,\text{Å}$. It is evident that no fine-structure lines are present at this resolution. The profiles tend to be asymmetric, with a steeper rise to continuum on the blue side. Apparent differences from star to star are probably caused by discrete clouds with differing radial velocities in the lines of sight, the presence of which is indicated by structure in the narrow gas phase interstellar lines of (e.g.) Na I in the same spectra (Snell and Vanden Bout 1981; Westerlund and Krelowski 1988). For example, the structure evident in figure 5.7 for χ^2 Orionis is very probably the result of discrete Doppler components with a velocity spread of $\sim 40\,\text{km s}^{-1}$. It is not clear that any of the observed star to star profile variations are real in the sense of reflecting *intrinsic* differences in DIB shapes. The general asymmetry of the $\lambda 5780$ profile is well established as an intrinsic property of the feature, and similar effects have been observed in other DIBs, including $\lambda 5797$ and $\lambda 6614$ (Wu 1972; Welter and Savage 1977; Herbig and Soderblom 1982). The $\lambda 6203$ feature has a complex profile with a sharp core flanked by asymmetric red and blue wings (Herbig 1975; Smith *et al* 1981).

The $\lambda 6196$ feature is the sharpest of all the diffuse bands. The intrinsic width (FWHM) is approximately $0.4\,\text{Å}$ after allowance for Doppler motions in discrete clouds. Like $\lambda 5780$, it has underlying absorption from a much broader feature ($\lambda 6177$). At a resolution of 0.05–$0.10\,\text{Å}$, $\lambda 6196$ is symmetric to within observational error (Smith *et al* 1981; Herbig and Soderblom 1982), in contrast to $\lambda 5780$. There is an absence of fine structure and wings, and the central wavelength is stable (after correction for cloud mo-

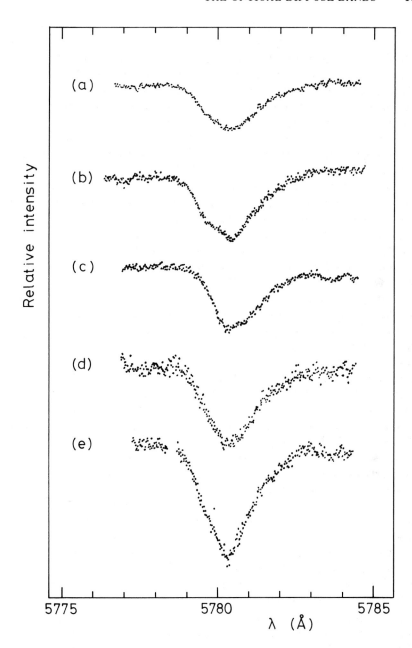

Figure 5.7   Profiles of the $\lambda 5780$ diffuse interstellar band at a spectral resolution of 0.07 Å: (a) σ Sco; (b) χ^2 Ori; (c) ν Cep; (d) HD 21389; (e) HD 187982. The central wavelength is standardized to that of σ Sco, on the assumption that differences present in the original spectra are the result of Doppler shifts due to cloud motions. (Adapted from Snell and Vanden Bout 1981.)

tion) to better than ±0.1 Å. Other features displaying symmetric profiles include λ6379 and λ7224 (Welter and Savage 1977; Herbig and Soderblom 1982). Thus, the features with well-studied profiles appear to fall into two distinct groups: those with symmetric profiles (λ6196, λ6379, λ7224) and those with marked asymmetries characterized by a red wing (λ5780, λ5797, λ6613). It seems likely that this result contains important clues as to the identity of the carrier(s).

The absence of polarization enhancement in the DIBs, noted in §5.2.1, is consistent with the profile information. Features produced by solid-state transitions in the classical-sized grain population would be expected to exhibit changes in both profile shape and central wavelength with grain size, and emission wings would be expected for radii $> 0.1 \mu$m (Savage 1976; see figure 5.8 below). No variation in DIB profiles occurs when lines of sight with different values of R_V and $\lambda_{\max}$ are compared (Snow *et al* 1982), and the lack of any detectable intrinsic drift in central wavelength argues against an association with grains responsible for the optical extinction and polarization, as previously argued for the λ2175 feature. The presence of emission wings has been advocated from time to time (e.g. York 1971) but the evidence is not convincing (Herbig and Soderblom 1982). If the DIB carrier is a component of the dust, then the grains responsible are likely to be small and/or unaligned.

5.2.4   Origins: Small Grains or Large Molecules?

The observed properties of the diffuse band spectrum appear to favour an origin either in particles smaller than those responsible for the visual extinction and polarization, or in gas phase molecules. Before discussing specific mechanisms, we begin by placing an observational constraint on the abundance of the carrier. Expressing equation (2.15) in terms of $W_\lambda = (\lambda^2/c) W_\nu$, the column density of the absorber is given in m^{-2} by

$$N_{\rm X} = 1.14 \times 10^{24} \frac{W_\lambda}{f \lambda^2} \tag{5.8}$$

where W_λ and λ are in Å. Considering the strongest feature (λ4430), the mean production efficiency is (Herbig 1975)

$$\frac{W_\lambda(4430)}{E_{B-V}} \simeq 2.3 \text{ Å mag}^{-1}, \tag{5.9}$$

and thus

$$\frac{N(4430)}{N_{\rm H}} \simeq \frac{2.3 \times 10^{-9}}{f} \tag{5.10}$$

by analogy with equation (5.7). This result provides a useful test for proposed carriers, discussed below.

The intrinsic width of the diffuse features in comparison to interstellar lines produced by simple gas phase molecules such as CH, CH^+ and CN is readily explained if they are produced by some component of the dust, as the absorption lines of atoms and molecules embedded in solids are broadened by the electric field of the crystal lattice. Solid-state mechanisms which have been suggested include plasma oscillations in metallic particles, and various absorptions associated with impurities or surface defects in the grain lattice. Plasma oscillations, considered in §5.1.2 above as a mechanism for the $\lambda 2175$ feature, were suggested for the diffuse bands by Unsöld (1964). Features with symmetric profiles are produced in the small-particle limit, and the variety of observed band widths could be attributed to solids with different electrical conductivities. However, each of the observed features is required to originate in a distinct grain type, and no specific proposal based on plasma oscillations has been made for any interstellar feature other than $\lambda 2175$.

Absorptions associated with 'impurity centres' have been discussed more widely. This term is used here to describe any foreign atom or ion, vacancy, or other defect in a crystal lattice which produces optical absorption. The impurity spectra of solids are discussed in detail by Rebane (1970), and models for impurity absorptions in small particles have been formulated by van de Hulst (1957), Savage (1976), Purcell and Shapiro (1977) and Shapiro and Holcomb (1986a). Of the many possibilities, most may be ruled out on the basis of band width or abundance considerations. Features up to a few tens of Ångstroms wide are very *narrow* by solid-state standards, and most impurity centre absorptions are typically an order of magnitude wider than this in the visible. Some bands produced by Fe ions in silicates or iron oxides are as narrow as 20–30 Å and may thus be capable of explaining some of the broader diffuse features (Huffman 1970; Manning 1976). These bands are numerous in terrestrial minerals, and it is not therefore surprising that some wavelength coincidences are found with the DIB spectrum. However, they are intrinsically very weak in silicates: for example, the andradite garnet proposed by Manning has $f \sim 3.5 \times 10^{-6}$, and the required abundance of Fe^{3+} impurity centres is $\sim 7 \times 10^{-4}$ (equation (5.10)), which exceeds the solar abundance of Fe by a factor of ~ 20. Huffman (1977) has pointed out that the absorptions associated with intrinsic Fe^{3+} ions in oxides such as Fe_2O_3, Fe_3O_4 and $MgFe_2O_4$ have f-values which are two or three orders of magnitude higher than those in silicates, and are thus viable on abundance grounds, but the optical properties of these oxides are not well known as their opacity hinders laboratory measurements.

A mechanism capable of producing bands as narrow as 0.5–5 Å (FWHM) is pure electronic (no-phonon) absorption at impurity centres (Rebane 1970). The narrowness arises because the vibrational state does not change, i.e. the transition is decoupled from vibrations in the crystal lattice. Pure electronic transitions have been advocated as a mechanism for the $\lambda 5780$

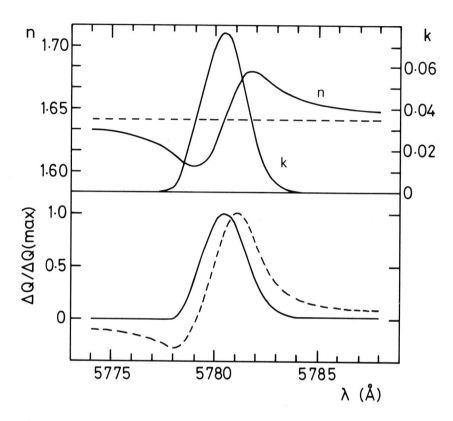

Figure 5.8 A model for the λ5780 feature based on no-phonon absorptions due to impurity centres in a low-temperature dielectric lattice (Savage 1976). The optical constants.n and k are plotted against wavelength in the upper frame; the dashed line indicating the 'continuum' value of n. The lower frame compares the resulting absorption profiles (ΔQ_{ext} normalized to unity at maximum absorption) calculated for two values of grain radius ($a = 0.05\,\mu$m, solid curve; $a = 0.125\,\mu$m, dashed curve).

diffuse band by Wu (1972) and Savage (1976), although no specific material was proposed. In bulk solids, Voigt profiles are produced, and on this basis it is possible to calculate generalized optical constants for an impurity centre absorption for use in Mie theory calculations (Savage 1976). Typical results are illustrated in figure 5.8, assuming that the host material is dielectric with refractive index appropriate to a silicate. The profile becomes symmetric for sufficiently small grains ($a \leq 0.05\,\mu$m), as illustrated by the continuous curve in figure 5.8, resulting in a good fit to symmetric diffuse bands such as λ6379 (Welter and Savage 1977). For larger grains, the peak of the feature shifts to the red and asymmetry is introduced (figure 5.8, dashed curve). This asymmetry has the same sense as that observed for features such as λ5780 (figure 5.7), giving the appearance of a symmetrical 'core' with a long-wavelength wing; a fit to λ5780 in HD 183143 for $a = 0.075\,\mu$m is shown in figure 5.9. Another interesting property of pure electronic transitions is that they may be accompanied by transitions involving phonons, producing broad vibrational sidebands which accompany the narrow no-phonon lines. It is conceivable that the pairing of sharp and broad features noted in §5.2.1 is produced in this way (see Wu 1972).

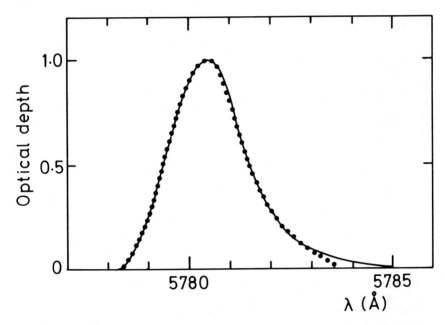

Figure 5.9 An attempt to model the observed λ5780 feature using an impurity centre model (after Savage 1976). The calculated profile for a constant grain radius of $0.075\,\mu$m (curve) is compared with smoothed observational data for HD 183143 (points), in each case normalized to unit optical depth at the band centre.

The impurity model was further developed by Shapiro and Holcomb (1986a, b), who present a more sophisticated treatment of the electromagnetic interaction between host grain and impurity centre. Although results are encouraging, the sensitivity of the profile shape to grain dimensions above the small-particle limit remains a significant problem: this places tight constraints on the size distribution of the grains required to match the asymmetry in features such as λ5780, although mantle growth can be accommodated if the mantle is impurity-free (Shapiro and Holcomb 1986b). If the host is a silicate, a further difficulty arises in explaining the presence of polarization structure in the infrared silicate feature and its apparent absence in at least some of the diffuse bands. These problems are alleviated for smaller particles, but only the symmetric features can then be fitted. Duley (1975) discusses factors which influence the dependence of profile on particle size. Surface effects become increasingly important for progressively smaller particles, and impurity centres located near the surface of a grain may experience electric and magnetic fields different from those enclosed inside, resulting in modifications to the spectrum. Surface stresses can affect the lattice spacing in small particles, which may result in a shift in wavelength with grain size. Duley suggests that the pairing of sharp and broad diffuse features corresponds to the same absorption in large and small grains respectively, but this requires that a bimodal size distribution is rigorously maintained in all directions.

Turning our attention to gas phase carriers, a first requirement is to identify a mechanism capable of producing features which are sufficiently broad. Although interstellar lines arising in isolated atoms and molecules are generally sharp, diffuse features may be produced by preionization or predissociation transitions, in which the carrier is temporarily excited to a discrete energy level above its first ionization or dissociation potential, or by molecular absorptions containing unresolved rotational structure. Preionization of negative ions such as C^-, N^- and O^- was suggested by Herzberg (1955) on the basis of low ionization energy and high f-values. Specific identifications have been proposed for a few of the diffuse bands, including $\lambda4430$ (H^-), $\lambda5780$ and $\lambda5797$ (O^-) (Wu 1972; Rudkjobing 1969). Preionization is capable of producing asymmetric profiles in either sense, displaying characteristically broad absorption wings (see Fano 1961 for detailed discussion). A comparison with interstellar profiles has been carried out by Savage (1976) and Welter and Savage (1977), who concluded that the core and wings of the features cannot be matched simultaneously. The occurrence of convincing asymmetries in only one sense in the diffuse band spectrum is further evidence against preionization as a viable mechanism.

The case for a molecular origin of the diffuse bands has been reviewed by Smith, Snow and York (1977). Predissociation in small molecules (3–4 atoms) with low dissociation energies was advocated by Herzberg (1955, 1967), but numerous investigations of cosmically abundant simple molecules have failed to identify the carrier (Douglas 1977). Another difficulty with predissociation is that, as each absorbed photon destroys the molecule, an efficient formation process is required to restore the population. Many potential carriers may be excluded on the basis of their short lifetimes against photodissociation in low-density regions, bearing in mind that DIBs are detected in lines of sight where $N(H_2)$ is extremely low (Meyer 1983). However, this problem is alleviated considerably for large polyatomic molecules, which can survive the fracture of a single bond and may be capable of absorbing ultraviolet photons in an internal excitation mode without undergoing dissociation. Molecules containing ten or more atoms have been shown to exist in dense interstellar clouds, and gas phase reaction schemes provide a basis for understanding their origin (e.g. Mitchell *et al* 1978), but formation in low-density clouds is improbable in the extreme in view of the short lifetimes of each intermediate step against photodissociation. The large molecule hypothesis (Danks and Lambert 1976) thus requires a species which originates in dense regions (molecular clouds or cool circumstellar envelopes) and survives subsequent exposure to a low-density environment.

A number of types of polyatomic molecule can give absorption bands which may appear structureless at currently available resolutions. Smith *et al* (1977) advocate quasi-linear, quasi-planar or ring molecules of intermediate molecular weight, composed of H, C, N and O atoms, which produce optical features with oscillator strengths typically $f \sim 0.05$. This gives a minimum abundance limit of $\sim 5 \times 10^{-8}$ (equation (5.10)). There is thus no shortage of the required chemical elements for such a model, but, of the observed interstellar gas phase molecules, only simple diatomics such as H_2, CO and CH typically exceed this level of abundance. The possibility of an origin for the DIBs in polycyclic aromatic hydrocarbons has been discussed by a number of authors (e.g. van der Zwet and Allamandola 1985; Leǵer and d'Hendecourt 1985). Sufficiently large PAHs are stable in the diffuse ISM for the reasons discussed above, and can undergo thermal fluctuations without sublimation. They have first and second ionization potentials typically $\sim 7\,\text{eV}$ and $\sim 20\,\text{eV}$, respectively, and are thus expected to be singly ionized under normal interstellar conditions. Some laboratory data are available, which show that a range of fairly strong ($f \sim 0.1$) absorptions are present in the visible. PAHs are postulated to contain some 5–10% of the available carbon, and thus easily satisfy abundance constraints. However, no *specific* identification with a particular PAH or group of PAHs has been proposed. Leach (1991) concludes from an examination of profile shapes for a number of PAH absorptions that they are inconsistent with the observed DIB profiles. As PAHs generally show stronger absorption bands in the ultraviolet than in the visible, the absence of additional interstellar features between $\lambda 2175$ and $\lambda 4430$ in stellar spectra is further evidence against these particles as carriers for the DIBs.

Although it is not possible to distinguish unambiguously between molecular and solid-state mechanisms for the diffuse bands on the basis of currently available data, the weight of evidence appears to support an origin in some component of the grain material. I conclude this discussion by suggesting future lines of research which may point to a solution of this seemingly intractable problem. Observationally, high-quality spectra for stars with anomalous DIB strengths may prove crucial. The primary goal is to establish with certainty whether all the diffuse bands have a common origin or whether distinct families exist, and this may best be achieved by studying the entire spectrum in stars where some bands are known to be weak or strong with respect to reddening. If distinct families are proven to exist, the question of polarization structure in the DIBs should be re-examined, as its absence in some features would not imply its absence in others. A related question is the degree of common behaviour between DIBs within putative families and the λ2175 feature and FUV extinction rise. Future profile studies should concentrate on stars with single or well-separated cloud components, determined by the velocity structure of the narrow atomic and molecular interstellar lines to a resolution better than the velocity dispersion of the individual clouds. The depletions of the gas phase elements should be determined and compared in lines of sight with normal and anomalous diffuse band strengths. Finally, the importance of laboratory investigations of the optical properties of candidate materials under simulated interstellar conditions cannot be too highly stressed. Although the observations mentioned above may serve to isolate the mechanism for the diffuse band absorptions, a specific identification is unlikely to emerge in the absence of complementary laboratory studies.

5.3   THE INFRARED ABSORPTION FEATURES

"It seems to me that in order to determine the composition of the dust, we must turn to the infrared part of the spectrum..."

J E Gaustad (1971)

5.3.1   The Observed Features: an Overview

Dust-related absorption features arise at infrared wavelengths characteristic of vibrational transitions in the grain lattice. The frequency of a molecular vibration is determined by the masses of the vibrating atoms, the molecular geometry, and the forces holding the atoms in their equilibrium positions. For example, the vibrations of a diatomic molecule containing atoms of masses m_1 and m_2 may be represented by a harmonic oscillator with a fundamental vibrational frequency

$$\nu_F = \frac{1}{2\pi}\left\{\frac{k}{\mu}\right\}^{\frac{1}{2}} \tag{5.11}$$

where k is the force constant of the chemical bond between the two atoms and $\mu = m_1 m_2/(m_1 + m_2)$ is the reduced mass. In the gas phase, rotational splitting of the vibrational energy levels leads to the production of molecular bands (e.g. Banwell 1983), but rotation is suppressed in the solid phase and the characteristic P and R branches of the gas phase spectrum are replaced by a broad, continuous spectral feature at a frequency close to ν_F. Taking into account the masses of the most abundant elements likely to be present in dust grains (H, O, C, N, Si, ...) and the kinds of chemical bonds they form, the expected values of ν_F correspond to wavelengths in the range 2–$25\,\mu$m. It should be noted that not all molecular vibrations actually produce spectral features: a prerequisite is that the dipole moment oscillates during the vibration; thus, the vibrations of all homonuclear diatomic molecules (such as O_2 and N_2) and the centrosymmetric vibrations of symmetric molecules cannot be studied by infrared spectroscopy. Table 5.3 lists vibrational modes which do give rise to absorption in a number of solids of potential astrophysical interest.

The infrared spectrum of interstellar dust has been observed between 2 and $25\,\mu$m using a variety of observing facilities, including ground-based and airborne telescopes, and IRAS. At the time of writing only the region from 4.1 to $4.5\,\mu$m remains totally unexplored: this is rendered unobservable from ground-based and airborne platforms by strong telluric CO_2 absorption. The principal observed features are listed in table 5.4†, together with proposed identifications, a guide to relative strengths, and a selection of key references. Most features are identified, at least to the extent of being attributed to a particular chemical bond (e.g. table 5.3), although the exact nature of the host molecule or molecular mix is uncertain in a number of cases. Infrared spectroscopy nevertheless provides the most direct diagnostic method available for investigating grain composition. The incidence of observational data on the various features is uneven: those at $3.0\,\mu$m and $9.7\,\mu$m (the principal 'ice' and 'silicate' absorptions) have been observed in the largest number of sources. The features are grouped in table 5.4 according to whether they occur in lines of sight which predominantly intercept (a) the diffuse interstellar medium, or (b) molecular clouds. Group (a) may be regarded as a subset of group (b) as its members have counterparts at similar wavelengths in group (b).

The distinction between 'diffuse' (atomic) and 'dense' (molecular) environments (§1.2) is fundamental to our discussion of the infrared absorption features, but it should be noted that it is not always a straightforward matter to make an *a priori* selection of sources for observation which provide independent samples. Studies of diffuse clouds, in particular, are hampered by the paucity of known suitable candidates which are both sufficiently bright and sufficiently extinguished to give measurable optical depth in the features. An ideal source would lie close to the galactic plane at a distance great enough to ensure a large column density of dust, accumulated through inclusion of many H I clouds (and no H_2 clouds) in the line of sight.

---

† Only well-established interstellar features are included; more extensive listings (which include some unconfirmed features) may be found in the reviews of Whittet (1987, 1988) and Roche (1989b).

Table 5.3   Molecular vibrational modes giving rise to absorption in some solids of astrophysical interest. Wavelength data are from Allamandola and Sandford (1988) for H_2O and other ices; Day (1979, 1981) for magnesium and iron silicates; and Whittet *et al* (1990) for silicon carbide.

Molecule	Mode	λ (μm)
H_2O	O–H stretch	3.05
	H–O–H bend	6.0
	libration	13.3
NH_3	N–H stretch	2.96
	umbrella	9.35
CH_4	C–H stretch	3.32
	C–H deformation	7.69
CO	C–O stretch	4.67
CO_2	C–O stretch	4.27
	O–C–O bend	15.3
CH_3OH	O–H stretch	3.08
	C–H stretch	3.35
	C–H stretch	3.53
	O–H bend, C–H deformation	6.89
	C–O stretch	9.75
$MgSiO_3$	Si–O stretch	9.7
	O–Si–O bend	19.0
Mg_2SiO_4	Si–O stretch	10.0
	O–Si–O bend	19.5
$FeSiO_3$	Si–O stretch	9.5
	O–Si–O bend	20.0
Fe_2SiO_4	Si–O stretch	9.8
	O–Si–O bend	20.0
SiC	Si–C stretch	11.2

The galactic centre has long been regarded as a good approximation to this ideal, but recent studies suggest that some proportion of the extinction actually arises in a molecular cloud (McFadzean *et al* 1989). Molecular clouds present different problems of sampling. Contamination by diffuse-cloud absorption can be largely avoided by selecting nearby clouds away from the galactic plane. However, the brightest infrared sources in molecular clouds are generally embedded protostellar or pre-main-sequence objects, and the radiative flux and winds they emit may influence the interstellar environment around them. A good example is W33A-IR, which has been the

Table 5.4   The principal observed dust-related absorption features in the infrared. Mean relative strengths are expressed in terms of the observed peak optical depth of each feature, normalized relative to that of 9.7 μm for diffuse clouds (a) and 3.0 μm for molecular clouds (b).

a. Diffuse clouds

λ (μm)	Identification	$\tau_\lambda/\tau_{9.7}$	References[†]
3.0	H_2O?	0.0–0.2	1, 2
3.4	HAC?	0.06	1, 2, 3
9.7	Silicate	1.0	4, 5
18.5	Silicate	0.4	4, 6

b. Molecular clouds

λ (μm)	Identification	$\tau_\lambda/\tau_{3.0}$	References[†]
2.85	?	0.2–0.5	7
3.05	H_2O	1.0	7, 8, 9
3.2–3.6	Hydrocarbons?	0.15–0.35	7, 8
4.62	'XCN'	0.0–0.2	10, 11
4.67	CO	0.0–1.2	10, 11, 12
6.0	H_2O	0.20	8, 13
6.85	Hydrocarbons?	0.20	8, 13
9.7	Silicate	0.4–5.0	8, 9
18.5	Silicate	?	14

† References: 1. Butchart *et al* (1986); 2. McFadzean *et al* (1989); 3. Adamson *et al* (1990); 4. Roche (1988); 5. Roche and Aitken (1984a); 6. McCarthy *et al* (1980); 7. Smith *et al* (1989); 8. Willner *et al* (1982); 9. Whittet *et al* (1988); 10. Lacy *et al* (1984); 11. Larson *et al* (1985); 12. Geballe (1986); 13. Tielens *et al* (1984); 14. Aitken *et al* (1989).

subject of intense investigation because of its brightness and the presence of exceptionally deep infrared absorption features (Roche 1989b and references therein); evidence for accelerated grain evolution in the vicinity of this source is provided by observations of the 4.62 μm feature (strong in W33A and weak or absent elsewhere), suggestive of local annealing of grain material (Lacy *et al* 1984). We return to this topic in §7.2.

5.3.2   Diffuse Clouds

The line of sight to the cluster of infrared sources (IRS) associated with the galactic centre (GC) in Sagittarius (Becklin and Neugebauer 1975) is

obscured by some 30 magnitudes of visual extinction accumulated along a
7–8 kpc path-length (Roche 1988). This cluster contains individual sources
which are believed to be luminous, cool stars on the basis of 2–4 μm spectroscopy (e.g. Wollman, Smith and Larson 1982), and which should thus
provide suitable background sources in which to measure interstellar absorptions. In this section, results for the galactic centre are discussed and
compared with those for reddened stars closer to the solar system, notably the highly reddened B-type supergiant Cyg OB2 no. 12. Peak optical
depths (τ_λ) for various features discussed below are listed in table 5.5 for
GC-IRS7 and Cyg OB2 no. 12.

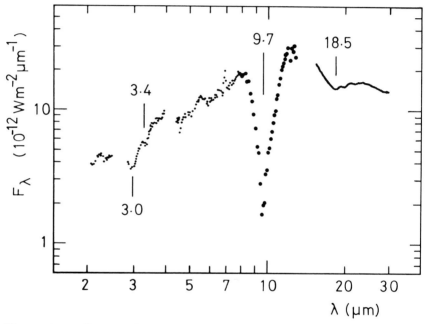

Figure 5.10   Integrated infrared spectrum of the galactic centre source Sgr A
from 2 to 30 μm (adapted from Roche 1988). This spectrum is characteristic of
dust in diffuse clouds. Dust features discussed in the text are labelled by their
wavelengths in μm. Absorptions at 4.6 and 2.3 μm are identified with fundamental and overtone bandheads, respectively, in interstellar and photospheric gas
phase CO.

The spectrum of the galactic centre from 2 to 25 μm is illustrated in
figure 5.10, as observed through apertures large enough to include the flux
from all the individual point sources (Roche 1988 and references therein).
The spectrum is dominated by deep 9.7 μm absorption, with weaker features occurring at 3.0, 3.4 and 18.5 μm. Strong absorption at 9.7 μm is also
seen towards some external galactic nuclei (Roche *et al* 1986a, 1991). The

9.7 and 18.5 μm absorptions are attributed to stretching and bending resonances in silicates (table 5.3). Corresponding peaks have been shown to occur in spectropolarimetric data (§5.3.4), an important result as it demonstrates that silicates are included in the aligned component of the grains in the diffuse ISM (Chapter 4). Of the remaining features observed to date towards the galactic centre, that at 3.4 μm appears to be associated with carbonaceous material and that at 3.0 μm is most likely due to water-ice. A number of other features discernible in molecular cloud sources (table 5.4b), notably those at 6.0 and 6.85 μm, are not detected in diffuse clouds.

Of the features seen in diffuse clouds, only that at 9.7 μm is strong enough to be easily observable in optically selected reddened stars with more modest extinctions than the galactic centre group. Roche and Aitken (1984a) have studied this feature in stars with A_V in the range 3–15 mag. They conclude that the profile shape shows no significant variations from star to star and is closely similar to the emission profile seen in the dust shells of many cool, late-type stars (§7.1). Details of the profile shape provide information on the nature of the silicate material. Highly crystalline silicates (as in many terrestrial igneous rocks) produce structure which is incompatible with the smoothness of the observed profile, whereas amorphous, disordered, hydrated or layer-lattice silicates appear capable of matching the data more closely (Zaikowski *et al* 1975; Krätschmer and Huffman 1979; Day 1979, 1981; Butchart and Whittet 1983; Dorschner and Henning 1986; Dorschner *et al* 1988). The possibility that interstellar silicates are hydrated is open to observational test, as an O–H resonance in such materials is predicted to occur in the 2.6–2.9 μm region. This part of the spectrum is unobservable from the ground, and available data from the Kuiper Airborne Observatory (Knacke *et al* 1985) do not place very rigorous constraints on the abundance of hydrated silicates in the ISM.

Table 5.5 Optical depths, visual extinctions and distances compared for the lines of sight to CG-IRS7 and Cyg OB2 no. 12. (Data from Adamson *et al* 1990; McFadzean *et al* 1989; Roche and Aitken 1985; Whittet *et al* 1990.)

Parameter	GC-IRS7	Cyg OB2 no. 12
$\tau_{3.0}$	0.65	< 0.02
$\tau_{3.4}$	0.25	0.03
$\tau_{9.7}$	3.6	0.58
$\tau_{11.2}$	< 0.1	< 0.08
A_V	30	10.2
d (kpc)	7.5	1.8

The peak optical depth in the 9.7 μm feature ($\tau_{9.7}$) correlates closely with visual extinction for lines of sight towards stars within 3 kpc of the Sun. Figure 5.11 shows a plot of $\tau_{9.7}$ versus A_V. For the sample of 'nearby' stars (which includes Cyg OB2 no. 12), a close linear correlation is apparent; an unconstrained least-squares fit passes through the origin to within error limits, and the slope provides an estimate of the ratio $A_V/\tau_{9.7}$:

$$A_V/\tau_{9.7} = 18.5 \pm 1.0. \tag{5.12}$$

Towards the galactic centre, the strength of the 9.7 μm feature is enhanced relative to A_V by a factor of about 2 compared with the solar neighbour-hood ($A_V/\tau_{9.7} = 9 \pm 1$), as illustrated in figure 5.11. The value of $A_V/\tau_{9.7}$ predicted for pure silicate dust is critically dependent on the size distribu-tion (which determines the contribution of scattering to A_V). Laboratory-generated silicate smokes discussed by Stephens (1980) containing grains with radii $a \sim 0.01\,\mu$m have $A_V/\tau_{9.7} \sim 1$, but Mie calculations for more realistic size distributions producing fits to the visible extinction curve have $A_V/\tau_{9.7} \sim 10$ (Gillett *et al* 1975a). The most likely explanation for regional variations in the observed $A_V/\tau_{9.7}$ ratio is differences in the contributions of O-rich and C-rich dust to A_V, with silicate dust dominant towards the galactic centre and with carbonaceous dust producing 50% or more of A_V in the solar neighbourhood (see Adamson *et al* 1990).

The average density of silicate dust in a column of length L required to account for the observed strength of the 9.7 μm feature may be estimated from the relation

$$\rho = \frac{\tau_\lambda}{\kappa_\lambda\,L} \tag{5.13}$$

where κ_λ is the mass absorption coefficient of the absorber at the wave-length of peak absorption. For grains of radius a and specific density s, κ_λ is related to the absorption efficiency by

$$\kappa_\lambda = \frac{3Q_{\text{abs}}}{4as} \tag{5.14}$$

and may thus be deduced from Mie theory computations, if the optical con-stants are known, or measured directly for laboratory-generated smokes. The available data suggest that a value of $\kappa_{9.7} \simeq 300\,\text{m}^2\,\text{kg}^{-1}$ is typical of amorphous silicates (Dorschner and Henning 1986). Combining equa-tion (5.12) with the average value of A_V/L in the galactic plane near the Sun (1.8 mag kpc^{-1}; equation (1.7)), we obtain $\tau_{9.7}/L \simeq 0.097\,\text{kpc}^{-1} \simeq 3.2 \times 10^{-21}\,\text{m}^{-1}$, and thus equation (5.13) gives

$$\rho_d(\text{sil.}) \simeq 1.1 \times 10^{-23}\ \text{kg m}^{-3}. \tag{5.15}$$

This result is consistent with the value deduced independently from the abundances and depletions of the constituent elements (§2.4.4, equa-tion (2.21)). The strength of the observed 9.7 μm absorption feature per

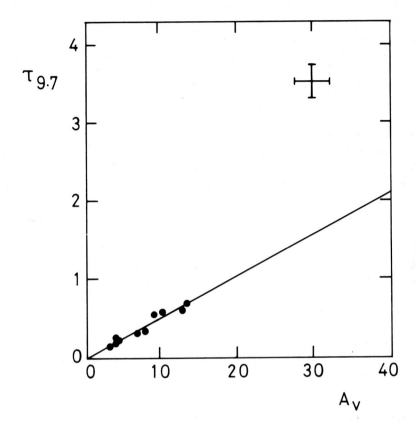

Figure 5.11   Plot of the optical depth in the 9.7 μm silicate feature against visual extinction (Whittet 1987 and references therein). Points represent stars in the solar neighbourhood, fitted by equation (5.12). The cross represents the locus occupied by galactic centre sources.

unit extinction in the solar neighbourhood thus requires that essentially all of the available Si atoms are tied up in silicate grains. For the galactic centre, values of $L \simeq 7.7\,\mathrm{kpc}$ and $\tau_{9.7} \simeq 3.6$ in equation (5.13) imply an enhancement in the mean silicate density by a factor of about 5 compared with the solar neighbourhood.

One resonance listed in table 5.3 which is surprisingly absent in interstellar spectra is that of silicon carbide (SiC) at 11.2 μm. This feature is widely observed in emission in C-rich stellar atmospheres (§7.2) and dust containing SiC is therefore presumably ejected into the interstellar medium. SiC particles of probable extra-solar-system origin have been isolated in C-type meteorites (§7.3), providing further circumstantial evidence for the existence of SiC at some level in interstellar dust. Its non-detection is re-

markable, as the intrinsic strength of the feature associated with the Si–C
bond is greater than that of the corresponding Si–O feature in silicates, observed in many lines of sight (Whittet, Duley and Martin 1990). Figure 5.12
shows 7.7–13.5 μm spectra of 10 individual sources in the galactic-centre
group after correction for the presence of foreground silicate absorption.
Most of the spectra in figure 5.12 exhibit residual silicate *emission*, which
is likely to be intrinsic to the circumstellar environments of the underlying late-type stars (§7.1.2). The position and width of the expected SiC
feature are marked in the figure, assuming it to be an inversion of the corresponding circumstellar feature (a good approximation for silicates). It is
clear that no appreciable SiC absorption is present. The upper limit on the
optical depth ($\tau_{11.2}$) is a factor of nearly 40 less than that of the silicate
feature towards the same sources (table 5.5).

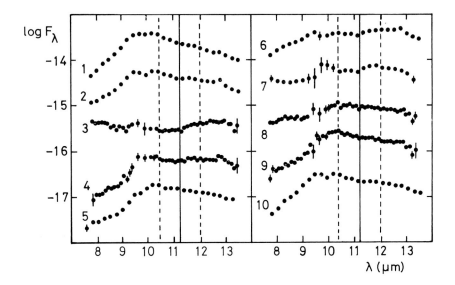

Figure 5.12   A search for interstellar SiC. The 7.5–13.5 μm spectra of galactic
centre sources (observed by Roche and Aitken 1985 at a resolution of 0.23 μm)
are displayed after correction for foreground silicate absorption.   The central
wavelength and width (FWHM) of the predicted SiC feature are marked by solid
and dashed lines (Whittet *et al* 1990).

We may use this result to estimate the relative abundance of Si in silicate
and silicon carbide dust in diffuse clouds. The column density of Si atoms
contained within grain species X (where X represents SiC or silicates) is
given by

$$N_{\mathrm{Si}}(X) = \frac{p(X)\tau_\lambda}{28 m_\mathrm{H} \kappa_\lambda} \tag{5.16}$$

where $p(X)$ is the proportion by mass of Si in X. For silicon carbide, $p(SiC) = 0.7$, whilst for silicates we may estimate $p(sil.) \simeq 0.2$ (appropriate to a mixture of magnesium and iron silicates with formulae $MSiO_3$ and M_2SiO_4, where M = Mg or Fe). Substituting for $p(X)$ in equation (5.16), and using a value $\kappa_{11.2} \simeq 660\,m^2\,kg^{-1}$ for the mass absorption coefficient of SiC (Whittet *et al* 1990), the ratio of Si atoms in SiC to Si atoms in silicates is given by

$$\frac{N_{Si}(SiC)}{N_{Si}(sil.)} \simeq 1.6\,\frac{\tau_{11.2}}{\tau_{9.7}}. \tag{5.17}$$

Using observed optical depths for GC-IRS7 from table 5.5, the number of Si atoms which can reside in interstellar SiC particles is thus < 5% of those in silicates (Cyg OB2 no. 12 sets a less rigorous limit). The scarcity of SiC in the interstellar medium is in marked contrast to its apparent ubiquity in carbon-rich circumstellar shells (§7.1) but consistent with its rarity in meteorites (§7.3).

The absorption feature at $3.4\,\mu m$ may provide an important clue to the nature of the carbon-rich dust component in diffuse clouds. This feature is observed in various galactic-centre sources (McFadzean *et al* 1989) and is also detected at a low level in other lines of sight with more modest extinctions (Tapia *et al* 1989; Adamson *et al* 1990). The feature and its associated substructure are widely attributed to a blend of interstellar absorptions due to C–H resonances: the wavelength of peak absorption corresponds to the asymmetric C–H stretch in aliphatic surface groups on hydrogenated amorphous carbon (HAC; Duley and Williams 1981, 1983), and also to aliphatic C–H absorption in organic hydrocarbons (Sandford *et al* 1991) produced in the laboratory by ultraviolet photolysis or proton irradiation of ices (e.g. Sagan and Khare 1979; Moore and Donn 1982; Greenberg 1989). The profile observed in GC-IRS7 is illustrated in figure 5.13 and compared with laboratory data for HAC. Adamson *et al* (1990) show that the $3.4\,\mu m$ feature is present in Cyg OB2 no. 12 with essentially the same profile as in GC-IRS7, but its central depth is a factor of ~ 3 weaker per unit A_V (table 5.5). This may reflect the optical properties of the absorber as a function of hydrogenation: carbonaceous dust with a low degree of hydrogenation inevitably produces a weak $3.4\,\mu m$ feature, and is strongly absorbing in the visible; as hydrogenation increases, the strength of the $3.4\,\mu m$ feature is enhanced due to the greater number density of C–H bonds, and, simultaneously, the visible absorption is reduced as the carbon becomes more polymeric. Systematic variations in the apparent strength of $\tau_{9.7}$ per unit A_V would thus also result. The observed intercorrelation of $\tau_{3.4}$, $\tau_{9.7}$ and A_V may be understood if there is a greater degree of hydrogenation towards the galactic centre compared with Cyg OB2 no. 12 (Adamson *et al* 1990).

In principle, the contribution of the $3.4\,\mu m$ absorber to the total mass density of dust may be estimated from the mass absorption coefficient us-

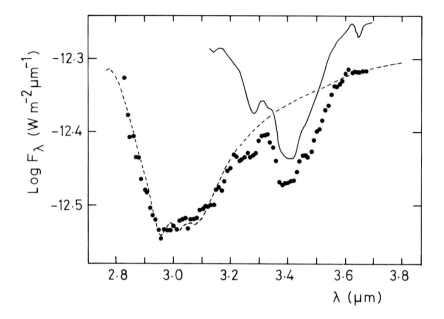

Figure 5.13   Spectrum of Galactic Centre source IRS7, illustrating the profiles of the 3.0 and 3.4 μm features. Observational data (Butchart *et al* 1986, points) are compared with laboratory data (Schutte and Greenberg 1988) for an ice mixture (dashed curve) and amorphous carbon (solid curve).

ing equation (5.14). Unfortunately, $\kappa_{3.4}$ is rather imprecisely known for most candidates. On the basis of the discussion in §2.4, upper limits on the density of organic matter and hydrogenated carbon in the solar neighbourhood are $\sim 7 \times 10^{-24}\,\mathrm{kg\,m^{-3}}$ (equation (2.22)) and $\sim 4 \times 10^{-24}\,\mathrm{kg\,m^{-3}}$ (table 2.2), respectively. In the line of sight to Cyg OB2 no. 12, $\tau_{3.4}/L \sim 5.4 \times 10^{-22}\,\mathrm{m^{-1}}$ (from table 5.5) and making use of equation (5.14), we deduce lower limits of $\kappa_{3.4} > 77\,\mathrm{m^2\,kg^{-1}}$ for organics and $\kappa_{3.4} > 130\,\mathrm{m^2\,kg^{-1}}$ for HAC. The former value is consistent with available laboratory data for organic refractory residues, which suggest $\kappa_{3.4} \sim 60 \pm 20\,\mathrm{m^2\,kg^{-1}}$ (unpublished data quoted by Whittet 1988). The laboratory study of Ogmen and Duley (1988) suggests a somewhat lower value ($\kappa_{3.4} \sim 20\,\mathrm{m^2\,kg^{-1}}$) for HAC, but this result is very uncertain and depends critically on the specific mass density and the number density of C–H bonds, both of which are strongly dependent on the nature of the sample.

It seems likely that the 3.0 μm feature observed towards the galactic centre is caused by O–H stretching vibrations in water-ice, and that its presence is due to the chance alignment of a molecular cloud in the line of sight contributing a proportion of the total extinction (see McFadzean *et al* 1989 for further discussion and references). Figure 5.13 demonstrates that the observed profile in GC IRS7 is consistent with ice absorption. Observations of the nucleus of the edge-on spiral galaxy NGC 4565 lend support to the view that $3.0\,\mu m$ absorption is not a common feature of the diffuse interstellar media of galactic disks (Adamson and Whittet 1990). Towards highly reddened OB stars in the solar neighbourhood, such as Cyg OB2 no. 12, the $3.0\,\mu m$ feature is undetected at a level a factor of 10 less (in terms of $\tau_{3.0}/A_V$) than that observed in IRS7. It is thus implausible that $3.0\,\mu m$ absorption is produced by the same grain material as $3.4\,\mu m$ absorption, which tends to argue against an identification with organic refractory material (in which both O–H and C–H bonds are expected to occur).

In summary, infrared spectroscopy of lines of sight which sample only low-density regions support the presence of amorphous silicates and a carbon-rich component which may be hydrogenated amorphous carbon, organic refractory residues, or some combination of these substances. Abundance requirements suggest that these materials contribute essentially all of the grain mass in the diffuse interstellar medium in the solar neighbourhood, and the strengths of the principal absorption signatures at 3.4 and $9.7\,\mu m$ are consistent with materials with mass absorption coefficients $\kappa_{3.4} \sim 100\,m^2\,kg^{-1}$ and $\kappa_{9.7} \sim 300\,m^2\,kg^{-1}$. Significant absorption at $3.0\,\mu m$ occurs only in those lines of sight which intersect a dense molecular cloud, and this feature thus provides a reliable diagnostic test for molecular cloud material.

5.3.3   Molecular Clouds

The characteristic spectral signatures of grain material observed in molecular clouds are those at 3.05, 4.67, 6.0, 6.85 and $9.7\,\mu m$ (table 5.4), illustrated in figure 5.14 for a typical embedded 'protostellar' source (NGC 7538E). Of this group, the $9.7\,\mu m$ silicate feature is the only one which is also prominent in the line of sight to the galactic centre (compare figures 5.14 and 5.10); the others may be assigned to vibrational modes in various ices (table 5.3), and the observations are thus qualitatively consistent with a grain model in which refractory cores containing silicates (present in all types of cloud) act as substrates for the condensation of volatile molecular mantles in the more shielded environments of dense clouds. The mantles are composed predominantly of water-ice producing stretching and bending mode features at 3.0 and $6.0\,\mu m$. The ratio $\tau_{3.0}/\tau_{9.7}$ is environmentally sensitive, measuring the contribution of mantle material to the total grain mass. The profile of the $3.0\,\mu m$ feature and its strength relative to other volatile grain signatures are consistent with mantles containing typically $\sim 60\%$ by mass of H_2O overall (e.g. Tielens and Hagen

1982). The mass and detailed composition of the mantle as a function of location and time depend intimately on the prevailing physical conditions, and it must be borne in mind that these conditions are strongly influenced by the formation and evolution of stars within the clouds. In many cases, as already noted, these embedded objects provide the continuum against which the interstellar absorptions are observed.

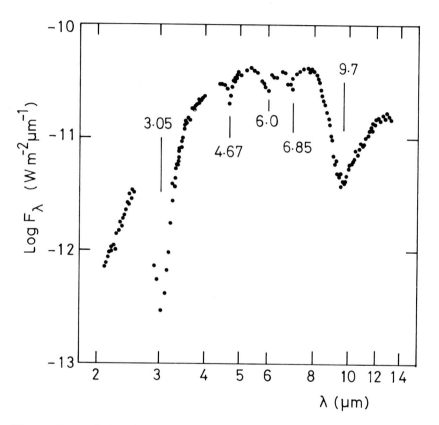

Figure 5.14   Infrared spectrum of NGC 7538E, a typical molecular cloud source, from 2 to 14 μm (Willner *et al* 1979). Dust-related features discussed in the text are labelled by their wavelengths in μm.

The problem of local disturbances associated with embedded objects can be avoided for the nearer molecular clouds by careful selection of background field stars, using techniques described by Elias (1978a, b). The Taurus dark-cloud complex, in particular, is an ideal 'laboratory' in which to study grain mantles in a relatively pristine state (Whittet *et al* 1988, 1989), as it is quiescent, with an absence of shocks or luminous internal sources of ultraviolet radiation. Figure 5.15 plots $\tau_{3.0}$ against A_V for field stars behind the Taurus cloud. A close linear correlation is evident, described by the equation

$$\tau_{3.0} = q\,(A_V - A_0) \tag{5.18}$$

(Whittet *et al* 1988) where $q = 0.093 \pm 0.001$ is the slope, and the intercept $A_0 = 3.3 \pm 0.1$ represents the threshold extinction for detection of the ice feature in this cloud. The value of this threshold undoubtedly varies from cloud to cloud; for example, it appears to be considerably larger $(A_0 \sim 12)$ in ρ Oph (Tanaka *et al* 1990). Figure 5.15 may be contrasted with figure 5.11, which indicates a null intercept for silicates. The difference may be understood if we assume that the presence or absence of ice mantles on the grains is governed by the ambient radiation field. In the outer layers of the Taurus cloud, mantles are destroyed (or their growth inhibited) by photolysis or sublimation, whereas the shielded inner region provides a more hospitable environment for mantle growth and survival.

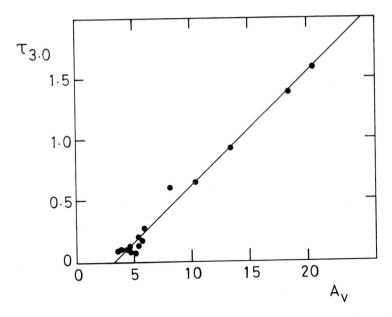

Figure 5.15 Plot of optical depth in the 3.0 μm ice feature against visual extinction for background field stars in the region of the Taurus dark cloud. The straight line is the least-squares fit (equation (5.19)). (Based on data from Whittet *et al* 1988.)

We may use these results to estimate the abundance of H_2O molecules in grain mantles in the Taurus cloud. The column density N of an absorber producing a feature of peak optical depth τ_{max} is given by (Allamandola and Sandford 1988)

$$N = \frac{\gamma \, \tau_{max}}{\mathcal{A}} \qquad (5.19)$$

where γ is the full-width half maximum, expressed in terms of wavenumbers, and $\mathcal{A}$ is the integrated absorption cross-section per molecule. For water-ice, $\mathcal{A} \simeq 2 \times 10^{-18}$ m per molecule (Allamandola and Sandford 1988), and thus equation (5.19) becomes

$$N(H_2O) \simeq 5 \times 10^{23} \, \gamma_{3.0} \, \tau_{3.0} \qquad (5.20)$$

with $N(H_2O)$ in m^{-2} and $\gamma_{3.0}$ in μm^{-1}. Assuming the normal gas to reddening ratio (§1.2.1), the column density of hydrogen for molecular cloud material above the ice threshold is

$$N_H \simeq 1.9 \times 10^{25} \, (A_V - A_0) \qquad (5.21)$$

and combining equations (5.18), (5.20) and (5.21) gives

$$\frac{N(H_2O)}{N_H} \simeq 0.026 \, q \, \gamma_{3.0}$$
$$\simeq 8.6 \times 10^{-5} \qquad (5.22)$$

where a value of $\gamma_{3.0} \simeq 0.036 \, \mu m^{-1}$ has been assumed. For comparison, the solar abundance of oxygen is $N_O / N_H \simeq 8.5 \times 10^{-4}$, so only $\sim 10\%$ of the O is tied up in H_2O on grains in the Taurus cloud.

The profile of the observed $3.0 \, \mu m$ feature is broader than that expected for pure ice absorption. In particular, the observed feature has a prominent long-wavelength wing extending from 3.2 to $3.6 \, \mu m$. It has been suggested (e.g. Léger et al 1983) that the wing is due to scattering by unusually large ($a \sim 1 \, \mu m$) core/mantle grains, but detailed modelling appears to rule out this possibility (Smith et al 1989). It is now widely accepted that the feature is a blend due to resonances in H_2O and other molecules contained within the grain mantles. Figures 5.16 and 5.17 compare the observed $3 \, \mu m$ profiles in two sources with similar central depths but which sample different interstellar environments (Smith et al 1989). In each case, a model based on Mie theory calculations for silicate cores coated with pure H_2O-ice mantles is also shown, and the lower frame plots the residual curve ($\tau_{obs} - \tau_{fit}$). Mon R2 IRS3 (figure 5.16) is an example of an embedded protostellar source, whereas Eias 16 (figure 5.17) is a field star seen through the Taurus cloud. The fit to each spectrum is optimized with respect to the temperature of the ice, which affects the profile shape. It is evident from inspection that the fit to Elias 16 is much better than the fit to Mon R2 IRS3. In both cases there is significant residual absorption in

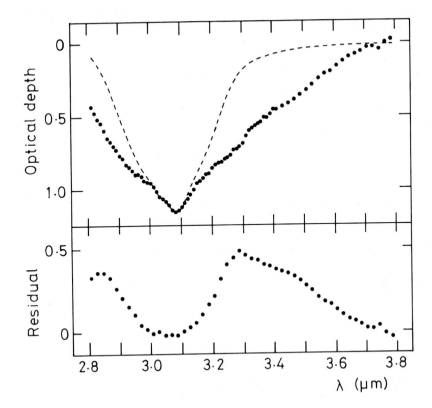

Figure 5.16 The profile of the $3\,\mu$m ice feature in Mon R2 IRS3 (points) compared with that calculated for water-ice deposited on silicate cores (dashed curve). The fit is optimized for a mixture of ices at temperatures 23 K, 77 K and 150 K. The lower frame shows the residual curve obtained by subtracting the calculated optical depth from the observed optical depth at each wavelength (see Smith *et al* 1989 for further discussion).

the long-wavelength wing; in Mon R2 IRS3, there is also significant excess absorption shortward of the main feature. The origin of these adjacent absorptions is not clear. The long-wavelength wing has been attributed to group resonances in H_2O–NH_3 ice mixtures (van de Bult *et al* 1985); however, the presence of NH_3 should lead to sharp structure in the ice feature at $2.96\,\mu$m due to N–H stretching (table 5.3), and this has not been observed to a low level which seems to exclude a substantial abundance of NH_3 in grain mantles (Smith *et al* 1989). C–H stretching vibrations in hydrocarbons may contribute absorption in the 3.2–$3.6\,\mu$m region; if this is the case, then the long-wavelength wing is likely to be a blend of many distinct absorptions arising in both aromatic and aliphatic groups, and the

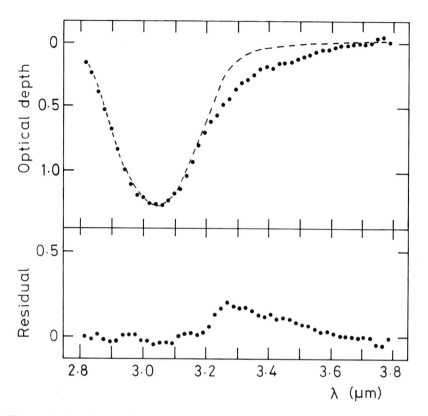

Figure 5.17   The profile of the 3 μm ice feature in Elias 16 (Smith *et al* 1989). As in figure 5.16, the fit (dashed curve) is calculated for a core/mantle grain model, in this case using optical constants measured at a single temperature of 23 K for the mantle. The lower frame shows the residual curve.

smoothness of the observed profile is somewhat surprising.

The possibility that solid CO might be a vital constituent of grain mantles was first suggested by Duley (1974), a prediction confirmed observationally some 10 years later. The 4.67 μm C–O stretch feature of solid CO has subsequently been observed in a number of lines of sight (see Whittet and Duley 1991 for a review). The absorption profile is continuous at high spectral resolution (Larson et al 1985), confirming a solid-state origin. All sources found to have the CO feature also show 3.0 μm absorption, but the converse is not true (see Geballe 1986), which is expected in view of the difference in sublimation temperatures for CO and H_2O ices. In the Taurus cloud, for example, the optical depth of the CO feature behaves in a similar way to that of the H_2O feature (equation (5.19)) but with a higher threshold extinction (Whittet *et al* 1989). Because of its status as the most abundant gas phase molecule to exhibit an observable rotational spectrum, the abundance and depletion of interstellar CO are topics of considerable significance and debate (e.g. van Dishoeck and Black 1987). Millimetre-wave observations are widely used to map the distribution of molecular material in the ISM, and column densities of CO (and hence H_2) will be seriously underestimated if a significant fraction of the CO resides in the solid phase. In addition to its importance as a tracer of molecular material, CO is vital to the production of many polyatomic molecules by gas phase reaction schemes, and the freeze-out of CO is thus a limiting factor on chemical models for the gas as well as the dust. A comparison of solid-state CO column densities deduced from the $4.67\,\mu$m feature using equation (5.19) with infrared and mm-wave measurements of gas phase $N(CO)$ for the same lines of sight suggest that an appreciable, if not dominant, fraction of the total CO abundance (up to $\sim 40\%$) may be depleted onto grains in quiescent molecular clouds (Whittet and Duley 1991). Observations of the $4.67\,\mu$m feature also provide a powerful means of studying the nature and evolution of the grain mantles: laboratory studies of ice mixtures containing CO demonstrate that the precise position, width and profile shape of the feature are sensitive to the composition of the molecular mix containing the CO, and to the occurrence of annealing. We return to this topic in Chapter 7 (§7.2).

The absorption features at $6.0\,\mu$m and $6.85\,\mu$m correspond in wavelength to the O–H bend and C–H deformation modes, respectively, in molecular ices containing hydrocarbons or alcohols. Good fits to the profiles observed in W33A can be obtained using laboratory data for H_2O and CH_3OH (methanol) respectively (Tielens and Allamandola 1987b). However, the appearance of the absorptions in this region of the spectrum changes markedly from source to source (Tielens et al 1984); profile variations in the $6.85\,\mu$m absorption, in particular, suggest that it is not a single feature but a blend of two or more, with varying relative strengths in different lines of sight. It is also clear from other evidence that methanol cannot be the only, or even perhaps the major, contributor to the $6.85\,\mu$m feature, because it should produce an absorption of comparable optical depth at $3.53\,\mu$m (the C–H stretch), but the feature observed at this wavelength in W33A is an order of magnitude weaker (Baas et al 1988). It is reasonable to attribute the $6.85\,\mu$m feature to a blend of molecular absorptions in a variety of carriers, as a number of vibrational modes in (e.g.) hydrocarbons occur in this spectral region, and changes in their relative abundances could account for the profile variations. However, observations at higher resolution than currently available will be needed to provide unique identifications. Alternative assignments of the 6.0 and the $6.85\,\mu$m features to water of hydration and carbonates, respectively, in mineral grains (e.g. Sandford and Walker 1985) appear untenable, as these absorptions are uncorrelated with the $9.7\,\mu$m silicate band (Willner et al 1982) and do not appear outside of molecular clouds.

The profile of the silicate absorption feature at $9.7\,\mu m$ in molecular clouds is significantly broader that that observed in the diffuse ISM and in circumstellar shells (Whittet *et al* 1988; Roche and Aitken 1984a,b). Although the implications of this result have yet to be fully assimilated, it appears to indicate real structural or compositional differences between the silicates in the two environments. The growth of mantles may be expected to affect the profile shape (e.g. by introducing additional long-wavelength absorption due to libration in the H_2O component of the mantles). However, this cannot be responsible for the effect noted above, as the width of the $9.7\,\mu m$ feature does not correlate with the depth of the $3.0\,\mu m$ feature. Also, the the $9.7\,\mu m$ profile is well fitted by an inversion of the 'Trapezium' silicate emission feature which arises in molecular cloud dust recently heated by the young OB stars embedded in the Orion Nebula (§6.3.1). These grains are unlikely to have retained ice mantles, but they have apparently retained the property which leads to the feature being broadened.

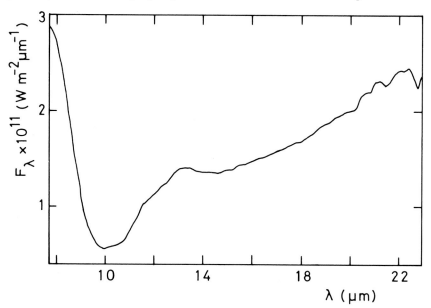

Figure 5.18   Infrared spectrum of Mon R2 IRS3 in the wavelength range 8 to $23\,\mu m$, obtained with the IRAS low-resolution spectrometer (Whittet and Walker 1991).

The assignment of features at 3.0 and $6.0\,\mu m$ to water-ice leads to the prediction that the libration (hindered rotation) feature should also be observable in molecular clouds, with a strength relative to the $3.0\,\mu m$ feature of $\sim 15\%$. Its apparent absence has proved puzzling. Laboratory studies (Kitta and Krätschmer 1983) show that the feature is centred at $11.5\,\mu m$ in pure crystalline ice, and at $12.5\,\mu$m in pure amorphous ice. In sources with deep $3.0\,\mu$m absorption, such a feature should be easily seen blended with the silicate feature, but convincing detections have been reported only for cool circumstellar shells around evolved stars (§7.1). When other cosmically abundant molecules are added to a laboratory H_2O matrix to simulate impure ices occurring in molecular clouds, the libration feature is broadened and its central position is shifted to longer wavelengths, typically 13–$14\,\mu$m (Tielens and Allamandola 1987b). Its absence in ground-based spectra (which cut off at $\sim 13\,\mu$m) is therefore explicable. Low-resolution spectra covering the whole of the 7–$23\,\mu$m region were obtained by IRAS, and an example is shown in figure 5.18. The prominent $9.7\,\mu$m silicate feature and shallower $18.5\,\mu$m feature have broad wings which may overlap such that no true continuum exists between them. Ice libration may be responsible for additional absorption at $\sim 12\,\mu$m and $\sim 14\,\mu$m in figure 5.18, but reliable profile information cannot be extracted. Cox (1989) has reported an apparently distinct libration feature at $13.6\,\mu$m in AFGL 961, but this spectrum appears to be exceptional (Whittet and Walker 1991). In view of these difficulties, the lack of an unequivocal detection of the libration feature in molecular clouds does not cast serious doubt on the correctness of the assignments of the 3.0 and $6.0\,\mu$m water-ice features.

5.3.4   Spectropolarimetry of the Infrared Features

Spectropolarimetry across absorption features provides an important diagnostic technique for investigating the optical and alignment properties of specific components of the dust (see Aitken 1989 for a review). Polarization maxima corresponding to the silicate features at $9.7\,\mu$m and $18.5\,\mu$m have been observed in a number of sources, including the galactic centre group (Capps and Knacke 1976; Aitken et al 1986) and various objects in molecular clouds (Dyck and Lonsdale 1981; Aitken et al 1988, 1989). Observations of the BN object in Orion are illustrated in figure 5.19. The observed wavelength of peak polarization is appreciably longer than that of peak absorption, consistent with theoretical models based on Mie theory (Aitken 1989). Significantly, the position angle of polarization within the features is the same as that in the near-infrared continuum. These results confirm that silicates within both diffuse and dense cloud environments share the general alignment properties of interstellar grains discussed in Chapter 4. In molecular clouds, polarization maxima corresponding to the $3.0\,\mu$m water-ice feature are also observed (Dyck and Lonsdale 1981; Hough et al 1988, 1989). An example appears in figure 4.8. Again, position angle is conserved with respect to adjacent continuum. These results are qualitatively consistent with a core/mantle model for grains in molecular clouds and indicate that mantle growth does not suppress the alignment mechanism.

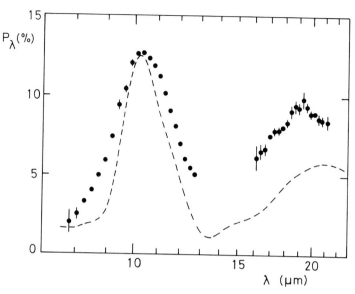

Figure 5.19 Comparison of the observed 7–22 µm polarization of the BN object (Aitken *et al* 1989, points) with a model assuming spheroidal grains of amorphous olivine (Day 1979, dashed curve).

Because polarization represents the difference between two extinction efficiencies (equation (4.2)) which are independently related to the optical constants n and k, spectropolarimetry across an absorption feature is highly sensitive to the optical properties of the grain material concerned. The ratio p_λ/τ_λ is determined primarily by the band strength of the resonance, and the amplitude of variation in p_λ/τ_λ across the 9.7 µm feature in BN requires $\kappa_\lambda \sim 300\,\mathrm{m^2\,kg^{-1}}$ (see Aitken 1989), consistent with amorphous silicates. However, a detailed fit to the 7–22 µm data cannot be obtained using available optical constants, as illustrated in figure 5.19 for the case of amorphous olivine. The 10 µm peak may be broadened by including a mixture of silicates, but the high level of observed polarization at 19 µm is inconsistent with any known silicate material or mixture (Aitken *et al* 1989). Again, the need for further laboratory studies is emphasized.

6

Continuum and Line Emission

"...the dust that is merely opaque in the visible is self-luminous in the infrared, and so in the midst of this optical darkness there has appeared a great infrared light."

N Woolf (1973)

Diffuse emission from interstellar dust was predicted by van de Hulst (1946) as a consequence of the presence of absorption: the balance of energy absorbed by the dust grains over the entire electromagnetic spectrum must re-emerge in the infrared. In a typical interstellar environment, a dust particle gains energy mainly by absorption of ultraviolet photons from the ambient interstellar radiation field. A steady state is established, such that the grain emits a power equal to that absorbed, at some temperature T_d which depends on its size and composition; van de Hulst showed that for classical dielectric spheres of radii $a \sim 0.1\,\mu m$, dust temperatures in the range 10–20 K are expected, and emission should thus occur in the far infrared[†], a prediction confirmed some 25 years later (Pipher 1973). Dust emission constitutes a significant fraction (~ 10–30%) of the total radiative output of the Galaxy and therefore has a major effect on the detailed energy balance of the galactic disk.

The flight of the Infrared Astronomical Satellite in 1983 and the parallel development of ground-based facilities for infrared and submillimetre astronomy represent major landmarks in our ability to study the diffuse emission from dust (see Cox and Mezger 1989 for a review). As an interstellar cloud is, in general, optically thin ($\tau_\lambda \ll 1$) at FIR wavelengths, observed flux densities sample emission at all depths in the cloud with equal efficiency. Observations of the diffuse emission thus provide valuable information on the spatial distribution of dust in the interstellar medium

---

[†] For convenience, the term 'far-infrared' (FIR) is taken here to mean the wavelength range 30–300 μm; similarly, 'near-infrared' (NIR), 'mid-infrared' (MIR) and 'submillimetre' are 1–5 μm, 5–30 μm and 300–1000 μm, respectively.

as well as on grain properties (Hildebrand 1983). Whereas the detection of thermal FIR emission from cool dust had been anticipated, the discovery of strong excess emission in the mid-infrared (MIR) was more surprising, apparently indicating the presence of a much hotter ($T_d \sim 100$–$500\,\text{K}$) dust component. Studies of near-infrared (NIR) continuum and line emission from reflection nebulae surrounding dust-embedded stars also indicate the occurrence of grain temperatures much greater than expected for classical particles in thermal equilibrium with their environment, leading to the development of models which include emission from very small grains subject to transient heating. The spectral emission features which generally accompany the NIR and MIR continuum emission provide clues to the nature of the emitting particles.

In this chapter, we begin in §6.1 by discussing the theoretical basis for emission of radiation by dust. Observations of continuum and line emission are described in §6.2 and §6.3. We are concerned primarily here with emission from interstellar dust heated by the ambient radiation field. The emissive properties of dust in circumstellar envelopes are discussed in Chapter 7.

6.1   THEORETICAL CONSIDERATIONS

6.1.1   Equilibrium Dust Temperatures

Interstellar grains exchange energy with their environment as a result of absorption and emission of radiation, collisions, and exothermic surface reactions such as hydrogen recombination. However, the equilibrium (steady-state) temperature is determined primarily by radiative processes except in clouds of the highest density (see Spitzer 1978, pp 191–193). We begin by considering a perfect spherical blackbody: its equilibrium temperature T_b is given by the Stefan–Boltzmann law

$$U = \frac{4\sigma}{c}\,T_b{}^4 \tag{6.1}$$

where U is the total energy density of the interstellar radiation field (ISRF). This has a mean value of $U \simeq 7 \times 10^{-14}\,\text{J}\,\text{m}^{-3}$ in diffuse H I clouds (e.g. Allen 1973), and substitution into equation (6.1) gives $T_b \sim 3.1\,\text{K}$, a result first deduced by Eddington (1926).

Now consider a spherical dust grain of radius $a \sim 0.1\,\mu\text{m}$. The power absorbed from the ISRF is

$$W_{\text{abs}} = c(\pi a^2) \int_0^\infty Q_{\text{abs}}(\lambda)\,u_\lambda\,\text{d}\lambda \tag{6.2}$$

where Q_{abs} is the efficiency factor for absorption by the grain, and u_λ is the energy density of the ISRF with respect to wavelength. Note that if the grain were composed of perfectly dielectric material, no energy would be absorbed ($Q_{\mathrm{abs}} = 0$); however, all real solids absorb to some extent, due either to their intrinsic properties or to the presence of impurities. The power radiated by the grain is

$$W_{\mathrm{rad}} = 4\pi(\pi a^2) \int_0^\infty Q_{\mathrm{em}}(\lambda)B_\lambda(T_{\mathrm{d}})\,\mathrm{d}\lambda \qquad (6.3)$$

where $Q_{\mathrm{em}}(\lambda)$ is the efficiency factor for emission from the grain (usually termed the emissivity), and

$$B_\lambda(T) = \frac{2hc^2}{\lambda^5}\frac{1}{\exp(hc/\lambda kT) - 1} \qquad (6.4)$$

is the Planck function. It follows from Kirchhoff's law that $Q_{\mathrm{abs}}(\lambda)$ and $Q_{\mathrm{em}}(\lambda)$ are, in fact, identical at a given wavelength, and we may replace them in equations (6.2) and (6.3) by a single function Q_λ. For equilibrium between rates of gain and loss of internal energy, we have $W_{\mathrm{abs}} = W_{\mathrm{rad}}$ and thus

$$\int_0^\infty Q_\lambda u_\lambda\,\mathrm{d}\lambda = \frac{4\pi}{c}\int_0^\infty Q_\lambda B_\lambda(T_{\mathrm{d}})\,\mathrm{d}\lambda. \qquad (6.5)$$

This equation may be used to deduce the temperature T_{d} of the grain if Q_λ can be determined. If we set $Q_\lambda = 1$ at all wavelengths, equation (6.5) leads to the Eddington result. Dust temperatures higher than T_{b} arise from the fact that most power is absorbed at wavelengths short compared with those at which most power is emitted, i.e. absorption occurs predominantly in the ultraviolet and emission occurs predominantly in the far infrared. Q_λ may be calculated from Mie theory (§3.1), and is found to have widely different values in the wavelength domains of interest: for weakly absorbing materials, $Q_{\mathrm{UV}} \sim 1$ and $Q_{\mathrm{FIR}} \ll 1$. In the FIR, $a \ll \lambda$ and we may use the small-particle approximation (equation (3.13)) to specify Q_λ. In general, Q_λ follows a power law in the FIR, i.e.

$$Q_{\mathrm{FIR}} \propto \lambda^{-\beta} \qquad (6.6)$$

for some index β which depends on the nature of the material. Theoretically, we expect $\beta = 2$ for metals and crystalline dielectric substances, and $\beta = 1$ for amorphous, layer-lattice materials (see Tielens and Allamandola 1987a and references therein). Considering a weak absorber with $\beta = 1$, van de Hulst (1946) showed that a dust temperature

$$T_{\mathrm{d}} \sim T_{\mathrm{s}}\, w^{\frac{1}{5}} \qquad (6.7)$$

is predicted, where the ISRF is represented by a blackbody of temperature $T_s = 10,000\,$K and dilution factor $w = 10^{-14}$, and thus $T_d \sim 15\,$K. Detailed calculations for specific materials (e.g. Greenberg 1971; Mathis *et al* 1983; Draine and Lee 1984) largely confirm the van de Hulst result for dielectric materials such as silicates in the low-density ISM. Strong absorbers such as graphite reach equilibrium at temperatures typically a factor of 2 higher than those of silicates (Mathis *et al* 1983). These results apply to the solar neighbourhood and are only weakly dependent on grain size for classical particles. It should be noted that the intensity of the ISRF is a function of galactocentric radius such that systematically higher grain temperatures are predicted in regions closer to the galactic centre (Mathis *et al* 1983).

Deep within a dense cloud heated only by the external ISRF, the ultraviolet and visible flux is severely attenuated and the grains are heated predominantly by absorption of infrared radiation. Models for the transport of radiation within dense clouds allow dust temperatures to be calculated as a function of optical depth (Leung 1975; Mathis *et al* 1983). In, for example, a spherical cloud with total visual extinction $A_V = 10$ through its centre, the estimated temperature of a silicate grain falls from the external value $T_d \simeq 15\,$K to a central value $T_d \simeq 7\,$K. Ices are expected to condense onto grains in such an environment, and this will influence their optical properties and hence their equilibrium temperatures. The effect is very small for silicates but more substantial for strong absorbers, lowering the absorption efficiency (and hence the temperature) compared with uncoated particles; differences in temperature between different grain types thus tend to be reduced.

The results discussed above apply to regions remote from individual stars. In the vicinity of an early-type star, dramatic enhancement of the ambient energy density at ultraviolet wavelengths naturally leads to much higher dust temperatures. Consider, for example, a particle situated in an H II region a distance $r = 0.5\,$pc from a star of radius $R_s = 1.1 \times 10^{12}\,$m and surface temperature $T_s = 40,000\,$K (equivalent to spectral type O6). The dilution factor is $w = (R_s/2r)^2 = 1.27 \times 10^{-9}$ (Greenberg 1978) and substitution into equation (6.7) gives $T_d \sim 650\,$K. Ultimately, the temperature a grain can reach is limited by its sublimation temperature.

6.1.2   FIR Continuum Emission from an Interstellar Cloud

Consider an idealized cloud containing $\mathcal{N}$ spherical dust grains of uniform size, composition and temperature. As in §6.1.1 above, we assume that the grains are classical spheres of radius $a \sim 0.1\,\mu$m, and that each grain is in thermal equilibrium with the ambient radiation field. A real interstellar cloud will, of course, contain grains with a range of sizes, but the FIR emissivity should be described by the small-particle approximation (equation (3.13)) over the entire size distribution for any reasonable grain model. Temperature effects in very small grains, resulting in additional flux at shorter wavelengths, are considered in §6.1.4 below. Assuming that the cloud is optically thin in the FIR, a flux density

$$F_\lambda = \mathcal{N} \left\{ \frac{\pi a^2}{d^2} \right\} Q_\lambda B_\lambda(T_d) \qquad (6.8)$$

is received, where d is the distance to the cloud. The wavelength at which the flux spectrum peaks, deduced by analogy with the Wien displacement law, is related to the temperature by

$$\lambda_{\text{peak}} \simeq 3000 \left\{ \frac{5}{\beta + 5} \right\} T_d^{-1} \qquad (6.9)$$

with λ_{peak} in μm. Setting $\beta = 0$ yields the normal form of Wien's law, but for interstellar grains we expect $1 < \beta < 2$ and thus $\lambda_{\text{peak}} \sim 2300/T_d$ (μm).

Firstly, consider the case of a quiescent dense cloud which contains no internal sources of luminosity. We noted in §6.1.1 that internal dust temperatures of 10 K or less are predicted, resulting in peak emission at wavelengths $> 200\,\mu$m. Because of the inherent properties of the Planck function (equation (6.4)), such cold grains emit negligible flux at $\lambda \leq 100\,\mu$m. However, a quiescent cloud will not be truly isothermal but will have an outer layer of warmer dust heated by the external ISRF to temperatures approaching those in unshielded regions; for $T_d \sim 23$ K (averaged over silicate and carbon grains), peak emission from the outer layer occurs near $100\,\mu$m. Thus, emission in (e.g.) the IRAS $100\,\mu$m passband depends only on the energy density of the ISRF and on the surface properties of the cloud, and clouds of different mass may thus have similar surface brightness.

Few interstellar clouds are truly devoid of embedded sources. Molecular clouds are sites of star formation, and a massive young star embedded in its parent cloud may generate intense local ultraviolet flux. Short-wavelength radiation cannot penetrate far into the cloud; absorption and re-emission by dust close to the source convert it into infrared radiation, to which the cloud is relatively transparent, and this tends to heat the dust elsewhere in the cloud to fairly uniform temperatures which depend on the total internal luminosity. The flux observed from an internally heated cloud may thus be used to deduce the mass of the cloud and to place constraints on grain properties, as discussed by Hildebrand (1983) and Thronson (1988). The volume of dust in the cloud is given by

$$V = \mathcal{N}v \qquad (6.10)$$

where $v = (4/3)\pi a^3$ is the volume of an individual grain. Substituting from equation (6.8) to eliminate $\mathcal{N}$, equation (6.10) becomes

$$V = \left\{ \frac{F_\lambda d^2}{\pi a^2 Q_\lambda B_\lambda(T_d)} \right\} v. \qquad (6.11)$$

If the grains are composed of material of density s, equation (6.11) may be written in terms of the total dust mass

$$M_d = Vs = \frac{4sF_\lambda d^2}{3B_\lambda(T_d)}\left\{\frac{a}{Q_\lambda}\right\}. \tag{6.12}$$

In the small-particle approximation, the quantity a/Q_λ is independent of a and depends only on the refractive index m at wavelength λ. The contribution of each grain to the observed flux spectrum (equation (6.8)) is therefore proportional to a^3, i.e. to its *volume*. Adopting a suitably weighted average of a/Q_λ, equation (6.12) may thus be used to estimate the total mass of dust in a cloud from the observed flux density without detailed knowledge of the grain size distribution.

Multiwaveband observations of an internally heated cloud in the FIR and submillimetre region allow the spectral index of the emissivity to be determined, and, in principle, this may be used to place constraints on grain models. For example, if the emission at these wavelengths were predominantly from graphite, as in the Draine and Lee (1984) model, we would expect $\beta \simeq 2$, whereas for amorphous carbon, $\beta \simeq 1$. Silicates typically have intermediate values of β (Day 1979, 1981) which may vary with particle size (Koike *et al* 1987) as well as crystallinity. Ideally, in clouds sufficiently warm that the Rayleigh–Jeans approximation is valid, the spectrum (equation (6.8)) of a cloud will follow a power law $F_\lambda \propto \lambda^{-(\beta+4)}$ independent of temperature†. However, $T_d > 170\,\text{K}$ is required for this to be true to an accuracy of ± 0.1 in β in the submillimetre region (Helou 1989). It is thus generally necessary to obtain an estimate of the temperature of a cloud in order to evaluate β, and the uncertainties are such that results do not provide a very sensitive discriminator between models for dust in the interstellar medium (see Helou 1989 and Whittet 1988 for further discussion and references). The technique places more significant constraints on dust in circumstellar shells, in which dust temperatures are generally higher and the composition more homogeneous (§7.1.2).

6.1.3   Effect of Particle Shape

Spherical grains are assumed in the preceding discussion; we briefly review here the implications of non-spherical shape (implied by observations of optical polarization) on the emissive properties of the dust. Greenberg and Shah (1971) investigated the effect of shape on calculations of dust temperatures, and showed that, in general, the temperatures of non-spherical

---

† Note that flux densities are often expressed in terms of frequency in the literature; the equivalent power-law dependence in the Rayleigh–Jeans approximation is $F_\nu \propto \nu^{\beta+2}$ where $Q_\nu \propto \nu^\beta$.

particles are somewhat lower that those of equivalent spheres. However, the difference is generally very small and is always less than 10% for dielectric spheroids and cylinders; only highly flattened or elongated metallic particles show more significant departures. The total volume of dust determined from the infrared flux density emitted by a cloud (equation (6.11)) depends on the ratio of volume to projected cross-sectional area (v/σ) for an individual grain. The value of v/σ remains within a factor of 2 of its maximum value (equal to that of a sphere of the same cross-section) for a wide range of grain shapes (Hildebrand 1983); serious discrepancies again arise only if the grains are highly flattened or elongated: unless the bulk of the dust is made up of very thin needles or flakes, shape will have little influence on the determination of dust mass.

The most pertinent property of a non-spherical grain is that, for significant elongation, the emitted flux is linearly polarized. The $\boldsymbol{E}$-vector of the emitted radiation has its maximum value in the plane containing the long axis of the grain and is thus perpendicular to the magnetic field for Davis–Greenstein alignment. For an ensemble of aligned grains, the degree of polarization P_{em} (%) of the emitted radiation may be defined by analogy with equation (4.5):

$$P_{em} = 100 \left\{ \frac{F_{max} - F_{min}}{F_{max} + F_{min}} \right\} \qquad (6.13)$$

where F_{max} and F_{min} are the maximum and minimum flux densities measured with respect to the rotation of some analysing element. At a given wavelength, this quantity is directly related to the efficiency P_λ with which the grains introduce polarization to transmitted radiation by differential extinction (Hildebrand 1988): assuming $P_\lambda \ll 100\%$,

$$P_{em}(\lambda) = -\frac{P_\lambda}{A_\lambda} \qquad (6.14)$$

where A_λ is the mean extinction. The negative sign in equation (6.14) arises because the position angles for emission and absorption are orthogonal. It should be noted that P_λ and A_λ become very small in the FIR where $P_{em}(\lambda)$ is measured. However, in situations where polarized emission in the FIR and polarized absorption at shorter wavelengths can be observed for the same region, the planes of polarization are predicted to be orthogonal if the two effects originate in the same grain population.

6.1.4   Effect of Particle Size

So far, we have assumed that dust temperatures are determined by the time-averaged rates of absorption and emission of energy (equations (6.2) and (6.3)) and that individual quantum events are unimportant. Two conditions must hold for this assumption to be valid: (i) the total kinetic energy content of a grain must be large compared with that received when a single energetic photon is absorbed; and (ii) the energy of the absorbed photon must be distributed throughout the full volume of the grain (i.e. the grain must reach internal thermal equilibrium) on a timescale short compared with that for photon emission. We assume, initially, that condition (ii) holds for all types of grain and examine the inevitable breakdown of condition (i) as progressively smaller particles are considered.

The thermal properties of solids are described by the theory of Debye (1912). For a material with Debye temperature Θ, the heat capacity at constant volume varies as $C_V \propto T^3$ in the low-temperature approximation ($T \ll \Theta$). Values of $\Theta \sim 500\,\text{K}$ are expected for typical candidate grain materials (Duley 1973; Greenberg and Hong 1974; Purcell 1976). Purcell showed that the internal heat energy due to lattice vibrations for a grain of radius a (μm) and temperature T_d (K) is given in electron volts by

$$H \simeq 17a^3 T_d{}^4 \tag{6.15}$$

assuming that the bulk heat capacity is applicable to small particles. (A more detailed treatment includes surface effects: see Aannestad and Kenyon 1979.) If the grain absorbs a single photon of energy ε_{ph}, the initial and final temperatures T_1 and T_2 are then related by

$$T_2 \simeq \left\{ T_1{}^4 + \frac{\varepsilon_{ph}}{17a^3} \right\}^{\frac{1}{4}} \tag{6.16}$$

with ε_{ph} in eV. For illustration, consider the amplitude ($\Delta T = T_2 - T_1$) of the increase in temperature induced by the absorption of individual $10\,\text{eV}$ photons as a function of grain size. An initial temperature $T_1 = 15\,\text{K}$ is assumed. For classical grains ($a > 0.05\,\mu\text{m}$), ΔT is negligible ($< 0.5\,\text{K}$), but ΔT increases rapidly for smaller radii, e.g. to 13 K for $a = 0.01\,\mu\text{m}$ and to 32 K for $a = 0.005\,\mu\text{m}$. The heated grain subsequently cools by emission of infrared radiation. Very small grains (VSGs) may thus undergo a sequence of temperature fluctuations or 'spikes' with a frequency dependent on the arrival rate of energetic photons, as shown schematically in figure 6.1 for $a = 0.005\,\mu\text{m}$.

At dust temperatures greater than those considered above, equation (6.15) is no longer appropriate. For $T > \Theta$, the heat capacity becomes insensitive to temperature and is given by $C_V = 3Nk$ where $3N$ is the number of degrees of freedom in the grain. The amplitude of the temperature spike produced by a photon of energy ε_{ph} (eV) is then

$$\Delta T \simeq 4 \times 10^3 \, \varepsilon_{ph}/N. \tag{6.17}$$

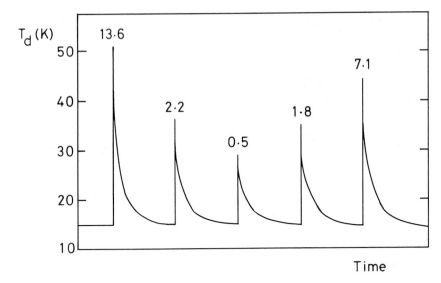

Figure 6.1 A schematic representation of temperature fluctuations in a small grain (radius $a = 0.005\,\mu$m) produced by periodic absorption of individual photons. Each temperature spike is labelled by the photon energy in electon volts, and the time interval between spikes is one hour. (Adapted from Aannestad and Kenyon 1979.)

For a VSG of radius $0.001\,\mu$m, $N \sim 40$ and a spike of amplitude $\Delta T \sim 1000\,$K is predicted for a 10 eV photon (Sellgren 1984).

We deduce above that the amplitudes of temperature spikes in large grains of homogeneous composition and structure are negligible. Only in grains of radii $a < 0.03\,\mu$m need small-particle effects normally be considered. However, real interstellar grains are perhaps unlikely to behave as homogeneous solids in which vibrational energy is rapidly transmitted throughout the lattice. Grains formed by an accretion process may contain many subunits in poor thermal contact with one another. Amorphous carbon, in particular, typically exhibits such a structure (Duley and Williams 1988), leading to the possibility that PAH-like clusters within the framework of a larger amorphous carbon grain may undergo some degree of stochastic heating.

A cloud containing a distribution of particle sizes from molecular dimensions up to classical submicron-sized grains is expected to emit a complex infrared spectrum. VSGs exposed to the ISRF contribute continuum flux peaking at wavelengths $2 < \lambda < 100\,\mu$m (dependent on their time-averaged temperature), and, if specific vibrational resonances are excited, discrete line emission may also be seen (§6.3.2). The temperatures of true classical grains are predicted to be much more uniform with peak emission emerging in the FIR.

6.2 CONTINUUM EMISSION FROM THE GALACTIC DISK

6.2.1 Morphology

Diffuse far-infrared emission from the galactic disk was first detected in the early 1970s (Pipher 1973) but could not be studied in detail until the launch of IRAS in 1983. A major discovery of the IRAS survey was the detection of large-scale filamentary emission in the 60 and 100 μm passbands (the 'cirrus'), arising not only from the plane of the Milky Way but also from regions of intermediate and high galactic latitude (Low *et al* 1984). The distribution of compact 100 μm cirrus clouds is illustrated in figure 1.5. The 100 μm intensity of the high-latitude cirrus is generally well correlated with other tracers of interstellar matter, including H I line emission and visual extinction from star counts (Terebey and Fich 1986; Laureijs *et al* 1987; Boulanger and Pérault 1988), indicating that it is emitted by individual diffuse clouds in the solar neighbourhood, typically as close as 100–200 pc. The cores of the brighter high-latitude cirrus clouds are also associated with CO emission (Weiland *et al* 1986; Boulanger and Pérault 1988). These individual clouds are superposed on a smooth component of galactic FIR emission which follows a cosec b law (Beichman 1987). Towards the galactic plane, the cirrus naturally merges into a continuous sheet of diffuse emission in which a typical line of sight intercepts many clouds.

6.2.2 Spectral Energy Distribution

The integrated spectral energy distribution of the emission from the galactic disk is shown in figure 6.2, combining IRAS and other data in the spectral range 4–900 μm. The spectrum shows a prominent, broad maximum centred near 100 μm (the FIR peak), and a secondary peak or plateau near 10 μm (the MIR excess). Both features are also seen in high-latitude galactic cirrus (Weiland *et al* 1986). The FIR peak is attributed to classical-sized grains in thermal equilibrium with the ISRF (§6.1.1) and the position of the peak implies temperatures $T_{\rm d} \sim$ 20–30 K (equation (6.9)), in reasonable agreement with the predictions of the graphite–silicate model (Draine and Lee 1984). The MIR excess implies substantially higher grain temperatures. A contribution to the integrated MIR flux from the galactic disk may come from warm circumstellar shells, but this is unlikely to account for more than $\sim$ 10% of the total. It is now widely accepted that the excess is predominantly caused by continuum and line emission from VSGs which are stochastically heated (§6.1.4) to mean temperatures typically in the range 100–500 K (Draine and Anderson 1985; Weiland *et al* 1986). Although the possibility of thermal fluctuations in VSGs had been discussed some years before the detection of the MIR excess, its presence and remarkable strength (accounting for $\sim 40\%$ of the total power radiated by dust at $\lambda < 120\,\mu$m according to Boulanger and Pérault 1988) was nevertheless a major surprise.

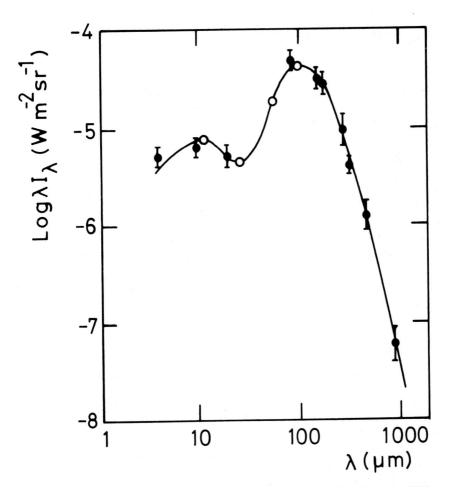

Figure 6.2   The spectral energy distribution of dust emission from the Milky Way between galactic longitudes 3° and 30°. Points with error bars: Pajot *et al* (1986) and references therein; open circles: IRAS data (Cox and Mezger 1987); continuous curve: model based on the superposition of a number of grain components (Cox and Mezger 1987).

Excess emission from VSGs is also apparent at shorter wavelengths in the environs of hot (B-type) stars. The NIR spectra of several optical reflection nebulae were investigated by Sellgren (1984), and an example is shown in figure 6.3. A peak occurring at 3.3 μm is attributed to C–H vibrational emission in PAHs (§6.3.2). This is superposed upon continuum emission which correlates closely in surface brightness with that of the visible nebula due to scattered light from the illuminating star. The continuum emission has a colour temperature ~ 1000 K and a spectral shape which shows no variation with distance from the star. This behaviour cannot be readily explained by scattering, fluorescence, free–free emission, or thermal emission from grains in equilibrium with the stellar radiation field, but is consistent with emission from VSGs subject to transient heating in the enhanced ultraviolet radiation field close to a hot star.

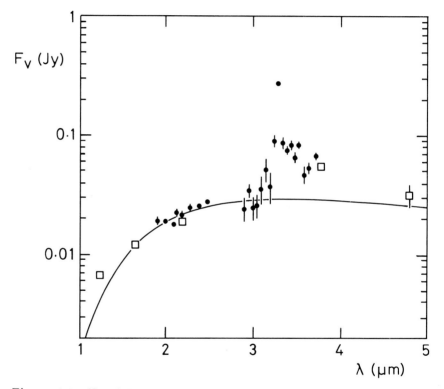

Figure 6.3   Near-infrared excess continuum and 3.3 μm line emission in the reflection nebula NGC 7023 (Sellgren 1984). Squares represent broadband photometry, circles are spectrophotometric data of resolution 0.03 μm. The solid curve represents a 1200 K gray body normalized to the observations at 1.65 μm.

Studies of external systems serve to illustrate the universality of the VSG phenomenon in spiral galaxies (Cox and Mezger 1987; Rice *et al* 1990). Figure 6.4 plots the integrated spectral energy distribution of M33 from ultraviolet wavelengths through to the submillimetre. As this Sc galaxy is a face-on system, the contribution of starlight (largely hidden hidden by dust in the case of the Milky Way) may also be seen. The points are compared with blackbodies of temperatures 3000 K (representing the coolest stars) and 30 K (representing a fit to the FIR peak attributed to 'classical' dust). The presence of excess emission from material at intermediate temperatures is clearly indicated by the points at 12 and 25 μm.

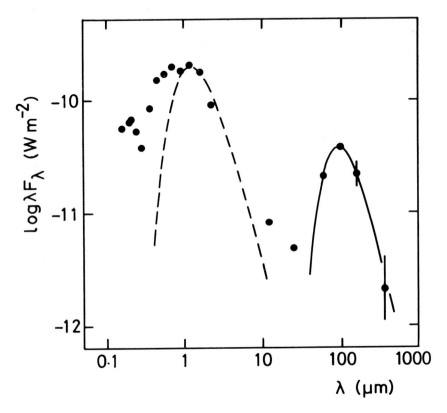

Figure 6.4   The integrated spectral energy distribution of the spiral galaxy M33 from 0.15 to 400 μm. Broad peaks near 1 μm and 100 μm are due to starlight and cool dust, respectively. The latter is fitted by a dust model with $T_d = 30$ K (smooth curve). A blackbody of temperature 3000 K (dashed curve) is shown to represent emission from cool stars. The elevation of the points at 12 and 25 μm indicates the existence of matter at an intermediate temperature. (Data are from Rice *et al* 1990 and references therein.)

A number of multi-component models may be constructed which account for the observed spectral energy distributions of the Milky Way and other galaxies. In the case of the Milky Way, the data are well matched by (e.g.) the model of Cox *et al* (1986), in which warm ($T_d \sim 30$–40 K) dust in extended, low-density H II regions is responsible for much of the FIR emission, and additional cool and hot components are superposed to give a detailed fit, as illustrated in figure 6.2. This model is not unique, and there is no consensus as to which phase of the ISM (§1.2.2) dominates the FIR emission from the Galaxy: neutral H I clouds and molecular clouds with embedded H II regions have also been widely discussed (see Sodroski *et al* 1989; Hauser *et al* 1984). In the solar neighbourhood, most of the emission appears to come from low-density phases of the ISM rather than from dense clouds associated with current star formation (Boulanger and Pérault 1988). Detailed mapping of the emission in nearby external galaxies such as M33, in which individual giant molecular clouds and H II regions can be resolved, may provide the key to understanding the origin of the large-scale emission from the Milky Way.

6.2.3   Dust to Gas Ratio

Observations of the FIR continuum emission may be combined with radio observations of gas phase emission lines to estimate the mass ratio of dust to gas. In some line of sight, assumed to be optically thin, the flux density observed within a beam of angular radius θ at wavelength λ may be conveniently expressed in terms of optical depth

$$\tau_\lambda = \frac{F_\lambda}{\pi \theta^2 B_\lambda(T_d)} \qquad (6.18)$$

(Hildebrand 1983). The optical depth is related to the dust density along a path-length L by equation (5.13); with reference also to equation (2.20), the dust to gas ratio is therefore

$$Z_d = 0.71 \frac{\rho_d}{\rho_H}$$
$$= 4.3 \times 10^{26} \frac{\tau_\lambda}{N_H \kappa_\lambda} \qquad (6.19)$$

where $N_H = \rho_H L / m_H$ is the total hydrogen column density in the same line of sight and κ_λ is the mass absorption coefficient of the grains. The atomic and molecular contributions to N_H may be quantified by means of 21 cm H I and 2.6 mm CO data, and their distributions compared with the diffuse 100 μm emission measured by IRAS (Sodroski *et al* 1987, 1989). Taking an average for the galactic plane, $\langle \tau_{100}/N_H \rangle \sim 3 \times 10^{-29}$ m^2/nucleon, yielding a value of $\langle Z_d \rangle \sim 0.003$ if a mass absorption coefficient of $\kappa_{100} \sim 4.1$ m^2 kg^{-1} appropriate to the MRN grain model is assumed. This result is a factor of 2–3 lower than that based on the extinction of starlight (equation (3.33)).

There are at least two significant sources of probable error associated with calculations of dust mass from FIR emission (see Draine 1990). Firstly,

κ_λ is poorly defined, and serious discrepancies exist between values measured directly in the laboratory (Tanabé et al 1983) and values implied by Mie theory calculations (Draine and Lee 1984). Secondly, the IRAS $100\,\mu$m band may sample emission predominantly from a warm component of the dust which constitutes a relatively minor fraction of the total mass. Dust temperatures $T_d \simeq 24\,$K are estimated from the IRAS data for H I clouds (Sodroski et al 1989), consistent with the MRN grain model in which most of the $100\,\mu$m emission is from graphite (Draine and Lee 1984). Silicates and other weakly absorbing grains, for which temperatures $T_d < 20\,$K are predicted, contribute almost all of their radiative energy at wavelengths beyond $100\,\mu$m. It is thus predictable that this method will tend to underestimate Z_d.

6.2.4   The Cold-Dust Problem

The contribution of molecular clouds to the total mass of dust in the Milky Way and other galaxies is controversial and difficult to estimate from observations of dust emission alone. As discussed in §6.1.2, the FIR flux from a molecular cloud devoid of internal sources of luminosity tends to be dominated by emission from the warm outer layers. Observations extending into the submillimetre are needed to sample the emission from cold internal dust with $T_d < 15\,$K. In general, the 100–$1000\,\mu$m emission from a molecular cloud will consist of a component due to cold dust with λ_{peak} typically in the range 150–$300\,\mu$m, superposed on the tail of the emission due to warmer dust with $\lambda_{\mathrm{peak}} \leq 100\,\mu$m (e.g. Mathis et al 1983). In principle, the total mass of dust may be estimated by modelling the spectral energy distribution (e.g. Draine 1990) if the temperatures of the various components can be determined reliably. However, because the amplitude of the emission at λ_{peak} declines with T_d, the observed submillimetre flux is rather insensitive to the total mass of cold dust (Helou 1989; Eales et al 1989; Draine 1990). A factor of 2 error in flux can typically lead to an order of magnitude error in the implied dust mass. In general, gas phase CO line emission provides a better tracer of molecular material than cold-dust emission.

6.2.5   Polarization and Alignment

Observations of polarization in the FIR and submillimetre continuum emission from molecular clouds provide information on the alignment of dust in the coldest, densest regions of the interstellar medium. The $270\,\mu$m polarization from OMC–1 has been measured (Hildebrand et al 1984; Dragovan 1986) and found to be orthogonal in position angle to that at $20\,\mu$m. This result can be understood (§6.1.3) if the same population of aligned grains absorb at the shorter wavelength and emit at the longer. The fact that the grains are, indeed, significantly aligned in dense clouds provides some discrimination between alignment mechanisms (§4.4). The temperature of the grains is much more closely coupled to that of the gas than is the case in diffuse regions, and the disorientating effects are stronger. An alignment mechanism which is marginal in diffuse clouds will therefore be ineffective in dense clouds (Hildebrand 1988). As discussed in §4.4, alignment does not seem to be a marginal process. The observations OMC-1 tend to favour superparamagnetic alignment over suprathermal paramagnetic alignment (Mathis 1986), as it seems unlikely that the latter mechanism will be effective in a region of such high density.

6.3   SPECTRAL EMISSION FEATURES

If characteristic vibrational modes within a grain lattice are excited, the corresponding infrared spectral features may appear in emission. This situation commonly arises in circumstellar shells and in interstellar matter near individual stars. However, few of the dust features seen in absorption in the interstellar medium (§5.3, table 5.4) have direct counterparts in emission: volatile solids detected in absorption in molecular clouds are generally destroyed in situations which would give rise to emission features. Also, populations of very small grains which are optically thin in absorption in the infrared may nevertheless be seen in emission if the relevant energy levels are efficiently 'pumped' by absorption at shorter wavelengths. The principal emission signatures attributed to cosmic dust arise in silicates and hydrocarbons.

6.3.1   Silicates

Silicates are sufficiently robust to survive under a variety of physical conditions. Emission features corresponding to the stretching and bending mode vibrations at wavelengths of 9.7 and $18.5\,\mu$m (table 5.3) are seen in H II regions and in the circumstellar envelopes of both young and evolved stars. In principle, direct comparisons can therefore be made between profiles observed in emission and absorption, and these may provide insight into the properties of silicates as a function of environment. However, a complication arises in that we may observe superposed circumstellar emission and foreground absorption components in the same line of sight. This typically occurs towards young stars embedded in molecular clouds, such as the BN object in Orion (see below), and towards cool evolved stars with

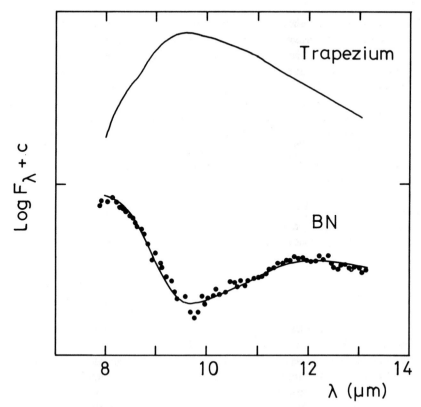

Figure 6.5   The Trapezium emissivity function (upper curve) compared with the spectrum of the BN object (points) in the region of the 9.7 μm silicate feature. The Trapezium curve represents $Q_\lambda \propto F_\lambda / B_\lambda(T_d)$, assuming a temperature of 250 K for the underlying continuum emission from dust. The BN spectrum is fitted with a model which assumes circumstellar emission and foreground extinction by dust with emissivity identical to that observed in the Trapezium (see Gillett *et al* 1975b).

intrinsic dust emission (§7.1), including those associated with the galactic centre used to probe diffuse-cloud extinction (§5.3.2; see figure 5.12).

Figure 6.5 plots the 8–13 μm emissivity curve deduced from observations of the Trapezium cluster in the Orion Nebula (Gillett *et al* 1975b). The curve is smooth and devoid of fine structure, implying that the silicates are not significantly annealed. Also shown in figure 6.5 is the spectrum of the BN object, which is deeply embedded in the molecular cloud behind the Trapezium cluster. A simple model for the spectrum of BN could be constructed by assuming blackbody continuum emission from the source, modified by foreground extinction by a slab of cold dust which absorbs according to the Trapezium emissivity curve. Such a model is appropriate to the case of a field star observed through a molecular cloud (Whittet *et al* 1988), but for BN a better fit to the data (illustrated in figure 6.5) is obtained if the model is refined to include circumstellar silicate band emission as well as continuum emission from the source, again subject to appropriate foreground absorption. Similar results are found for other BN-like objects. It appears that the Trapezium emissivity curve provides a good representation of the silicate profile in both absorption and emission in molecular clouds and associated H II regions. This conclusion is supported by observations of background field stars towards quiescent dark clouds (Whittet *et al* 1988) as well as embedded objects such as BN. It is remarkable that the same profile is seen in both quiescent and disturbed regions. The H II region N44A in the Large Magellanic Cloud also shows Trapezium-like emission (Roche *et al* 1987), supporting the existence of 'normal' silicate dust in that galaxy. The Trapezium curve is less successful in reproducing the absorption profile observed in low-density regions of the Milky Way (§5.3.2), where an emissivity curve deduced from silicate emission in the circumstellar shells of evolved stars (§7.1.2) appears to be more appropriate.

The silicate features observed in H II regions and circumstellar shells may be understood in terms of classical-sized ($a \sim 0.1\,\mu m$) grains which are nevertheless small compared with the wavelengths of the features. If much smaller silicate grains are abundant in the ISM, the silicate features should be prominent in emission whenever the grains are exposed to energetic radiation, as occurs in lines of sight where near- and mid-infrared continuum emission is observed (§6.2.2). Désert *et al* (1986) calculate 7–14 μm emission spectra for a graphite–silicate grain model, including both small and large grains, and compare their results with representative observational data for the reflection nebula NGC 2023 and the galaxy M82. The 9.7 μm feature is not of detectable strength in these sources. Désert *et al* conclude that the mass fraction of silicates which can be accommodated in grains small enough to undergo significant transient heating is < 1%. It appears that the VSG population is predominantly carbon-rich.

6.3.2   The PAH Spectrum

The occurrence of a series of narrow emission features with principal wavelengths at 3.3, 6.2, 7.7, 8.6 and 11.3 μm and widths (FWHM) in the range 0.03–0.5 μm was first reported by Gillett, Forrest and Merrill (1973). A wealth of observational data now exist for these features (see Bregman 1989 for a review). A statistical analysis suggests that they belong to a common 'generic' spectrum (Cohen *et al* 1986, 1989), although some variations do occur in their relative strengths. They are frequently referred to in the literature as the 'unidentified infrared' (UIR) features, although they are now widely attributed to polycyclic aromatic hydrocarbons (see below). The 'PAH' features are commonly observed in the lines of sight to planetary nebulae, H II regions, and reflection nebulae. As planetary nebulae are remnants of evolved stars, they often differ in chemical composition from average interstellar material, and the strength of the PAH spectrum correlates with the abundance ratio of C to O (Cohen *et al* 1986; Roche 1989a), supporting identification with a C-rich carrier. However, it is clear that the features are not limited to C-rich environments and must be produced by a widespread ingredient of the ISM. The factor linking the various sources in which they are seen is the presence of an intense ultraviolet radiation field softer than the Lyman limit, such as occurs, for example, in the neutral zones just beyond the ionization fronts of H II regions surrounding OB stars, and in reflection nebulae around stars of somewhat later spectral type. Under such conditions, small grains will be stochastically heated as described in §6.1.4, reaching temperatures of up to $\sim 1000\,\mathrm{K}$.

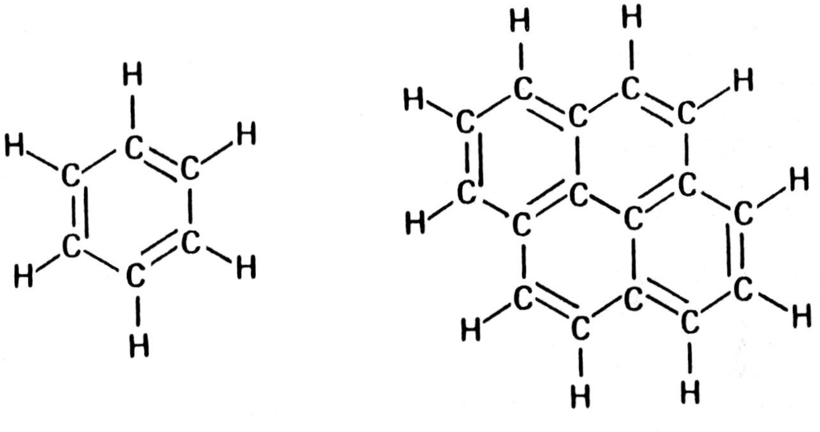

Figure 6.6   The molecular structure of benzene (left) and the 4-ringed polycyclic aromatic hydrocarbon pyrene (right).

Polycyclic aromatic hydrocarbons (PAHs) are a class of organic molecules composed of benzene rings, linked together in a plane, with H atoms or other radicals saturating the outer bonds of peripheral C atoms. The molecular structure of benzene and a simple 4-ringed PAH are illustrated in figure 6.6. The infrared spectra of PAHs are characterized by a number of resonances, of which the strongest are listed in table 6.1. Identification of the observed 'UIR' features with PAHs excited by absorption of ultraviolet photons (Duley and Williams 1981; Léger and Puget 1984) is based on wavelength correspondence (table 6.1) between observed and laboratory spectra. Figure 6.7 compares the observed spectrum for the line of sight towards a reflection nebula (the Red Rectangle) with a laboratory spectrum for a representative PAH (coronene). It is of interest to note that the observed emission features do *not* correspond to resonances commonly seen in absorption in the ISM (table 5.4). This indicates that PAHs, although pervasive, are not major contributors to the mass density of interstellar grains. The emission spectrum can be understood if some 5–10% of the total carbon abundance of the ISM is in PAHs (e.g. Allamandola *et al* 1989), a result which is not in serious conflict with limits placed by ultraviolet spectroscopy (§5.1.2).

Table 6.1   Wavelengths (μm) and assignments of the principal spectral features in polycyclic aromatic hydrocarbons (Allamandola *et al* 1989).

λ_{obs}	λ_{lab}	Assignments
3.29	3.29	C–H stretch ($v = 1 \to 0$)
6.2	6.2	C–C stretch
7.7	7.6–8.0	C–C stretch
8.6	8.6–8.8	C–H in-plane bend
11.3	11.2–12.7	C–H out-of-plane bend for peripheral or adjacent H atoms

Figure 6.8 illustrates the infrared spectra of several individual PAHs at 850 K (Léger and d'Hendecourt 1988). It is reasonable to assume that the ISM will actually contain a variety of PAHs, although the most stable (symmetrical and condensed) forms will naturally tend to be favoured†. Individual PAHs in a given environment will be subject to different degrees of excitation dependent on their absorption cross-sections and molecular weights, and on whether they are free or bound into some larger structure. All of these factors will affect the appearance of the emergent spectrum. The individual spectra shown in figure 6.8 have a number of common features, but also some differences. The features at 3.3 and 6.2 μm are stable in position and will arise in any mixture of aromatic molecules with appropriate excitation; they are generally accompanied by a weaker feature at 5.2 μm which has also been observed (Allamandola *et al* 1989). Spectral variations from one PAH molecule to another are more evident at the longer wavelengths. The 7.6–8.0 μm region contains a blend of several strong C–C

---

† An example of a PAH mixture composed of the most stable molecular forms is the soot produced by high-temperature combustion of hydrocarbons (Allamandola *et al* 1985), as in automobile exhaust.

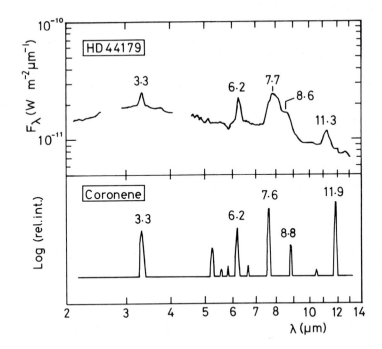

Figure 6.7   The infrared spectrum of HD 44179 in the Red Rectangle (Russell *et al* 1978, upper frame) compared with that for coronene (Léger and Puget 1984). The principal spectral features are labelled by their wavelengths in μm.

stretching bands which may combine to account for the relatively broad 7.7 μm feature in astronomical spectra (e.g. figure 6.7). Out-of-plane C–H bending modes occur in the 11–13 μm region. This part of the spectrum is particularly sensitive to the geometry and degree of hydrogenation of the PAH molecule; the fact that the principal interstellar feature in this region occurs at 11.3 μm (rather than 12 μm, as in coronene) implies emission by isolated, peripheral H atoms (Allamandola *et al* 1989) which would be favoured in partially hydrogenated species.

Studies of PAH features in the vicinity of prominent ionization fronts such as the Orion Bar are particularly valuable in constraining the properties of the carriers (Roche *et al* 1989a; Geballe *et al* 1989b). Figure 6.9 shows the 11–13 μm spectrum for a 6 arcsec aperture centred directly on the Orion Bar. The 11.3 μm PAH feature is particularly strong. It is accompanied by a weaker feature centred at 12.7 μm (blended with nebular [Ne II] emission at 12.8 μm), superposed on an emission plateau which extends between the two features. Additional spectra taken at either side of the ionization front show that the intensities of the [Ne II] and 11.3 μm fea-

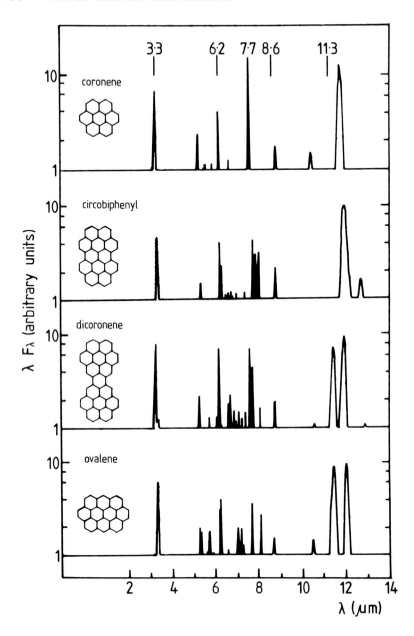

Figure 6.8 Comparison of the predicted emission spectra of several PAHs at 850 K (Léger and d'Hendecourt 1988): from the top, coronene, circobiphenyl, dicoronene and ovalene. The structure of each PAH is shown to the left of the spectrum. The positions of the principal interstellar features are labelled above the upper spectrum.

tures are anticorrelated, as expected if the former arises in the H II region and the latter in the neutral zone just outside it. In contrast, the intensity of the $12.7\,\mu$m feature correlates closely with that of $11.3\,\mu$m, implying that it belongs to the PAH family. A feature near $12.7\,\mu$m is, indeed, characterisitic of many PAHs (e.g. figure 6.8) and its detection is not unexpected. The plateau may be a quasi-continuum formed by many overlapping emissions in a variety of molecules, but the existence of only two well-defined features in this spectral range suggests that a relatively small number of species are dominant.

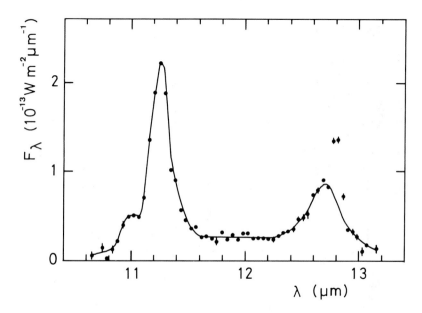

Figure 6.9   The 11–13 μm spectrum of the Orion Bar (points) taken through a 6 arcsec beam at a resolution of 0.09 μm (Roche *et al* 1989a). The profile of the PAH emissions (comprising the features at 11.3 and 12.7 μm and the associated plateau) is represented by a smooth curve. Additional emission at 12.8 μm is due to a [Ne II] line in the H II region.

Dust-related emission features in the 3–4 μm region of the spectrum have been studied at somewhat higher spectral resolution and in a larger number of sources. An example is illustrated in figure 6.10. The strong 3.3 μm feature (actually centred at 3.29 μm) dominates the spectrum, and is accompanied by a weaker feature centred at 3.40 μm, observed in many lines of sight. Further structure is apparent in the 3.3–3.6 μm region, with weak features at 3.46, 3.51 and 3.56 μm superposed on an underlying plateau (Allen *et al* 1982; de Muizon *et al* 1986). Comparison of these features in different sources suggests that a hierarchy governs their occurrence: con-

sidering the sequence (i) 3.29 μm, (ii) 3.40 μm, (iii) 3.3–3.6 μm plateau, and
(iv) the weaker features, each successive group requires the presence of all
the previous ones in order to be seen. The principal feature is identified
with the $v = 1 \rightarrow 0$ aromatic C–H stretch. Additional emissions at longer
wavelength may occur due to excitation of higher levels (Barker *et al* 1987),
and their intensities relative to the principle feature should vary with environment: spectra taken at several positions around the Orion Bar show
that the 3.40 and 3.51 μm features do vary in intensity with respect to the
principal feature in a manner consistent with an origin in the aromatic
$v = 2 \rightarrow 1$ and $v = 3 \rightarrow 2$ transitions, respectively (Geballe *et al* 1989b).
Other 3.3–3.6 μm features may be associated with aliphatic subgroups at-
tached to PAHs (Duley and Williams 1981; de Muizon *et al* 1986, 1990).
As in the case of the broad plateau longward of the 11.3 μm feature, a
plausible explanation of the 3.3–3.6 μm plateau is a quasi-continuum due
to overlapping features; these may perhaps arise in a variety of aliphatic
sidegroups.

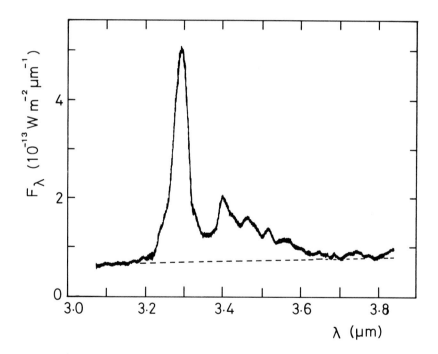

Figure 6.10   The 3.1–3.8 μm spectrum of IRAS source 21282+5050 (de Muizon
et al 1986) at a resolution of 0.008 μm. The dashed line represents the assumed
continuum underlying the emission features.

In addition to the normal family of infrared emission features associated with PAHs, other features have been observed in the $3-4\,\mu$m region which may have a different origin. The dominant 'anomalous' feature, centred at $3.53\,\mu$m, was first detected in HD 97048, a pre-main-sequence Be star which illuminates a reflection nebula in a dark cloud (Blades and Whittet 1980). The wavelengths of the features seen in HD 97048 (3.53, 3.43 and $3.29\,\mu$m) are similar to those in sources with 'normal' $3-4\,\mu$m emission spectra (e.g. figure 6.10), but the relative strengths are reversed with $3.53\,\mu$m very strong compared with $3.29\,\mu$m. Another difference is the fact that the $3.53\,\mu$m feature in HD 97048 is not spatially extended in the nebula but appears to be formed very close to the star (Roche *et al* 1986b). Despite an extensive search, this phenomenon has been detected with certainty in only two other objects (Whittet *et al* 1984; Geballe *et al* 1989a), suggesting that it requires specialized physical conditions to occur and perhaps represents a transient phase in the evolution of Be/Ae shells. A logical assignment would be C–H stretching in circumstellar dust-based molecules: good profile agreement is found with laboratory data for formaldehyde (H_2CO) in a cold matrix (Baas *et al* 1983), but the presence of H_2CO on dust in an environment where temperatures $T_d > 1000$ K are expected (Roche *et al* 1986b) is highly implausible. Schutte *et al* (1990) argue that C–C overtone and combination bands in highly excited PAHs or amorphous carbon particles offer a more likely explanation.

PAH features are observed in the nuclei of many external galaxies (see Roche 1989c for a review). They are particularly prominent in systems with nuclear H II regions indicative of recent OB star formation. Indeed, the occurrence of PAH features appears to be a useful discriminator between starburst galaxies (dominated by rapid star formation) and active galaxies (dominated by compact, energetic nuclei). PAH emission occurs in the spectra of starburst galaxies because many luminous H II regions are included in a typical beam. Mizutani *et al* (1989) demonstrate the existence of a correlation between the intensity of the $3.3\,\mu$m feature and that of the Brγ line of hydrogen in several galaxies, and argue that the former may be used as a quantitative tracer of star formation activity. In contrast, PAH features are generally absent from the spectra of active galaxies, suggesting that the carriers are destroyed by hard ultraviolet radiation in the vicinity of the nuclear source (Aitken and Roche 1985; Désert and Dennefeld 1988). It should be noted that the sample of galaxies for which spectra are available is strongly biased towards those that are brightest in the infrared, and little information exists for more normal, quiescent systems. PAH emission (at $3.3\,\mu$m) from the nuclear region of our own Galaxy was reported by Gatley (1984), a result which appears to have been largely overlooked by workers in the field. This contradicts a suggestion (Désert and Dennefeld 1988) that the supposed absence of PAH emission in the galactic centre implies nuclear activity.

6.3.3   Extended Red Emission

Many solids emit visible luminescence when exposed to energetic radiation, prompting a search for luminescence from interstellar grains. A broad emission band centred between 0.6 and 0.8 μm has been detected in a number of reflection nebulae surrounding B- and A-type stars (Witt 1988 and references therein). An example of the observed profile is shown in figure 6.11 (upper frame). This feature, termed the extended red emission (ERE), is superposed on the flux due to scattering, and typically contributes up to 30% of the total surface brightness of the nebula near its peak wavelength. The ERE is attributed to luminescence from dust excited by absorption of soft ultraviolet photons from the nearby star. Nebulae emitting ERE also display PAH features in their infrared spectra, suggesting a link between the two phenomena. However, only neutral PAHs are expected to show luminescence, and those occurring in reflection nebulae must surely be ionized (Witt and Schild 1988). The material responsible for the ERE appears to be securely identified with amorphous carbon.

Laboratory investigations demonstrate that hydrogenated amorphous carbon films exhibit a broad luminescence band in the spectral range 0.5–1.0 μm (Watanabe *et al* 1982) with a profile similar to that of the observed ERE. The position of peak emission varies as a function of the degree of hydrogenation of the laboratory samples, as illustrated in figure 6.11 (lower frame). This result provides a natural explanation for the fact that the observed peak wavelength of the ERE profile shows spatial variations both within individual nebulae and from one nebula to another (Witt and Boroson 1990). Hot amorphous carbon grains exposed to intense ultraviolet radiation normally have a low hydrogen content, producing luminescence which peaks at $\lambda > 0.7\,\mu$m and has relatively low intensity. A comparison of the spatial distribution of the ERE with that of fluorescent molecular hydrogen strongly suggests that luminescence is stimulated most efficiently in H_2 photodissociation zones where the carbon becomes rehydrogenated. The average value of the observed peak ERE wavelength in reflection nebulae is close to that expected for maximum yield. Amorphous carbons subject to extreme rehydrogenation tend ultimately to become polymeric, exhibiting a stable luminescence band which peaks at $\lambda \sim 0.56\,\mu$m (Wagner and Lautenschlager 1986). Duley and Whittet (1990) have pointed out the resemblance of this feature to very broadband structure in the interstellar extinction curve (§3.2.3).

Most of the nebulae in which the ERE has been studied arise due to the presence of a young star embedded in an interstellar cloud associated with recent star formation. The Red Rectangle provides an intriguing exception: this nebula is carbon-rich and appears to have formed as a result of mass loss from an evolved star. The nebula is illuminated by the B9–A0 III star HD 44179, and there has been some discussion in the literature as to

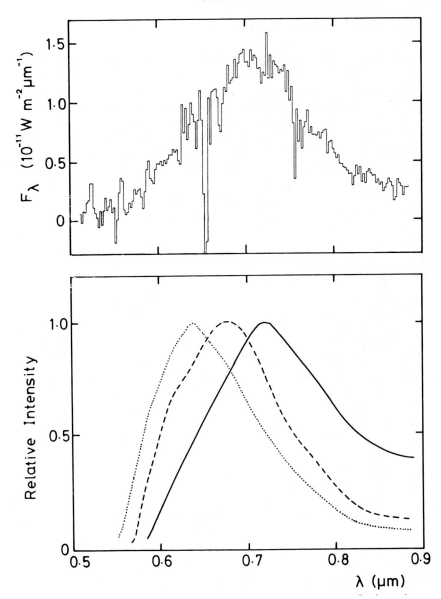

Figure 6.11 The profile of extended red emission observed in the reflection nebula NGC 2327, displayed after subtraction of the component due to scattering (Witt 1988, upper frame). This is compared with normalized photoluminescence spectra of hydrogenated amorphous carbon films (Watanabe *et al* 1982, lower frame). The solid, dashed and dotted curves in the laboratory spectra correspond to samples with increasing degrees of hydrogenation. The highest luminescence yield occurs for peak emission at $\lambda \sim 0.68\,\mu$m (dashed curve).

whether this object is also the source of the nebula itself or whether it is located by chance within or near the envelope of an evolved star enshrouded in self-generated dust (see Witt and Boroson 1990). Recent spectroscopic data show that HD 44179 is extremely metal-deficient (Waters, personal communication) and thus in an advanced evolutionary state, consistent with the occurrence of mass loss possibly associated with the formation of a planetary nebula. The ERE in the Red Rectangle pervades the entire nebula with an intensity as much as 25 times higher than that typically found in nebulae around young stars, although the peak occurs at a similar wavelength. PAH features are also seen (figure 6.7). The Red Rectangle thus provides direct spectroscopic evidence for the coexistence of PAH molecules and amorphous carbon dust in C-rich stellar ejecta.

7

The Origin and Evolution of Interstellar Grains

The evolution of interstellar matter is intimately linked with that of the stars within it. Stars may be associated with dust at almost any phase of their lifecycle from protostellar youth to cataclysmic old age. Star formation in our Galaxy currently occurs deep within dense molecular clouds. Dusty circumstellar shells or disks are ubiquitous to stars in the earliest stages of their evolution, emitting copious infrared radiation and providing raw materials for the accretion of planetary systems. Subsequently, much of the placental material is returned to the interstellar medium, but, in a number of cases, unaccreted remnants of protostellar disks survive around mature main-sequence stars. After leaving the main sequence, stars may evolve to become significant sources of new grain material. The cool atmospheres of red giants and supergiants provide environments conducive to the rapid nucleation and growth of refractory dust grains; these particles are accelerated by radiation pressure to become a component of the outflows associated with such stars. Dust may also be produced in cataclysmic events such as nova eruptions and supernovae explosions. The shock waves generated by supernovae are a major cause of grain destruction in the interstellar medium, resulting in sputtering by high-velocity gas and shattering by grain–grain collisions.

In this chapter, we trace the lifecycle of cosmic dust, illustrated schematically in figure 7.1. We begin (§7.1) with a discussion of grain formation in the atmospheres of evolved stars and the subsequent mass-loss processes which lead to the ejection of 'stardust' into the interstellar medium. We then consider the constructive and destructive processes which affect the dust in contrasting interstellar environments (§7.2). In the final section (§7.3), we assess the nature and significance of dust in the environments of newly formed stars, and examine the relationship between interstellar grains and diffuse matter in the Solar System.

193

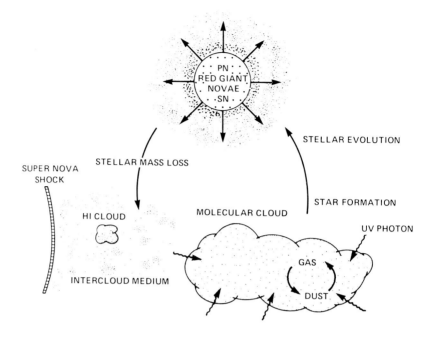

Figure 7.1   Schematic representation of the lifecycle of cosmic dust. Grains originating in the atmospheres and outflows of evolved stars (red giants, planetary nebulae, novae and supernovae) are ejected into the interstellar medium, where they are exposed to ultraviolet irradiation and to destruction by shocks. In molecular clouds, ambient conditions favour the growth of volatile mantles on the grains. The onset of star formation leads to the dissipation of molecular clouds. (From Tielens and Allamandola 1987b; reprinted by permission of Kluwer Academic Publishers.)

7.1   DUST IN THE EJECTA OF EVOLVED STARS

"...a complex interplay of gas dynamics, chemical kinetics and radiative transfer."

M Jura (1989)

The nucleation of dust grains in the atmospheres and outflows of evolved stars has been discussed for many years (e.g. Hoyle and Wickramasinghe 1962; Gilman 1969; Salpeter 1977), and this process is now widely accepted to be a major source of interstellar grains (e.g. Bode 1988; Jura 1989; Gehrz 1989). In this section, we begin (§7.1.1) by examining the physical conditions and chemical reactions which control the formation of dust in circumstellar environments. Observational constraints on the nature and quantity of dust entering the ISM are reviewed in §7.1.2 and §7.1.3.

7.1.1   The Formation of Dust in Stellar Atmospheres

The most important stellar sources of dust are luminous post-main-sequence stars with photospheric temperatures between 2000 K and 3000 K, i.e. objects which lie in the upper right-hand region of the Hertzsprung–Russell (HR) diagram. This group includes giants, supergiants and asymptotic giant branch (AGB) stars; for convenience, we refer to them collectively as 'red giants'. Close to the photospheres of such stars, the temperature is generally too high for solids to exist and all material is in the gas phase. Indeed, some red giants have chromospheres with temperatures as high as $\sim 10,000$ K (Gail and Sedlmayr 1987). The inner boundary of the dust shell is specified by the radial distance r_1 at which the gas temperature falls below the condensation temperature T_C of the dust. For a red giant of radius R_s forming grains with $T_C = 1000$ K, Bode (1988) estimates that $r_1 \sim 10 R_s$. The outer boundary of the shell is specified by the distance r_2 at which the density and temperature of the circumstellar material become comparable with those of average interstellar matter, which may typically be $10^4 R_s$. It is easily shown that the number density n of the gas in a stellar wind declines due to dilution with radial distance r from the star such that $n \propto r^{-p}$, with $p = 2$ for uniform expansion at constant velocity. Grain formation is thus expected to be most rapid in the zone immediately beyond the inner radius r_1: here, densities are typically $n \sim 10^{19}$ m^{-3}, many orders of magnitude greater than those characteristic of interstellar clouds. It is this combination of physical conditions, high densities coupled with temperatures comparable with the condensation temperatures of many solids, which renders red giant atmospheres particularly favourable sites for rapid grain formation. It should be borne in mind that evolved stars often have pulsational instabilites, leading to periodic variations in the physical conditions at a given radial distance, and these may induce cyclic phases of grain nucleation, growth and ejection.

The growth of solid particles in an initially pure gas that undergoes cooling has been discussed in terms of classical nucleation theory (e.g. Salpeter 1974), originally formulated to describe the condensation of liquid droplets in the Earth's atmosphere. Condensation of some species X occurs when its partial pressure in the gas exceeds the vapour pressure of X in the condensed phase, and is most efficient at temperatures appreciably below the nominal condensation temperature. We refer to an individual unit of X in the gas phase (an atom or molecule) as a monomer. Random encounters between monomers lead to the formation of clusters; for equilibrium, the number density of clusters containing i monomers is

$$n_i = n_1 \exp(-\Delta E_i / kT) \tag{7.1}$$

(e.g. Dyson and Williams 1980) where n_1 is the number density of individual monomers and ΔE_i is the thermodynamic free energy associated with the formation of the cluster. For small clusters, ΔE_i increases with i, and thus n_i decreases. Above some critical size, however, the addition of further monomers is energetically favourable; the clusters become stable and grow rapidly to some maximum size limited by the availability of monomers. Classical nucleation is thus a two-step process: (i) the formation of clusters of critical size, and (ii) the growth of these clusters into macroscopic grains or droplets. However, in a red giant wind, where pressures are typically at least a factor of 10^6 less than in the terrestrial atmosphere, the timescale for nucleation may be longer than the timescale in which the physical conditions undergo appreciable change, and condensation is not then an equilibrium process (Donn and Nuth 1985). Solids are held together by strong valence bonds rather than by the weak polarization forces which bind liquid droplets, and the formation of grains thus involves chemical reactions as well as physical clustering (Salpeter 1974). Whereas classical theory describes the nucleation and growth of particles which are chemically identical to the original monomers, chemical reactions integral to the nucleation process lead to the formation of solids for which no equivalent monomer is available in the gas (Gail and Sedlmayr 1986).

The atmospheres of red giants contain both atomic and molelular gas, and the abundances and physical properties of the dominant species regulate the condensation of solids. We assume that all relevant monomers are neutral, i.e., there is no significant source of ionizing radiation: in cases where this is not so (e.g. where there is local ultraviolet emission from a chromosphere or an early-type binary companion), grain formation will tend to be inhibited because positively charged monomers repel each other. CO is the most abundant of all molecules containing heavy elements. It forms readily and is stable in the gas phase in stellar atmospheres at temperatures below about 3000 K. The dissociation energy of CO (11.1 eV) is sufficiently high that the bond cannot readily be broken in simple gas phase reactions. Thus, despite its status as a radical, CO is chemically rather unreactive. It is also very volatile and cannot itself condense into solids in circumstellar environments. The abundance of CO in the gas is thus limited only by the abundances of the constituent atoms. This has dramatic consequences for the chemistry of dust condensation: whichever element out of C and O is less abundant will be almost entirely locked up in gas phase CO molecules and hence unavailable to form solids. The abundance ratio $N(\mathrm{C})/N(\mathrm{O})$ (hereafter abbreviated C/O) is thus crucial in determining the composition of the dust.

Another molecule of some interest in the same context is N_2. Like CO, N_2 is physically stable and chemically unreactive, and is thus confined to the gas phase in conditions pertaining in circumstellar shells. Essentially all of the nitrogen present in the gas is therefore expected to be locked up in N_2 (irrespective of C/O ratio), and this abundant element is thus unavailable to form solids.

Red giants which show observational evidence for the presence of dust (see §7.1.2 below) may be placed into two broad categories: those with approximately solar abundances (normal M-type stars with $C/O < 1$) and those enriched in carbon to the extent that $C/O > 1$. As discussed in §2.1, carbon enrichment occurs when ^{12}C, the product of helium burning, is dredged up to the surface. In the following discussion, it is assumed that a C-rich stellar atmosphere can be represented by a gas in which the abundance of ^{12}C is enhanced, but which otherwise has solar composition. In general, this is likely to be a reasonable approximation for non-binary carbon stars in the solar neighbourhood of the galactic disk, but it should be noted that some stars with carbon enrichment are also hydrogen deficient, and some are metal poor. For solar composition, the C/O ratio is 0.43 (table 2.1) and we thus expect all of the C and approximately half of the O to be tied up in CO in M stars, leaving the remaining O free to become bonded into molecules which may condense into solids. In a carbon star ($C/O > 1$), the rôles are reversed. Note that if the degree of C enhancement were such that C/O were precisely unity, both C and O would be tied up fully in gaseous CO and would effectively block each other from inclusion in solids.

The physical conditions prevailing in the atmosphere of a red giant of solar composition resemble those thought to have occurred in the early Solar System, and the solids predicted to condense are therefore broadly similar (Salpeter 1977; Barshay and Lewis 1976). The relevant temperature–pressure (T, P) phase diagram is illustrated in figure 7.2. CO is stable in the gas in all regions of the diagram above the dashed curve. Consider a pocket of gas which is steadily cooled, such that its locus in the T, P diagram follows the direction of the curved arrow in figure 7.2. The most abundant monomers which lead to the production of solids are expected to be Fe, Mg, SiO and H_2O. Initially, at $T > 1500\,K$, these remain in the gas, and only rare metals such as tungsten and refractory oxides such as corundum (Al_2O_3) and perovskite ($CaTiO_3$) are stable in the solid phase. The relevant monomers are of such low abundance that condensation of these materials can generally be ignored, with the possible exception of corundum, which may form via nucleation of AlO clusters. Any such high-temperature condensates which do appear may provide nucleation centres for the deposition of more abundant solids at lower temperatures (Onaka *et al* 1989). The major condensation phase occurs at temperatures $1200 \rightarrow 800\,K$ and involves the formation of magnesium silicates and metallic iron. The initial stage of silicate formation may involve the nucleation of SiO clusters which chemisorb Mg and H_2O, resulting in the growth of linked SiO_3 chains with attached Mg cations (enstatite, $MgSiO_3$) or individual SiO_4 tetrahedra joined by cations (fosterite, Mg_2SiO_4). Note that neither of these molecular structures can exist in the gas phase: they *must* form chemically within the clusters.

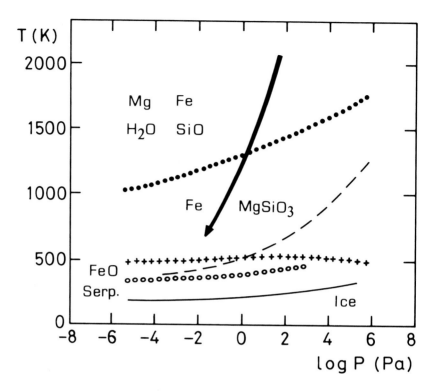

Figure 7.2   Temperature–pressure phase diagram illustrating stability zones of major solids in an atmosphere of solar composition (adapted from Salpeter 1974, 1977; Barshay and Lewis 1976). Above the dashed curve, gas phase CO is stable and essentially all C is locked up in this molecule. The most abundant gas phase reactants which lead to the production of solids are Fe, Mg, SiO and H_2O. The curved arrow indicates the variation in physical conditions which may occur in the outflow of a typical red giant. Magnesium silicates and solid Fe condense below the upper curve (● ● ●). At much lower temperatures, Fe is fully oxidized to FeO (below curve marked + + +) and may then be absorbed into silicates. Hydrous silicates such as serpentine are stable below the curve marked o o o. Finally, water-ice condenses below the continuous curve.

At $T \sim 700\,\mathrm{K}$, essentially all the metallic elements are likely to have condensed into solids in some form or other. As the temperature falls further, iron becomes increasingly oxidized to FeO (and sulphidized to FeS) until no pure metallic phase remains below $\sim 500\,\mathrm{K}$ (curve labelled FeO in figure 7.2). FeO may be absorbed into hydrous ferromagnesium silicates at somewhat lower temperatures. Finally, water-ice may condense as the temperature drops below $\sim 200\,\mathrm{K}$. A macroscopic grain emerging from such an atmosphere seems likely to have a layered structure, containing silicates of differing Mg/Fe ratios and degrees of hydration, perhaps deposited on an oxide 'seed' nucleus and coated with a thin surface layer of ice.

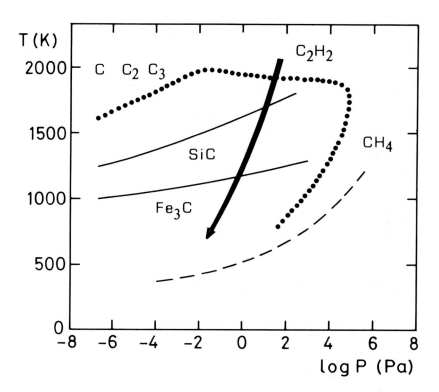

Figure 7.3   Temperature–pressure phase diagram illustrating stability zones of solids in a carbon-rich atmosphere (adapted from Salpeter 1974, 1977; Martin 1978). Solar abundances are assumed except that the abundance of C is enhanced to exceed that of O by 10%. Above the dashed line, gas phase CO is stable and essentially all O is locked up in this molecule. Other important gas phase carriers of C are labelled outside the dotted curve. The curved arrow indicates the change in physical conditions associated with a typical outflow from a red giant. Solid carbon is stable in the region enclosed by the dotted curve. Condensation curves for SiC and Fe_3C are also shown.

The equivalent phase diagram for a carbon-rich atmosphere is shown in figure 7.3. C is assumed to be enhanced such that its abundance exceeds that of O by 10%. (For qualitative discussion, the actual degree of enhancement is not critical provided that C/O > 1.) As before, CO is stable in the gas above the dashed curve. Solid carbon is stable in the area enclosed by the dotted curve. The shape of this curve arises because different carbon-bearing monomers (other than CO) predominate in different regions of the diagram (Salpeter 1974), as marked in figure 7.3. At pressures prevailing in red giant atmospheres, acetylene (C_2H_2) is the dominant form. The kinetic processes which lead to the production of carbon dust in the winds of red giants appear to be closely analogous to soot production by combustion of hydrocarbons (Frenklach and Feigelson 1989). The basic unit of solid carbon (ignoring exotic forms such as diamond) is the hexagonal ring (see figure 6.6). However, the molecular structure of the available monomer (acetylene) is $H-C{\equiv}C-H$, i.e. it is the simplest example of a saturated *linear* molecule involving carbon bonding with alternate single and triple bonds. In order to produce aromatic hydrocarbons, it is necessary to replace the triple bond with a double bond. The same atoms present in acetylene could be rearranged to form the radical $C=CH_2$, which has two unpaired electrons. Such a metamorphosis may be brought about collisionally, involving, for example, the removal (abstraction) of an H atom:

$$C_2H_2 + H \rightarrow C_2H + H_2. \qquad (7.2)$$

The product ($C=CH$) constitutes a ring segment (figure 6.6), and the ring may be closed by chemical reactions which attach two further C_2H_2 molecules. Once formed, the ring is stable and can grow cyclically by abstraction of peripheral H atoms and attachment of further C_2H_2 units. This growth process may be represented symbolically by the pair of alternating reactions

$$C_nH_m + H \rightarrow C_nH_{m-1} + H_2 \qquad (7.3)$$

and

$$C_nH_{m-1} + C_2H_2 \rightarrow C_{n+2}H_m + H \qquad (7.4)$$

which lead to the construction of planar PAH mass molecules containing many rings (Frenklach and Feigelson 1989). The final structure of a macroscopic grain depends on how the PAHs are assembled. To form graphite, they must be stacked regularly in parallel sheets; an amorphous stucture of randomly grouped clusters seems more plausible and in agreement with experimental evidence on hydrocarbon smokes.

In a carbon-rich atmosphere which is severely deficient in hydrogen, the nucleation process is likely to differ in detail from that described above. C, C_2 and C_3, rather than C_2H_2, will be the primary gas phase carriers of condensible carbon, and the relevant condensation curve will resemble an extrapolation of the low-pressure segment of the dotted curve in figure 7.3. In these circumstances, C and $C=C$ monomers may assemble into ring clusters which accumulate into amorphous carbon grains with low hydrogen content.

The rôle of the metallic elements in the formation of dust in C-rich atmospheres appears to have received little attention. Unlike the situation in O-rich atmospheres, chemical reactions involving metals do not regulate the condensation of the primary solid phase. Nevertheless, a cooling gas containing a solar or near-solar endowment of metals must inevitably form metal-rich condensates. Abundant species such as Fe may condense as pure metals, as in the O-rich case, but silicates are prevented from forming because O remains trapped in CO. Some free C and S may be available to form carbides and sulphides, however. Phase transition curves for SiC and Fe_3C are shown in figure 7.3.

As a red giant evolves, it may undergo successive phases of mass loss, possibly involving distinct episodes of O-rich and C-rich grain formation. Ultimately, many such stars are destined to become planetary nebulae in which the outer layers are ejected and the hot core is exposed. The gas expands at velocities typically in the range 20–50 $km s^{-1}$, compared with $\sim 10 km s^{-1}$ for red giant winds, and previously ejected material may thus be swept up and reprocessed.

7.1.2 Observational Constraints on Circumstellar Grain Composition

The observational techniques used to explore the nature and composition of dust in the shells of evolved stars are far-infrared photometry of continuum emission, near- and mid-infrared spectroscopy of absorption and emission features, and studies of the ultraviolet extinction curve. These methods are fairly successful in identifying at least some of the grain materials. We review the principal results below. Note that whereas interstellar clouds contain mixtures of both C-rich and O-rich dust, the circumstellar dust surrounding a given star at a given epoch is generally confined to one of these types for reasons discussed above, and the results therefore provide an important comparison for studies of interstellar dust.

Infrared continuum emission greatly in excess of that expected from the Rayleigh–Jeans tail of a normal stellar photosphere is a definitive characteristic of stars with dust shells. Optical and near-infrared radiation from the photosphere is absorbed by the grains and re-emitted at longer wavelengths. The spectral shape of the far-infrared emission provides a useful diagnostic of grain composition on the basis of arguments presented in Chapter 6 (§6.1). The flux density emerging from an isothermal dust shell of temperature T_d is given by equation (6.8), i.e. $F_\lambda \propto Q_\lambda B_\lambda(T_d)$ where Q_λ is the grain emissivity and $B_\lambda(T_d)$ the Planck function. A real circumstellar shell will contain a range of grain temperatures resulting in a composite spectrum, the form of which depends on the wavelengths of peak emission (equation (6.9)) at the inner and outer boundaries of the shell and on the radial distribution of material in the shell (Bode and Evans 1983). However, at sufficiently long wavelengths, B_λ is described by the Rayleigh–Jeans approximation independent of T_d, and we have

$$\log F_\lambda = C - (\beta + 4)\log \lambda \qquad (7.5)$$

where $Q_\lambda \propto \lambda^{-\beta}$ and C is a constant. The spectral index β may thus be evaluated from the slope of the logarithmic FIR flux distribution.

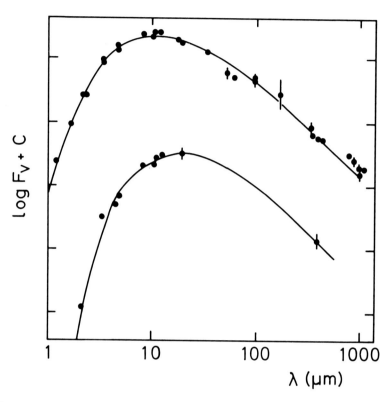

Figure 7.4   Spectral energy distributions in the infrared for two carbon stars, IRC+10216 (above) and CRL 3068. The observational data (points) are fitted with models which assume an emissivity function appropriate to amorphous carbon. (Sopka *et al* 1985.)

Far-infrared data are available for a number of cool, dust-ejecting stars. The well-known carbon star IRC+10216 is perhaps the best studied of all such objects (Campbell *et al* 1976; Sopka *et al* 1985; Le Bertre 1987; Martin and Rogers 1987). Figure 7.4 shows observed flux distributions for IRC+10216 and the similar but more highly obscured carbon star CRL 3068 (Sopka *et al* 1985). Both are well fitted by a model with $\beta = 1.2 \pm 0.2$ in the FIR, a result which implies that the emitting dust is amorphous carbon rather than crystalline graphite or silicon carbide (see Tanabé *et al* 1983). Graphite, in particular, is expected to obey a $Q_\lambda \propto \lambda^{-2}$ emissivity law. More generally, β-values close to unity are typical of dusty late-type stars irrespective of C/O ratio (Sopka *et al* 1985; Rowan-Robinson *et al* 1986;

Jura 1986), indicative of amorphous grain structure in both C-rich and O-rich environments. This is consistent with the predictions of a kinetic model for grain growth in stellar atmospheres (Donn and Nuth 1985).

A quantitative limit on the abundance of graphite in circumstellar shells may be deduced by means of a spectral feature expected to occur at $11.52\,\mu m$ due to a lattice resonance. If graphite is present, this feature should be seen in either absorption or emission, depending on the temperature and distribution of the dust (Draine 1984). It is intrinsically narrow (FWHM $\simeq 0.01\,\mu m$), demanding higher spectral resolution than is commonly used to study dust features. Its absence in IRC+10216 and other C-rich objects (Martin and Rogers 1987; Glasse et al 1986) leads to the conclusion that no more than about 5% of the solid carbon in their shells can be graphitic.

Table 7.1 Infrared spectral features observed in the circumstellar shells of evolved stars.

λ (μm)	Carrier	Abs/em?	ISM?	C/O?	Object type
9.7, 18.5	Silicates	a/e	Yes	< 1	M stars; PNe
11.2	SiC	e	No	> 1	C stars; PNe
30	MgS?	e	No	> 1	Extreme C stars; PNe
3.1, 6.0	H_2O ice	a	Yes	< 1	OH–IR stars
11.5	H_2O ice	a	No?	< 1	OH–IR stars
3.3, 6.2, 7.7, 8.6, 11.3	PAHs	e	Yes	> 1	Post AGB stars; PNe

Dust-related features which have been observed in the infrared spectra of evolved stars and planetary nebulae (PNe) are listed in table 7.1 (adapted from Whittet 1989). The columns indicate the wavelength, the proposed carrier, an indication of whether the feature is seen in absorption or emission, the presence or absence of a counterpart in the interstellar medium (ISM; see table 5.4), the C/O ratio of the source, and the object type. The family of emission features at wavelengths 3.3, 6.2, 7.7, 8.6 and $11.3\,\mu m$ are described and discussed in §6.3.2; they are assumed to be spectral signatures of PAH molecules which are vibrationally excited by soft ultraviolet photons. In the general ISM, they are seen whenever ambient dust is exposed to such irradiation, as typically occurs in reflection nebulae and beyond the ionization fronts of H II regions. In the circumstellar context, these features occur only in the spectra of C-rich planetary nebulae and their immediate precursors, the hot post-AGB stars; they do *not* occur in the spectra of physically similar objects which are O-rich (e.g. Roche 1989a). Their absence in C-rich red giants is explained by the lack of a source of exciting radiation.

Red giants have distinctive infrared spectral properties according to C/O ratio, as illustrated in figure 7.5. The 9.7 and 18.5 μm features associated with stretching and bending modes in silicates are seen in O-rich stars (Treffers and Cohen 1974; Forrest *et al* 1979), whereas these are almost always absent in C-rich stars and replaced by the 11.2 μm feature of silicon carbide (Treffers and Cohen 1974; Cohen 1984). Silicate or SiC emission features are also observed in planetary nebulae of the appropriate C/O ratio (Aitken *et al* 1979). The correlation between optically determined C/O abundance and infrared dust spectrum is extremely strong†. The dust features provide a useful diagnosis of C/O ratio in stars too heavily obscured to be studied optically, a technique which has been used extensively as a means of classifying IRAS low-resolution spectra (LRS).

The position, width, and profile shape of the infrared features observed in red giants provide information on grain size and structure. The SiC profile appears to be virtually constant from star to star (Little-Marenin 1986), suggesting that the absorbing grains are either in the small-particle limit or have closely similar properties in different lines of sight. The 9.7 μm silicate feature is more difficult to study because of uncertainty in the placement of the underlying continuum (note that the short-wavelength cut-off of the LRS at 7.5 μm is close to the edge of the feature), but real variations in width do seem to occur, with FWHM typically in the range 2.0–2.8 μm. The 9.7 μm profile observed in absorption in the diffuse interstellar medium closely resembles an inversion of the emission profile seen in those stars displaying the narrowest features (e.g. μ Cephei; Roche and Aitken 1984a). In other stars, broader 9.7 μm profiles more characteristic of the Trapezium emissivity function (§6.3.1) have been reported (Simpson and Rubin 1989). These variations may be related to the size distribution or thermal history of the grains (Papoular and Pegourie 1983; Nuth *et al* 1986), although it is surprising that no corresponding changes are apparent in the 18.5 μm feature (Simpson and Rubin 1989). Composition may also be important; for example, Onaka *et al* (1989) discuss the effect of absorption in Al_2O_3 on the 9.7 μm profile in Miras. However, what is most striking, overall, is the general absence of fine structure in the silicate features, a result which is consistent with amorphous rather than crystalline materials.

---

† A few exceptional cases exist, in which optically classified carbon stars show silicate features (Little-Marenin 1986); these seem most likely to be binary systems containing both O-rich and C-rich components (Bode 1988).

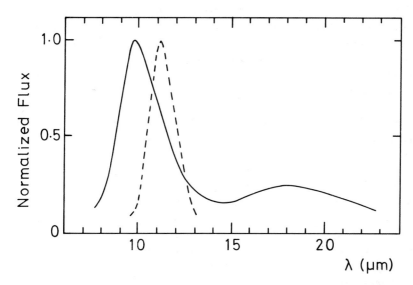

Figure 7.5   Smoothed mean profiles of 8–23 μm dust emission features observed in IRAS low-resolution spectra. Continuous curve: silicate emission in M-type stars with normal abundances (C/O < 1). Dashed curve: SiC emission in carbon stars (C/O > 1).

Spectral signatures of dust have also been observed in classical novae (see Bode 1989 for a review). A nova outburst is triggered when matter drawn from the surface of a cool main-sequence star undergoes thermonuclear ignition as it accretes onto the surface of a companion white dwarf. A number of novae subsequently develop strong excess continuum emission in the infrared (Gehrz 1988), generally attributed to condensation of carbon-rich dust in the outflow. The detection of 9.7 μm *silicate* emission in Nova Aquilae 1982 (Bode *et al* 1984) and Nova Vulpeculae 1984 (Gehrz *et al* 1986) was therefore surprising. Gehrz *et al* (1984) noted the apparent absence of the corresponding 18.5 μm emission in Nova Aquilae, but this result is critically dependent on the placement of the continuum and does not cast doubt of the silicate interpretation of the 9.7 μm excess (Bode 1989). The simultaneous presence of silicate and PAH features in the spectrum of Nova Centauri 1986 argues that the 'rule of exclusion' for O-rich and C-rich dust in stellar ejecta may not apply to novae (Bode 1989). The depletions of gas phase elements in Nova Aquilae (Snijders *et al* 1987) support the hypothesis that both types of grain material have condensed in its outflow.

A broad feature centred near 30 μm has been observed in a small sample of C-rich planetary nebulae and extreme carbon stars, and attributed to MgS, possibly as a mantle on more refractory grain cores (Goebel and Moseley 1985; Nuth *et al* 1985; Roche 1989a). Remarkably, in some objects (e.g. IC 418) as much as 25% of the total infrared flux is emitted in this feature. MgS is nevertheless unlikely to be a major component of the dust in carbon stars generally. For example, Martin and Rogers (1987) estimate that it contributes no more than a few per cent by mass of the dust in IRC+10216. A similar result is deduced for SiC on the basis of the strength of the 11.2 μm feature. Other possible condensates containing metallic elements, such as FeS and Fe$_3$C, have not been detected spectroscopically (Nuth *et al* 1985). None of these materials can account for the strong continuum emission from carbon stars, which, as previously discussed, is most probably due to amorphous carbon.

Absorption features attributed to water frosts have been observed in the spectra of several evolved stars (e.g. Soifer *et al* 1981; Eiroa *et al* 1983; Roche and Aitken 1984b; Geballe *et al* 1988; Smith *et al* 1988). The condensation of ice is not unexpected in the outflow of an O-rich star (§7.1.1). It is most likely to arise in dust shells which are optically thick, such that the outer layers are protected from photospheric radiation, as is the case towards infrared-bright evolved stars associated with OH maser emission (OH–IR stars). The spectrum of OH 231.8+4.2 is shown in figure 7.6. As expected for an optically thick shell, the 9.7 μm feature appears in absorption rather than emission. Corresponding data for the young molecular cloud source NGC 7538E appear in figure 5.14, and the comparison is informative. Both spectra show deep ice absoption centred at 3.1 μm (the O–H stretch) and a weaker bending mode absorption at 6.0 μm. Both also show deep silicate absorption at 9.7 μm, but in the OH–IR star this is blended with a feature centred at 11.5 μm and attributed to the libration mode in crystalline water-ice. Other features present in the molecular cloud (at 3.4, 4.67 and 6.85 μm) are very weak or absent in the OH–IR star. This is consistent with current views on the composition of ices in molecular clouds (§5.3.3): the carriers of these features are more volatile than water, and therefore do not condense in the stellar outflow. The OH–IR spectrum indicates the presence of pure H$_2$O frosts, whereas the ices in interstellar clouds include molecules involving C and N which condense at $T < 100$ K. As discussed in §5.3.3, the heterogeneous composition of the ices condensing in an interstellar environment provides a natural explanation for the absence of the 11.5 μm feature in ground-based spectra of molecular cloud sources. The evolved stars provide the 'control experiment' in which the ice is relatively pure.

The spectral dependence of circumstellar extinction can, in principle, provide information on both the composition and the size distribution of the particles. Traditionally, interstellar extinction curves are generated by the colour-difference method (§3.1.3), in which the spectral energy distributions of selected reddened stars (generally of normal early spectral type) are compared with intrinsically similar objects suffering little or no reddening. This technique is much less reliable for circumstellar extinction, as stars

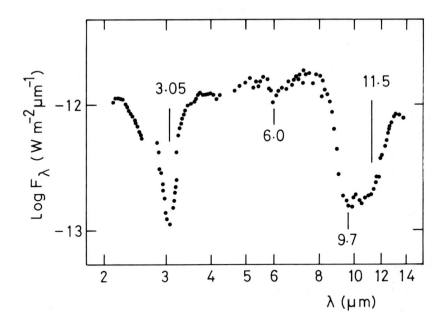

Figure 7.6   Infrared spectrum of the evolved star OH 231.8+4.2 (Soifer *et al* 1981). Spectral features discussed in the text are labelled in μm.

with circumstellar dust are generally intrinsically variable, with complex or anomalous spectra which may include emission lines. Also, practical considerations limit studies to objects which are relatively bright in the ultraviolet. Several techniques have been used to alleviate the problem of finding suitable comparisons. Intrinsic spectra may be reconstructed with reference to model atmospheres (e.g. Buss, Lamers and Snow 1989). In cases where stellar variability is caused entirely by varying degrees of circumstellar extinction, the comparison between spectra at bright and faint phases of the light curve provides an extinction curve without reference to another object (e.g. Hecht *et al* 1984). In serendipitous cases where a normal early-type star lies within or behind a late-type stellar atmosphere, the former may be used as a probe of the extinction in the latter using the standard technique (e.g. Snow *et al* 1987). In all cases, care must be taken to separate circumstellar and interstellar components.

Stars of the RCB class exhibit variability which is attributed to temporal changes in circumstellar dust opacity, and circumstellar extinction curves may thus be derived using the variable extinction method subject to matching the stellar pulsational phase at high and low obscuration. Results show the presence of a broad peak centred near 2500 Å, in place of the well-known 2175 Å feature which characterizes interstellar extinction curves (Hecht *et*

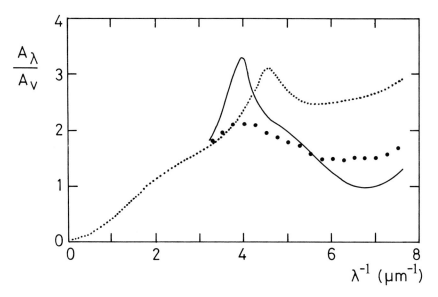

Figure 7.7   A comparison of the mean interstellar extinction curve (table 3.1; dotted curve) with extinction due to dust associated with the RCB star V348 Sgr (Drilling and Schönberner 1989; continuous curve) and the planetary nebula Abell 30 (Greenstein 1981; points).

al 1984; Evans *et al* 1985; Drilling and Schönberner 1989). Figure 7.7 compares data for the RCB star V348 Sgr, the C-rich planetary nebula Abell 30 and the average extinction curve for the diffuse ISM. Dramatic differences are apparent in the level of far-ultraviolet extinction as well as in the position and strength of the mid-ultraviolet peak. RCB stars are carbon-rich but also hydrogen-deficient, and the 2500 Å feature is attributed to glassy or amorphous carbon dust of low hydrogen content (Hecht *et al* 1984; see Stephens 1980 for relevant experimental data). Figure 7.8 shows a fit to the circumstellar extinction curve of the RCB star RY Sgr, deduced by means of Mie calculations for glassy carbon spheres of radii 0.015–0.050 μm which follow a power-law size distribution (Hecht *et al* 1984).

Ultraviolet extinction curves have also been derived for the carbon-rich post-AGB stars HR 4049 and HD 213985. Both are at high galactic latitude, minimizing the problem of interstellar contamination. In contrast to the RCB stars, HR 4049 is hydrogen-rich, and has PAH emission features in its infrared spectrum (Geballe *et al* 1989a). However, its ultraviolet extinction curve is featureless, devoid of peaks at either 2175 or 2500 Å, displaying instead a λ^{-1} dependence (Waters *et al* 1989; Buss *et al* 1989). The Red Rectangle, which may possibly be in a similar evolutionary state to HR 4049, also has PAH features (see figure 6.7) coupled with an absence of mid-ultraviolet extinction peaks (Sitko, Savage, and Meade 1981). These

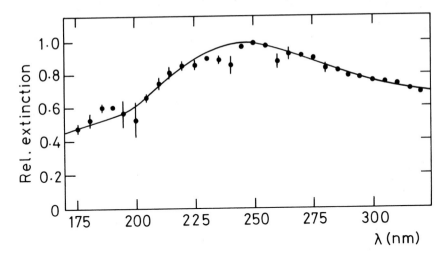

Figure 7.8   The profile of the mid-ultraviolet circumstellar absorption feature in the spectrum of the RCB star RY Sgr (Hecht *et al* 1984). Observational data (points) are fitted with a model based on Mie calculations for glassy carbon spheres with radii between 15 and 50 nm.

results give strong additional support to the view (§5.1.2) that PAHs are *not* carriers of the interstellar λ2175 feature. In contrast, HD 213985 combines an absence of PAH features in the infrared with a mid-ultraviolet peak near 2300 Å, intermediate in wavelength between the normal interstellar feature and the circumstellar feature in RCB stars (figure 7.7). This star may be the missing link, providing tentative support for the proposal that an evolutionary process converts carbon stardust into the carrier of the λ2175 feature in the interstellar medium. Its spectrum warrants investigation at higher signal-to-noise in both the ultraviolet and the infrared.

Circumstellar extinction towards O-rich supergiants has been investigated by the 'near-star' method (Snow *et al* 1987; Buss and Snow 1988). The data show no evidence for either the λ2175 bump or the far-ultraviolet rise. In the best-studied case (α Scorpii), the extinction declines steadily with wavenumber between 3 and 8 μm^{-1}. On the basis of Mie theory calculations, Seab and Snow (1989) conclude that the grain size distribution cuts off at a radius $a \sim 0.08\,\mu$m; i.e. there is an absence of grains significantly *smaller* than this value in the stellar wind.

Current observational constraints on the composition of dust in the shells of evolved stars may be summarized as follows. In O-rich stars, the data are consistent with the presence of a population of large ('classical') amorphous silicate grains. This explains the spectral index of the far-infrared emissivity, the occurrence and shape of the 9.7 and 18.5 μm spectral features, and the shape of the ultraviolet extinction curve. Water frosts also condense in the outflows of a subset of this stellar group, but no other material can currently be identified; if another abundant condensate is present, it must be essentially featureless between wavelengths of 0.1 and 30 μm. In C-rich stars, the predominant solid phase is deduced to be amorphous carbon rather than graphite. The former explains the properties of the infrared continuum emission and the mid-ultraviolet extinction. Silicon carbide emission at 11.2 μm is a ubiquitous feature of carbon stars, but SiC is not an abundant component of the dust. Finally, the emission features characteristic of PAHs are observed in post-AGB stars and planetary nebulae, and their absence in cool carbon stars is attributed to the lack of ultraviolet radiation.

7.1.3   Evolved Stars as Sources of Interstellar Grains

The existence of circumstellar dust does not in itself establish a connection between circumstellar and interstellar grains but the link is greatly strengthened by evidence that evolved stars with dusty envelopes generally undergo lengthy periods of continuous mass loss during their post-main-sequence evolution. Kinematic studies of stellar atmospheres and circumstellar gas have demonstrated that rapid mass-loss rates are common amongst the most luminous stars of all spectral types (see Dupree 1986 and Chiosi and Maeder 1986 for extensive reviews). All stars with atmospheres containing hot chromospheric and coronal gas lose mass to some degree in stellar winds driven by thermodynamic pressure. In the case of the Sun, this currently amounts to some $10^{-14} M_\odot \, yr^{-1}$, which is negligible. Empirically, the mass-loss rate for a thermally driven wind shows a power-law dependence on luminosity L; for stars of early and intermediate spectral type, this is given to an order of magnitude by the relation

$$\log \dot{M} \sim 1.6 \log(L/L_\odot) - 14 \qquad (7.6)$$

where $\dot{M} = dM/dt$ in $M_\odot \, yr^{-1}$; i.e. the typical rate for a supergiant with $L = 10^5 \, L_\odot$ is $10^{-6} \, M_\odot \, yr^{-1}$.

Cool stars tend to lose mass at rates which exceed those predicted purely on the basis of thermodynamic pressure. For a red giant of spectral type M5 III and luminosity $L = 10^3 \, L_\odot$, the measured rate is typically $\sim 10^{-7} \, M_\odot \, yr^{-1}$ (Dupree 1986 and references therein). Figure 7.9 shows a plot of mass-loss rate against spectral type for K- and M-type giants of luminosity classes II and III. A distinct correlation is seen: stars later than about M2 (i.e. stars with photospheres cooler than about 3000 K) tend to have the highest rates. This would be surprising for winds driven entirely by the pressure of hot gas; however, the highest mass-loss rates tend to occur for stars which generally show independent evidence for the presence of dust through observations of excess emission in the mid- and far-infrared. This provides circumstantial support for the view (e.g. Wannier *et al* 1990) that the presence of dust is the catalyst which induces high mass-loss rates in red giants.

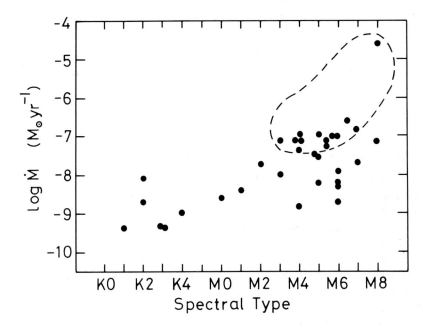

Figure 7.9   A plot of observed mass-loss rate against spectral type for red giants of luminosity classes II and III (Dupree 1986 and references therein). The locus of stars which show evidence for circumstellar envelopes, detected by means of continuum emission from the dust and line emission from gas phase CO and OH, is enclosed by the dashed curve.

Dust grains nucleating in stellar atmospheres are subject to outward acceleration due to radiation pressure. The rate at which a plane wave of intensity I carries linear momentum across a unit area normal to the direction of propagation is I/c. A grain which intercepts a portion of the wave experiences a net force of magnitude $(I/c)C_{pr}$, where C_{pr}, the cross-section for radiation pressure, defines the effective area over which the pressure is exerted (Martin 1978, p 137). We may also define an efficiency factor Q_{pr} for radiation pressure as the ratio of radiation pressure cross-section to geometrical cross-section (i.e. $Q_{pr} = C_{pr}/\pi a^2$ for a spherical grain of radius a). Q_{pr} depends on both the absorption and scattering characteristics of the grain, and is related to the corresponding efficiency factors (§3.1) by

$$Q_{pr} = Q_{abs} + \{1 - g(\theta)\}Q_{sca} \tag{7.7}$$

where $g(\theta)$ is the scattering asymmetry parameter (equation (3.16)). Consider a spherical grain of mass m_d and radius a situated in an optically thin shell at a radial distance r from the centre of a star of luminosity L and mass M. The outward force due to radiation pressure is

$$F_{pr} = \pi a^2 \langle Q_{pr} \rangle \left(\frac{L}{4\pi r^2 c} \right) \qquad (7.8)$$

where $\langle Q_{pr} \rangle$ is the average value of Q_{pr} with respect to wavelength over the stellar spectrum. The opposing force due to gravity is

$$F_{gr} = \frac{GMm_d}{r^2}. \qquad (7.9)$$

The ratio of these forces is of interest, as we require $F_{pr}/F_{gr} > 1$ for outward acceleration. Combining equations (7.8) and (7.9), and expressing m_d in terms of the specific density s of the grain material, we have

$$\frac{F_{pr}}{F_{gr}} = \frac{3L}{16\pi GMc} \left\{ \frac{\langle Q_{pr} \rangle}{as} \right\} \qquad (7.10)$$

which is independent of r and varies with grain properties as the term in brackets.

Q_{pr} may be calculated as a function of wavelength from Mie theory for a given grain model, and the average value estimated with respect to the expected spectral energy distribution for a given stellar type. Martin (1978) estimates $\langle Q_{pr} \rangle \simeq 0.18$ for graphite and 0.003 for silicates, assuming spheres of constant radius $a = 0.05\,\mu m$ in the atmosphere of a red giant of luminosity $10^4\,L_\odot$ and mass $4\,M_\odot$. For discussion, we may take the graphite value as representative of solid (amorphous) carbon. The force due to radiation pressure is typically much greater for absorbing grains than for dielectric grains of the same size, due to the dominant contribution of Q_{abs} to Q_{pr} (equation (7.7)). It may be shown that dust is accelerated from red giants of luminosity $L \geq 10^3\,L_\odot$ for a range of grain size and composition; in the above example, F_{pr} exceeds F_{gr} by a factor of ~ 2000 for carbon grains and by a factor of ~ 40 for silicate grains (equation (7.10)).

The ratio Q_{pr}/a determines the sizes to which grains are predicted to grow before they are expelled from a stellar atmosphere, and the dependence of this quantity on a is thus of interest. For absorbing grains, Q_{pr}/a is maximum for small sizes ($x = 2\pi a/\lambda \ll 1$), whereas for dielectric grains, Q_{pr}/a typically peaks at $x \sim 1$. It follows that silicate grains in an O-rich atmosphere will tend to grow to larger sizes than carbon grains in a C-rich atmosphere. Taking $\lambda \sim 1\,\mu m$ (the wavelength at which a red giant emits maximum flux), $a \sim 0.2\,\mu m$ for dielectrics: thus, silicates may grow to classical sizes before they are injected efficiently into the interstellar medium, in agreement with observational evidence.

The velocities of grains in a stellar atmosphere are limited by frictional drag exerted by the gas. Grains accelerated by radiation pressure thus impart momentum to the gas, which may be an important factor in driving mass loss. For dielectric particles, the terminal velocity depends on size such that large grains tend to overtake smaller ones, resulting in grain–grain collisions with impact velocities typically a few $km s^{-1}$, sufficient to cause fragmentation. Biermann and Harwit (1980) discuss the implications of such collisions for the size distribution of the particles. Multiple collisions lead to the imposition of a power-law size distribution, which is independent of the initial size distribution of the condensates provided that a range of sizes is present. On the basis of fragmentation theory (originally applied to asteroids), Biermann and Harwit argue that the emergent grains follow an $n(a) \propto a^{-3.5}$ law, consistent with that deduced for interstellar grains by fitting the extinction curve (§3.4).

Table 7.2   Estimates of integrated mass-loss rate in $M_{\odot} yr^{-1}$ for stars injecting dust into the interstellar medium in the disk of the Galaxy. The columns from left to right give (1) stellar type; (2) dust type (C-rich or O-rich); (3) total mass-loss rate; (4) mass-loss rate for dust ($\times 10^3$); and (5) percentage of the total mass of injected dust for all types of stars.

Stellar type	C or O	$[\dot{M}_G]$	$10^3 [\dot{M}_G]_d$	%
M giants	O	2.0	14	37
OH–IR stars	O	2.0	14	37
Carbon stars	C	1.0	7	19
M supergiants	O	0.2	1.4	4
Supernovae	O ?	0.1	0.7 ?	2 ?
Planetary nebulae	C, O	0.3	0.1	0.4
WC stars	C	0.01	0.07	0.2
Novae	C, O	0.003	0.02	0.05

In order to gain a quantitative picture of the injection of stardust into the interstellar medium as a whole, it is necessary to estimate space densities, mass-loss rates and dust to gas ratios for all types of star thought to contribute to the process. Table 7.2 shows typical results (based on Gehrz 1989 and references therein). The integrated mass-loss rates for the galactic disk $[\dot{M}_G]$ (column 3) are typically uncertain by a factor of 2 to 3. Dust to gas ratios are found to be in the range $0.004 < Z_d < 0.010$ for both O-rich and C-rich outflows (e.g. Jura 1989). In table 7.2, a value of $Z_d = 0.007$ has been used to estimate $[\dot{M}_G]_d = Z_d[\dot{M}_G]$ for all object types except planetary nebulae, where values $Z_d \sim 4 \times 10^{-4}$ are observed (Natta and Panagia 1981). The final column expresses $[\dot{M}_G]_d$ as a percentage of the total.

Whilst bearing in mind that the data are subject to considerable uncertainty, we may draw some general conclusions from the results in table 7.2. The contributions of novae, Wolf–Rayet (WC) stars and planetary nebulae to injected stardust are evidently very small compared with those of cool stars. The contribution of supernovae is more controversial. Supernovae inject $\sim 0.1 \, M_\odot \, yr^{-1}$ in total to the ISM, and there can be little doubt that this matter is enriched in heavy elements, a high proportion of which may ultimately become incorporated into solids. The detection of isotopic anomalies in meteorites strongly suggests a link between supernovae and presolar interstellar grains (§7.3.3), but there is a scarcity of *direct* observational evidence for the nucleation of dust in supernova ejecta. SN 1987a provided a particularly good opportunity to examine the development of a bright supernova; thermal infrared emission was detected, but may arise from pre-existing rather than newly formed dust (Roche *et al* 1989b; Dwek 1989; but see also Gehrz and Ney 1990). Observations of obscuration at visible wavelengths in the Crab Nebula (Fesen and Blair 1990) must also be interpreted with caution. A 'normal' dust to gas ratio of $Z_d = 0.007$ is assumed for supernova ejecta in table 7.2, but any value in the range $0 \leq Z_d \leq 0.01$ could not be deemed unreasonable. Similarly, it is not clear that planetary nebulae are sources of dust: the low measured values of Z_d noted above may result simply from the dilution of pre-existing dust and gas with purely gaseous ejecta, but Pottasch *et al* (1984) find that Z_d decreases with increasing age, suggesting that grains are being destroyed as a planetary nebula evolves.

We conclude that dust formation in stellar ejecta is dominated not by cataclysmic events but by the winds of cool stars (OH–IR stars, M giants and supergiants, and carbon stars), which together contribute some 90% or more of the total mass of stardust reaching the interstellar medium. This conclusion is valid even if the most optimistic estimates for supernovae are adopted. The balance between C-rich and O-rich stardust appears to be weighted distinctly towards the latter, carbon stars apparently contributing only about a fifth of the total (Gehrz 1989; Bode 1988; see table 7.2). It should be noted that carbon stars become more prevalent towards the outer regions of the galactic disk, as discussed in §2.3.2, and corresponding spatial variations are to be expected in dust composition as a function of galactocentric radius (see Thronson *et al* 1987). Jura and Kleinmann (1989) argue that the contributions of C-rich and O-rich stars are *comparable* in the solar neighbourhood.

How important is stardust as a source of refractory interstellar grains? A number of studies have concluded that grain destruction by shocks in the ISM (§7.2.4) is so efficient that *interstellar* sources of refractory dust must be invoked to explain the observed dust density and to maintain the high observed depletions of the heavy elements. The total injection rate for stardust, summing column 4 of table 7.2, is $\sim 0.04\,M_\odot\,\mathrm{yr}^{-1}$. Injection therefore contributes to the mass density of interstellar dust at a rate

$$\dot\rho_\mathrm{d} = \frac{[\dot{M_\mathrm{G}}]_\mathrm{d}}{V_\mathrm{G}} \sim 2 \times 10^{-32} \ \mathrm{kg\,m}^{-3}\,\mathrm{yr}^{-1} \qquad (7.11)$$

where V_G is the volume of the galactic disk (taken to have radius 15 kpc and thickness 200 pc). The timescale for injection of stardust into the ISM is thus

$$t_\mathrm{in} = \frac{\rho_\mathrm{d}}{\dot\rho_\mathrm{d}} \sim 1 \ \mathrm{Gyr} \qquad (7.12)$$

where the diffuse-cloud value of $\rho_\mathrm{d} \simeq 2 \times 10^{-23}\,\mathrm{kg\,m}^{-3}$ is assumed (§3.2.6). For equilibrium between injection and destruction, the mass fraction of interstellar dust originating in stars is

$$f_\mathrm{sd} = (1 + t_\mathrm{in}/t_\mathrm{sh})^{-1} \qquad (7.13)$$

where t_sh is the timescale for destruction in shocks. McKee (1989) estimates $t_\mathrm{sh} \sim 0.4\,\mathrm{Gyr}$, and equation (7.13) thus gives $f_\mathrm{sd} \sim 0.3$, i.e. stardust may account for some 30% of the mass of refractory interstellar dust at any given time.

7.2   THE EVOLUTION OF DUST IN THE INTERSTELLAR MEDIUM

"Once the newly formed grains are injected into the interstellar medium, they are subject to a variety of indignities..."

   C G Seab (1988)

   In previous chapters, a number of lines of evidence are discussed which illustrate that the properties of interstellar grains are not spatially uniform but vary in response to environment. As examples, the correlation of depletion with mean density (§2.4) and the occurrence of ice absorption features in dense clouds and their absence in diffuse clouds (§5.3) point to significant exchange of matter between gas and dust as a function of cloud density. The interaction of the dust with the ambient radiation field determines the grain temperature (§6.1), controlling evaporation rates and internal structure. Individual photons induce charge on the grains by the photoelectric effect, and may activate chemical evolution by dissociative production of radicals in molecular mantles. In this section, we begin (§7.2.1) by reviewing the adsorption process which leads to the attachment of gas phase atoms onto grain surfaces and the subsequent recombination of molecular hydrogen by surface catalysis. Important gas phase reactions which influence the chemical evolution of the interstellar medium are also briefly discussed. We then consider the growth of dust grains in dense clouds (§7.2.2), with emphasis on the composition and evolution of icy molecular mantles (§7.2.3). Finally, the energetic processing and destruction of dust is reviewed in §7.2.4.

7.2.1   Grain–Surface Reactions and Gas Phase Chemistry

Gas phase atoms and molecules impinging upon the surface of a solid may become attached to it by physical adsorption or by chemical bonding. The binding energies involved are markedly different in the two cases. Physical attachment is maintained by weak van der Waals forces with binding energies $< 0.1 \, \mathrm{eV}$, the surface acting merely as a passive substrate. The surfaces of real solids terminate in atoms or ions in states of low symmetry and low coordination which tend to be chemically reactive. Moreover, interstellar grain surfaces are likely to be highly disordered, containing numerous lattice defects and impurites which enhance their chemical reactivity (see, for example, pp 92–95 of Duley and Williams 1984). Amorphous carbon dust may have a high concentration of active surface sites due to the presence of C atoms with 'dangling' bonds, and in interstellar clouds these will tend to become saturated with hydrogen. Similarly, silicate and oxide grains will have active surfaces due to the presence of oxygen ions in low-coordination sites, and these will also tend to attach hydrogen. The binding energies for chemical adsorption are typically in the range 1 to $5 \, \mathrm{eV}$, i.e. at least an order of magnitude higher than those for physical adsorption.

Figure 7.10 illustrates potential energy diagrams for attachment of atoms and molecules onto a surface. Note that the forces for physical adsorption have longer range than those for chemical bonding, and two distinct potential wells therefore occur: a particle approaching the surface encounters, firstly, a shallow well associated with physical adsorption forces, followed by a deeper well associated with chemical bonding. Chemisorption of accreting atoms is limited only by the availability of active sites, and a surface in which all such sites are occupied may accrete further atoms into an outer layer by physical adsorption. The situation is different for accreting molecules, in that chemisorption is often inhibited by the presence of an activation energy barrier arising from the need to break or modify a molecular bond (Tielens and Allamandola 1987a). Chemisorption of molecules is thus generally unimportant in interstellar clouds.

The rate at which a dust grain adsorbs atoms or molecules from its environment may be deduced as a function of the kinetic temperature and density of the gas. Assuming a Maxwellian distribution of velocities, the

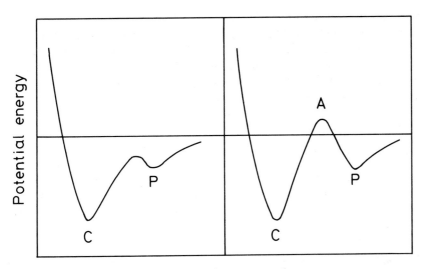

Figure 7.10 Schematic potential energy diagrams for the adsorption of an atom (left) and a molecule (right) onto the surface of a solid (adapted from Tielens and Allamandola 1987a). In each case, a particle approaching the surface (from right to left) encounters a relatively shallow negative potential well (P) associated with physical adsorption forces, and a deeper well (C) associated with chemical bonding. In the molecular case, chemical bonding is inhibited by an activation energy barrier (A).

mean speed of a particle of mass m is

$$\langle v \rangle = \left(\frac{8kT_{\mathrm{g}}}{\pi m} \right)^{\frac{1}{2}} \tag{7.14}$$

(Spitzer 1978), where T_{g} is the gas temperature. In diffuse interstellar clouds, we are concerned primarily with the adsorption of atomic hydrogen, and assuming $T_{\mathrm{g}} \simeq 80\,\mathrm{K}$ (table 1.1), equation (7.14) gives $\langle v_{\mathrm{H}} \rangle \simeq 1.4\,\mathrm{km\,s^{-1}}$. Kinetic motion leads to collisions between the gas atoms and the grain surface. Assuming the grain to be spherical of radius a, immersed in a gas of number density n, the number of collisions per second is $\pi a^2 n \langle v \rangle$. An impinging atom may be adsorbed or returned to the gas; the probability of adsorption is denoted by S, the sticking coefficient, which has a value in the range $0 \leq S \leq 1$. The number of atoms adsorbed per second is therefore $S\pi a^2 n \langle v \rangle$, and the timescale (mean free time) for adsorption events is the reciprocal of this quantity,

$$t_{\mathrm{ad}} = (S\pi a^2 n \langle v \rangle)^{-1}. \tag{7.15}$$

The sticking coefficient depends on the nature of the surface and on the temperatures of both the gas and the dust. In a hot gas, in which the mean kinetic energy of the impinging atoms exceeds the adsorption energy, collisions tend to be elastic with low probability of attachment ($S \ll 1$). However, theoretical calculations and experimental investigations strongly suggest that $S \simeq 1$ for chemical adsorption of atoms onto amorphous carbon or amorphous silicate grains in interstellar clouds (Tielens and Allamandola 1987a and references therein). With this value in equation (7.15) above, we estimate $t_{ad} \simeq 800\,s$ for classical ($a \simeq 0.1\,\mu m$) grains accreting H I in diffuse clouds of density $n \simeq 3 \times 10^7\,m^{-3}$.

The lifetime of an adsorbed atom against evaporation depends on the dust temperature T_d and the binding energy ε, and is given by

$$t_{ev} = t_0 \exp\left(\frac{\varepsilon}{kT_d}\right) \tag{7.16}$$

where t_0 is the oscillation time period of the adsorbed species in the direction perpendicular to the surface (typically $10^{-12}\,s$). For chemical binding ($\varepsilon \sim 3\,eV$), we have $\varepsilon \gg kT_d$ for any reasonable value of T_d, and t_{ev} is generally long compared with cloud lifetimes: i.e., a chemisorbed atom is essentially stable against thermal evaporation. For physical binding, t_{ev} is critically dependent on temperature: for (e.g.) $\varepsilon = 0.03\,eV$, values of $1.3 \times 10^3\,s$ and $1.1 \times 10^{-7}\,s$ are deduced from equation (7.16) for $T_d = 10\,K$ and $T_d = 30\,K$, respectively.

The accretion of atoms onto grains is of primary importance in interstellar chemistry because it provides the only plausible mechanism for the efficient production of molecular hydrogen (e.g. Duley and Williams 1984). This process may be written symbolically as

$$H + H + grain \rightarrow H_2 + grain. \tag{7.17}$$

The H atoms cannot recombine directly by two-body encounters in the gas phase because there is no feasible method of losing binding energy. The surface of a grain provides both a substrate on which the atoms become readily trapped and a heat reservoir which absorbs excess energy. Three distinct steps are involved: (i) the attachment of atoms to the surface, as described above; (ii) the migration of an atom from one surface site to another until a second atom is encountered; and (iii) recombination, leading to the ejection of the resulting H_2 molecule from the grain surface.

The mobility of surface atoms is discussed by (e.g.) Tielens and Allamandola (1987a). The surface binding energy of an adsorbed species varies with position due to the discrete nature of the lattice. The potential barrier against diffusion to an adjacent site is typically 30% of the binding energy for physical adsorption and 50–100% of the binding energy for chemical adsorption. The migration of H atoms involves quantum mechanical tun-nelling, which occurs on a timescale of 3×10^{-9} s for physical adsorption. This is generally short compared with the timescale for evaporation at low temperatures ($T_d < 30$ K). Thus, migrating physically adsorbed atoms rapidly seek out and fill any vacant sites for chemisorption on the grain. H_2 formation is most likely to involve an encounter between a migrating physically adsorbed atom and a trapped chemisorbed atom. Binding en-ergy of 4.48 eV is released on recombination, some fraction of which will be imparted to the newly formed molecule as rotational excitation and translational motion, any excess being absorbed by the grain as lattice vibrations.

With the availability of molecular hydrogen from the dust, a number of important gas phase reactions become viable as sources of heavier molecules known to exist in interstellar clouds. A detailed discussion of gas phase chemical models is beyond the scope of this book, but it is appropriate to summarize the processes leading to the formation of species which may subsequently condense onto the dust. For comprehensive accounts of gas phase chemistry, see, for example, Herbst (1987), Herbst and Leung (1986), Leung et al (1984), Duley and Williams (1984), Prasad and Huntress 1980, and Mitchell et al (1978).

Under normal interstellar conditions, only binary (two-body) reactions need be considered, as the probability of three-body reactions in the gas phase is extremely low. In general, only exothermic reactions are likely to occur at a significant rate in quiescent clouds; certain weakly endothermic reactions are thought to be driven by shocks, but these need not concern us here. Consider the formation of some molecule M by the schematic reaction

$$A + B \rightarrow M + N. \tag{7.18}$$

The reactants A and B may be atomic or molecular, neutral or ionic. The additional product N absorbs binding energy from the molecule M; N may be atomic, molecular or, in some instances, a photon. Reaction (7.18) will have some rate coefficient k_r associated with it which limits the abundance of M. In diffuse clouds, molecules are destroyed predominantly by photodis-sociation, and if the mean lifetime of molecule M against photodissociation, Δt, is short compared with cloud lifetimes, then the number density of M is given in terms of the number densities of A and B by the equilibrium equation

$$n_M = k_r n_A n_B \Delta t. \tag{7.19}$$

If the reactants bear no charge, the value of k_r is typically 10^{-17} m^3 s^{-1} (e.g. Duley and Williams 1984). However, if one of the reactants is ionized, much higher rates can occur, with k_r enhanced typically by a factor of 100. This arises because the electric field of the ion polarizes the neutral species, resulting in a net attractive Coulombic force. Exothermic ion–molecule reactions thus play an important, and, indeed, dominant rôle in current models for gas phase chemistry, and such models are generally quite successful in reproducing the observed abundances. A supply of ions is maintained by radiative and cosmic-ray ionization, with the latter dominating in dense clouds shielded from the external radiation field. It should be noted that in dense clouds, photodissociation timescales may become long compared with timescales for dynamical evolution, and molecular abundances may not reach a steady state.

Models predict that water molecules form in interstellar clouds by means of a sequence of gas phase reactions between oxygen-bearing ions and molecular hydrogen. O^+ is assumed to be present, as a result of either direct photoionization or charge transfer reactions such as

$$O + H^+ \rightarrow O^+ + H. \tag{7.20}$$

O^+ is processed to H_3O^+ by the ion–molecule sequence:

$$O^+ + H_2 \rightarrow OH^+ + H$$
$$OH^+ + H_2 \rightarrow H_2O^+ + H$$
$$H_2O^+ + H_2 \rightarrow H_3O^+ + H \tag{7.21}$$

and, finally, H_2O results from the recombination of H_3O^+:

$$H_3O^+ + e \rightarrow H_2O + H. \tag{7.22}$$

An analogous set of reactions produces NH_3. Carbon chemistry appears to be initiated by radiative association

$$C^+ + H_2 \rightarrow CH_2^+ + h\nu \tag{7.23}$$

as formation of CH^+ is endothermic. CH_2^+ may then be processed to CH_4 by reactions analogous to (7.21) and (7.22). The resulting abundances of saturated molecules such as H_2O, NH_3 and CH_4 depend on the rate coefficients of each step of the formation sequence, and the dissociation and recombination cross-sections of the products. Over a wide range of cloud densities and physical conditions, these molecules have relatively low abundances and account for only a minor fraction of the available O, N and C (e.g. Herbst and Leung 1986).

By far the most abundant gas phase molecule after H_2 is CO, which is formed in diffuse clouds (van Dishoeck and Black 1987) by the ion–molecule reaction

$$C^+ + OH \rightarrow CO + H^+ \tag{7.24}$$

and by the alternative sequence

$$C^+ + OH \rightarrow CO^+ + H$$
$$CO^+ + H_2 \rightarrow HCO^+ + H$$
$$HCO^+ + e \rightarrow CO + H. \tag{7.25}$$

In dense clouds, CO is produced directly by the neutral exchange reaction

$$CH + O \rightarrow CO + H \tag{7.26}$$

and by recombination of HCO^+ formed by the reactions

$$C^+ + H_2O \rightarrow HCO^+ + H \tag{7.27}$$

and

$$CH_3^+ + O \rightarrow HCO^+ + H_2. \tag{7.28}$$

Once created, CO is difficult to destroy because of its relatively large binding energy, as previously noted in our discussion of cool stellar atmospheres (§7.1.1). In diffuse clouds, CO is destroyed principally by photodissociation due to photons with energy in the range 11.1–13.6 eV. In dense clouds, the most important sink for gas phase CO is probably condensation onto grains.

Table 7.3   Predicted gas phase abundances for atomic and molecular species containing the C, N, O group of elements in a cloud of intermediate density, based on the model of Herbst and Leung (1986). Abundances are expressed as a fraction of $N_H = N(HI) + 2N(H_2)$.

Species	N_X/N_H	Species	N_X/N_H
C	2.1×10^{-5}	CN	1.5×10^{-7}
C^+	1.2×10^{-5}	CH_2	1.9×10^{-7}
N	2.1×10^{-5}	NH_2	1.6×10^{-9}
O	1.4×10^{-4}	H_2O	1.0×10^{-8}
CH	3.7×10^{-7}	CO_2	3.2×10^{-11}
NH	1.9×10^{-9}	H_2CO	6.0×10^{-11}
OH	3.0×10^{-8}	NH_3	4.7×10^{-12}
C_2	4.9×10^{-7}	CH_3	4.2×10^{-10}
CO	3.9×10^{-5}	CH_4	7.0×10^{-9}

Table 7.3 lists abundances for 18 atomic and molecular C-, N- and O-bearing species, calculated for a cloud of density $n = 2 \times 10^9 \, \mathrm{m}^{-3}$, temperature $T_g = 50 \, \mathrm{K}$, and visual extinction $A_V = 2$ mag (Herbst and Leung 1986). These parameters were chosen to represent the intermediate, transitional state between a diffuse cloud and a dense cloud, and results may thus provide a reasonable guide to the abundances of atoms and molecules available to condense onto grains as the density reaches a critical value for the onset of rapid mantle growth. It is notable that, with the single exception of CO, all molecular species listed have very low abundances ($< 5 \times 10^{-7}$) compared with those of atomic C, N and O.

7.2.2   Mechanisms for Grain Growth

Grains may increase their average size and mass in interstellar clouds by two distinct mechanisms, (i) coagulation, as a result of low-velocity grain–grain collisions, and (ii) mantle growth, in which gas phase atoms and molecules are adsorbed on to the surface. In either case, the collision rate, and hence the rate of growth, increases with cloud density. It is highly likely that both mechanisms occur in the interstellar medium: the two main observational indicators of grain growth, extinction curve variations and the presence of certain infrared spectral signatures, appear to arise independently and are attributed to coagulation and mantle growth, respectively (see below). Enhanced adsorption rates in interstellar clouds are also suggested by the correlations of observed depletion indices for various metals with mean density (§2.4.3; e.g. figure 2.8). However, the most refractory elements generally have high fractional depletions even in relatively low-density regions, and there is little scope for significant growth by further adsorption of these elements. It is clear from consideration of the abundances and depletions of the elements capable of forming solids that grain mantles resulting from adsorption from the gas must be composed of compounds predominantly involving C, N and O, and that complete adsorption of these elements would increase the dust to gas ratio typically by a factor of about 2.

The timescale for grain–grain collisions is

$$t_{col} = (n_d \pi a^2 \langle v_d \rangle)^{-1}$$
$$\sim (10^{-25} n_H \langle v_d \rangle)^{-1} \tag{7.29}$$

where n_d and $\langle v_d \rangle$ are the number density and average velocity of the dust grains (see Jura 1980). For a cloud of density $10^9 \, \mathrm{m}^{-3}$, and with $\langle v_d \rangle \sim 0.1 \, \mathrm{km \, s}^{-1}$ (Jura 1980), equation (7.29) gives $t_{col} \sim 3 \, \mathrm{Myr}$, and thus coagulation may be significant within cloud lifetimes. Results discussed in §3.3 imply that grain growth leading to enhanced values of the ratio of total to selective extinction R_V appears to involve changes in the optical properties of the dust in the blue–visible region of the spectrum, and these changes are not generally accompanied by spectral signatures of ice mantles (e.g. Whittet and Blades 1980). The ρ Oph cloud provides a good example. Stars with moderate values of visual extinction $(1 < A_V < 10)$ show significant $(\sim 40\%)$ enhancement in R_V compared with the diffuse ISM, whereas ice mantles are detected only towards sources with $A_V > 10$ (Tanaka et al 1990). It thus seems likely that the phenomenon of high R_V is associated with coagulation and not with mantle growth. Coagulation may result in a reduction of the relative number of small grains due to accretion onto larger ones, and this will affect the shape of the extinction curve in the blue–visible. Independent evidence to support this interpretation is provided by the ratio N_H/A_V observed towards ρ Oph, which is found to be enhanced by a factor of ~ 2 compared with the diffuse-cloud value. This effect is predicted by a coagulation model (Jura 1980) in which grains tend to 'hide' behind one another, reducing the effective opacity in the visible. In contrast, a model based on mantle growth would predict a reduction in N_H/A_V.

Whereas coagulation may modify the form of the grain size distribution function, we show below that the thickness of an adsorbed mantle is independent of the initial size; thus, accretion shifts the distribution function to larger sizes whilst preserving its functional form. Consider a spherical grain of radius a immersed in a gas of identical particles of mass m and number density n. The grain adsorbs particles at a rate $S\pi a^2 n\langle v\rangle$, where S is the sticking coefficient and $\langle v\rangle$ the mean particle speed (§7.2.1; see equation (7.15)). The rate of increase in the mass of the dust grain is thus

$$\frac{dm_d}{dt} = S\pi a^2 n \left(2.5 k T_g m\right)^{\frac{1}{2}} \qquad (7.30)$$

where T_g is the kinetic temperature of the gas. If adsorption leads to the formation of a mantle of density s, then the rate of growth of the grain is given by

$$\frac{da}{dt} = \frac{1}{4\pi a^2 s} \frac{dm_d}{dt}$$
$$= 0.4 S n s^{-1} (k T_g m)^{\frac{1}{2}} \qquad (7.31)$$

which is independent of a. Assuming that the quantities on the right-hand side of equation (7.31) are constant with respect to time, integration gives the radius at time t simply as

$$a(t) = a_0 + 0.4 S n s^{-1} (k T_g m)^{\frac{1}{2}} t \qquad (7.32)$$

where a_0 is the radius at $t = 0$, i.e. the core radius. In the absence of significant desorption, the timescale for accretion of a mantle of thickness

$\Delta a = a(t) - a_0$ is thus

$$t_{\rm m} = \frac{2.5 s \, \Delta a}{S n \, (k T_{\rm g} m)^{\frac{1}{2}}} \, . \tag{7.33}$$

We may use the result in equation (7.33) to investigate whether the formation of molecular mantles observed spectroscopically in dense clouds results from 'frosting' (i.e., the direct condensation of pre-existing molecules from the gas) or from surface reactions involving adsorbed atoms and radicals. Following Jones and Williams (1984), we take the Taurus dark cloud as a basis for discussion, noting that the spectral signature of water-ice is observed at visual extinctions as low as 3–4 magnitudes (§5.3.3), and that this can be interpreted in terms of the presence of thin ($\Delta a \simeq 0.02 \, \mu$m) mantles of amorphous ice of density $s \simeq 750 \, \text{kg m}^{-3}$ (Whittet *et al* 1983; Jones and Williams 1984). Assuming cloud parameters and gas phase abundances from the Herbst and Leung model (table 7.3), equation (7.33) predicts mantle accretion times of approximately 1 Myr, 5 Gyr and 10 Gyr if the process is limited by the abundance of O, OH and H_2O, respectively, assuming $S = 1$ for all species. Given a typical cloud age of 1–10 Myr, these calculations clearly show that the growth of ice mantles *cannot* result from frosting of gas phase H_2O. Only surface reactions involving adsorbed atomic oxygen appear to be capable of producing such mantles on a realistic timescale. Even in much denser clouds, where gas phase H_2O abundances of 5×10^{-6} are predicted (Herbst and Leung 1986), surface reactions may still be the dominant source of H_2O on grains.

7.2.3   The Nature and Evolution of Grain Mantles in Molecular Clouds

Icy mantles are ubiquitous to the molecular cloud environment. They accumulate in dense regions where the grains are cold and well protected from the destructive effects of the interstellar radiation field, and they contain a substantial fraction of the dust mass in such regions, regulating the abundances of species remaining in the gas. Moreover, many chemical reactions are possible on grain surfaces which cannot occur in the gas phase, and the evolution of the mantles thus governs the evolution of the cloud as a whole. Ultimately, the fate of the mantles in regions of recent star formation determines the chemical form of the CNO group of elements in protostellar shells and protoplanetary disks.

Models developed by d'Hendecourt, Allamandola and Greenberg (1985) show that the composition of grain mantles in an interstellar cloud is critically dependent on the abundance ratio of atomic to molecular hydrogen. As a diffuse H I cloud gradually collapses, the initial mantle material accumulating upon the refractory grain cores will be hydrogen-rich, formed by surface reactions between impinging H, C, N and O atoms. This initial condensate is predicted to be composed predominantly ($\sim 68\%$ by mass) of H_2O, but will also contain significant quantities of NH_3 and CH_4. However, once the cloud density has increased to the extent that essentially all the gas phase hydrogen is converted to H_2, surface reactions leading to the formation of H_2O, NH_3 and CH_4 are inhibited. A qualitatively different mantle material then grows on the surface of the H_2O-rich mantle. This secondary mantle is expected to be rich in non-polar molecules such as O_2 and N_2, and will also contain appreciable quantities of CO. Whereas O_2 is formed primarily by surface catalysis of atomic oxygen, CO is primarily accreted from the gas, whilst N_2 may be an intermediate case (d'Hendecourt *et al* 1985). A composite grain structure is thus predicted by these models, as illustrated schematically in figure 7.11.

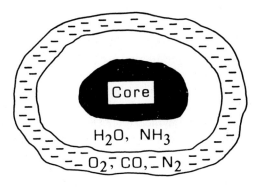

Figure 7.11   A schematic representation of grain structure in a dense molecular cloud. The core is composed of refractory materials such as silicates, amorphous carbon and organic residues. The inner mantle is composed largely of polar molecules which form by surface reactions in the presence of H I, and is dominated by H_2O. The outer mantle is composed predominantly of non-polar molecules such as O_2, N_2 and CO, which form in an H_2 environment.

Whilst the homonuclear molecules O_2 and N_2 are not available to direct spectroscopic observation in the infrared, detection of the feature near $4.67\,\mu$m has demonstrated the presence of CO in grain mantles (§5.3.3; see Whittet and Duley 1991 for an extensive review). In general, the profile of the observed CO feature is consistent with the presence of two distinct but overlapping absorptions which have different relative strengths in different lines of sight: (i) a sharp feature (FWHM $\sim 0.013\,\mu$m) centred at $4.674\,\mu$m, and (ii) a shallower, broader feature (FWHM $\sim 0.020\,\mu$m) centred at $4.681\,\mu$m, which appears as a long-wavelength wing in the profile of the sharp feature (see figure 7.12). These absorptions provide strong observational support for the view that mantle structure is, indeed, hierarchical (figure 7.11), with segregation of H-rich (polar) and H-poor (non-polar) ices. Laboratory studies of variations in the profile as a function of the composition of the mantle in which the CO molecules are embedded (e.g. Sandford *et al* 1988) are vital to the interpretation of the observations. Pure solid CO produces a feature near $4.675\,\mu$m which is significantly narrower (FWHM $\sim 0.005\,\mu$m) than that generally seen in the interstellar medium (but see Kerr *et al* 1991 for an exception). Impurities in solid CO lead to shifts in the position of the feature and to broadening, and, significantly, the scale of these changes depends on the dipole moment of the mixture. All laboratory mixtures in which the dominant constituent is H_2O produce a feature centred at $4.681\,\mu$m with FWHM $\sim 0.02\,\mu$m, i.e., they match the properties of the broad component of the observed profile but not those of the narrow component, as illustrated in figure 7.12. A further characteristic of cold mixtures dominated by water-ice is the presence of a sideband centred near $4.64\,\mu$m, which has no counterpart in the spectrum of any source observed to date, placing further constraints on the abundance of CO in a mixture dominated by H_2O. The observed properties of the solid CO profile appear to be consistent with an origin for the broad and narrow components in polar and non-polar materials, respectively. In particular, ices containing a high proportion of H_2O mixed with the CO *cannot* explain the narrow feature, and its existence provides the first direct observational evidence for the presence of grain mantles in which H_2O is not the dominant constituent. Of the laboratory mixtures studied by Sandford *et al* (1988), that which corresponds most closely to the observations is CO_2:CO ($\lambda \simeq 4.674\,\mu$m, FWHM $\simeq 0.013\,\mu$m).

Many sources with solid CO detections are highly obscured objects characterized by strong infrared excess emission, and are presumably young stars still enshrouded in their placental molecular clouds. Such sources may have a marked effect on the nature of the dust in their vicinity through local enhancement of the radiation field. Warm-up leading to evaporation will clearly tend to remove the volatile outer layer of the mantles more readily than the underlying water-dominated component. The strength of the CO feature relative to that of the $3.0\,\mu$m water-ice feature thus gives a measure of the degree of such processing. Observations of background field stars seen through quiescent dark-cloud material are valuable as a 'control experiment', providing the opportunity to study the behaviour of the mantles in an undisturbed state, unaffected by local sources of luminosity within the clouds (§5.3.3). Figure 7.13 plots $\tau_{4.67}$ against $\tau_{3.0}$ for sources in the Taurus dark cloud; $\tau_{4.67}$ is here taken to be the peak optical depth of the composite profile, normally dominated by the narrow feature. Field and embedded objects are distinguished by plotting symbol. The locus of the pre-main-sequence star HL Tau illustrates how an embedded object influences its environment: in this line of sight, the dust responsible for the optical extinction resides predominantly in a circumstellar disk

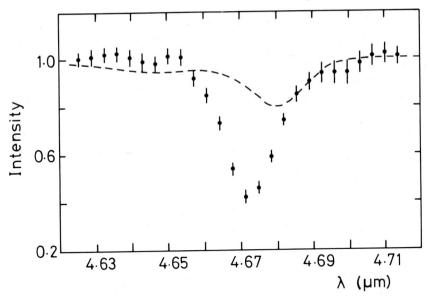

Figure 7.12   Comparison of the observed 4.67 μm CO profile in Elias 16 (Whittet *et al* 1985, points) with the laboratory profile for a 20:1 H_2O:CO mixture (Sandford *et al* 1988, dashed curve). The central depth of the feature in the laboratory spectrum is scaled to 30% of that of the observed feature.

(§7.3.2) maintained at too warm a temperature for solid CO to survive in the mantles, yet the H_2O-ice feature is quite strong. The remaining sources show a close correlation, and a least-squares fit gives

$$\tau_{4.67} = 0.69\,(\tau_{3.0} - 0.25) \tag{7.34}$$

(Whittet *et al* 1989), as plotted in figure 7.13. This correlation indicates that both CO and H_2O ice are ubiquitous in undisturbed regions of the dark cloud, and suggests that the strengths of both features increase with the total column density of dust. The existence of a threshold value of $\tau_{3.0}$ (the intercept in equation (7.34)) below which the CO feature is undetected reflects the different degrees of volatility of the inner and outer mantles. Figure 7.14 compares the correlation line for the Taurus cloud (equation (7.34)) with the distribution of a heterogeneous sample of 'protostellar' sources in the $\tau_{4.67}$ versus $\tau_{3.0}$ diagram. In contrast to the situation in the single dark cloud, there is much scatter in the points and, as expected, a general tendency for the CO feature to be weak relative to the H_2O feature. Indeed, it is somewhat surprising that two sources (NGC 7538 IRS9 and SVS 20) lie *above* the dark-cloud line. If we assume that Taurus represents the ideal static environment for CO condensation, then it may be that dynamic models for interstellar chemistry should be applied to these lines of sight, leading to cyclic wind-driven molecule formation and

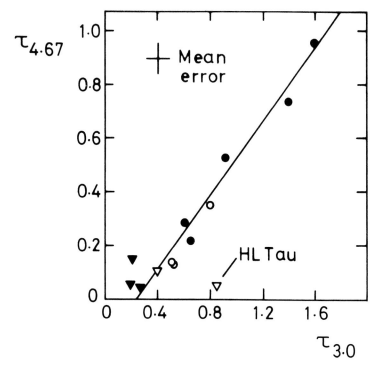

Figure 7.13   Plot of peak optical depth in the 4.67 μm solid CO feature ($\tau_{4.67}$) against that in the 3.0 μm water-ice feature ($\tau_{3.0}$) for stars in the region of the Taurus dark cloud (Whittet *et al* 1989). Filled symbols represent field stars seen through the cloud, open symbols represent embedded pre-main-sequence objects; upper limits are denoted by triangles. The straight line is the least-squares fit to field stars.

mantle growth (Charnley *et al* 1988).

The width, central wavelength and profile shape of the solid CO feature are sensitive to the thermal history as well as the composition of the molecular matrix containing the CO. When a species is deposited at a temperature much lower than its melting point, an amorphous rather than a crystalline structure is produced, containing a distribution of molecular environments which leads to broad, Gaussian line profiles. If the material is warmed, the molecules arrange themselves into more energetically favourable orientations, resulting in evolution of the line profiles towards the sharper features generally seen in crystalline solids. The effect of this annealing process on the infrared spectra of laboratory ice mixtures containing CO has been investigated by Sandford *et al* (1988) and Schmitt *et al* (1989). Whereas a pure CO frost would vaporize at ∼ 20 K, CO is retained in an H_2O-dominated matrix at much higher temperatures: for

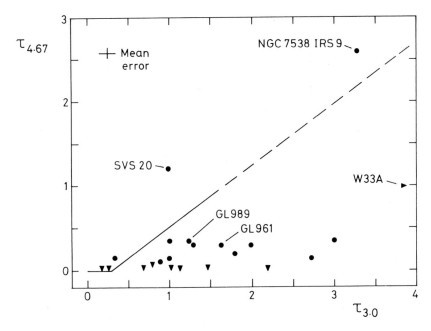

Figure 7.14   The $\tau_{4.67}$ versus $\tau_{3.0}$ diagram for 'protostellar' objects in various molecular clouds excluding Taurus (Whittet and Duley 1991). The line is the observed correlation for Taurus field stars (figure 7.13), dashed in the region of extrapolation). Triangles denote upper and lower limits.

example, Schmitt *et al* find that a $H_2O:CO$ mixture containing 25% CO at 10 K retains 7.5% CO when warmed to 100 K. These authors investigate the evolution of the profile shape with temperature. The most notable effect is that the strength of the sideband at $4.64\,\mu m$ relative to that of the main feature is reduced with increasing temperature, until it has almost disappeared at 100 K. Simultaneously, the position of the main feature shifts to slightly longer wavelength (from $4.682\,\mu m$ to $4.686\,\mu m$). The non-detection of the $4.64\,\mu m$ sideband in the spectra of luminous, embedded sources could thus be due to warming (although its absence in quiescent regions devoid of luminous stars, such as the Taurus clouds, cannot be explained in this way). Independent evidence for annealing of grain mantles is provided by profile studies of the $3.0\,\mu m$ water-ice feature (Smith, Sellgren and Tokunaga 1989). These authors find evidence for annealing at temperatures $T > 70\,K$ towards BN, Mon R2 IRS3, and AFGL 2591 (none of which have a CO detection), and find *no* evidence for annealing (implying $T < 30\,K$) towards Elias 16, AFGL 961, AFGL 989 and Mon R2 IRS2 (all of which have substantial solid CO features). On the basis of this sample, it appears that warm-up may not be a marginal process: when it occurs, it appears to proceed rapidly until essentially all the CO is desorbed. Combined investigations of the 4.67 and 3.0 μm dust features clearly have great potential for studying the thermal evolution of dust in the protostellar environment.

In some lines of sight, freeze-out of CO onto grains may be sufficiently high to have a drastic effect on the abundance of CO remaining in the gas. As noted in §5.3.3, this is an issue of considerable astrophysical significance. Ideally, the degree of CO depletion onto grains is measured by means of a direct comparison of solid phase and gas phase column densities in the same line of sight, deduced by absorption spectroscopy of a background source (see Whittet and Duley 1991). However, the high spectral resolution and sensitivity required to obtain reliable gas phase column densities limits this technique to a few bright protostellar sources such as W33A (Mitchell *et al* 1988) and NGC 2024 no.2 (Black and Willner 1984) with current facilities. In these lines of sight, the depletion of CO appears to be low, ∼ 5% or less. Towards a number of intrinsically similar objects (e.g. BN, OMC-1 IRc2, AFGL 2591), the solid CO feature is undetected at a level below that expected if a substantial fraction of the observed gas phase CO were depleted. However, as discussed above, these regions are very probably highly disturbed by the embedded sources within them, and are unlikely to be representative of the dark-cloud environment.

The first indication that substantial quantities of CO may be depleted in a dark cloud came from the observations of Whittet, Longmore and McFadzean (1985) for the Taurus cloud. No infrared observations of gas phase CO exist for these stars, but extensive millimetre-wave data are available. Figure 7.15 shows a comparison of gas and solid phase CO column densities plotted against A_V, using data from Frerking *et al* (1982) and Crutcher (1985) for the gas and from Whittet *et al* (1985, 1989) for the dust. Estimates of the former depend on observations of $^{12}C^{18}O$, $^{12}C^{17}O$ and $^{13}C^{18}O$, assuming terrestrial isotopic ratios. The plot shows that both solid and gas phase CO column densities tend to increase linearly with visual extinction above a threshold value of ∼ 5 mag, to within observational error. The depletion of CO is essentially constant at ∼ 30% for $A_V > 5$. These results are in conflict with models for grain growth which suggest that CO should become almost totally depleted in clouds with densities ∼ 10^9 m^{-3} and visual extinctions $A_V \sim 2$ (e.g. Léger 1983). The timescale for accretion is

$$t_{\rm acc} = (S n_{\rm d} \pi a^2 \langle v_{\rm CO} \rangle)^{-1} \tag{7.35}$$

where $\langle v_{\rm CO} \rangle$ is the mean thermal velocity of gas phase CO. Assuming $T_{\rm g}({\rm CO}) = 20$ K, we deduce $t_{\rm acc} \sim 3 \times 10^6 / S$ years (see also equation (7.29)), which is comparable with cloud lifetimes for $0.3 < S < 1$. The survival of grain mantles in such regions depends critically on the efficiency of appropriate desorption mechanisms such as photodesorption (Barlow 1978b) and impulsive heating by X-rays and cosmic rays (Léger, Jura and Omont

1985), as discussed in detail by Duley *et al* (1989b). If it is assumed that the refractory component of the grain material is amorphous, localized heating of subunits or islands within the grain structure due to the absorption of individual photons can act as an efficient desorption mechanism. The existence of threshold extinctions for detection of H_2O and CO ices in Taurus is then easily understood if the cloud is heated predominantly by the external interstellar radiation field. Deep within the cloud, cosmic rays are likely to be the most significant source of energy contributing to desorption from grain surfaces, and Duley *et al* argue that the ultraviolet field produced by cosmic-ray excitation of H_2 can account for the observed abundances of CO on grains and in the gas phase.

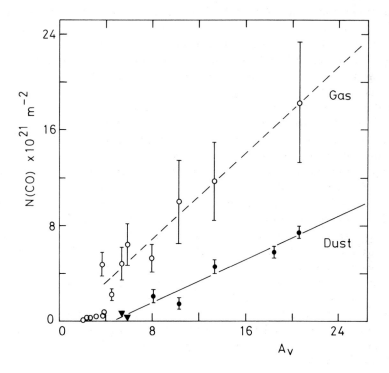

Figure 7.15   Plot of gas phase and solid-state CO column densities against visual extinction for field stars in the line of sight to the Taurus dark cloud (Whittet and Duley 1991 and references therein).

The presence of CO in grain mantles is suggestive of the possibility that carbon dioxide (CO_2) may also be a significant constituent of the mantles. As noted above, laboratory CO_2:CO mixtures give $4.67\,\mu m$ profiles which resemble the narrow component of the observed feature. Unlike CO, CO_2 is not predicted to have appreciable abundance in the gas phase in molecular clouds (e.g. table 7.3) and direct freeze-out cannot therefore lead to appreciable quantities of solid CO_2. However, CO_2 may form on the grains themselves, given the presence of CO, by means of the reaction

$$CO + O \rightarrow CO_2. \tag{7.36}$$

This reaction possesses an activation energy barrier (Grim and d'Hendecourt 1986) and will not normally occur at low temperatures, but it has been demonstrated that CO_2 is produced readily in laboratory interstellar ice analogues containing CO which are subjected to energetic processing such as ultraviolet irradiation or particle bombardment (e.g. d'Hendecourt *et al* 1986; Sandford *et al* 1988). Stretching and bending mode vibrations occur in CO_2 at wavelengths of approximately 4.27 and 15.2 μm, respectively (table 5.3; see Sandford and Allamandola 1990, for a review). Figure 7.16 illustrates the evolution of a laboratory H_2O:CO ice mixture subject to ultraviolet photolysis. Prior to irradiation, spectral features are seen at 3.1, 6.0 and 13 μm (corresponding to the H_2O stretch, bend and libration modes) and at 4.67 μm (corresponding to the CO stretch). After irradiation, the signatures of CO_2 also appear. The abundance of CO_2 in grain mantles is thus, in principle, a measure of the degree of CO processing. However, neither of the spectral regions containing CO_2 features can be studied with ground-based or airborne instruments because of the strength of telluric CO_2 absorption; only the 15 μm region, covered by the IRAS low-resolution spectrometer (LRS), has so far been observed from space. The detection of CO_2 was reported by d'Hendecourt and Jourdain de Muizon (1989), who found an absorption feature at 15.2 μm in LRS data for AFGL 961. It is logical to suppose that CO_2 formation might be activated in the vicinity of embedded sources such as AFGL 961, yet lines of sight towards other embedded sources displaying comparable optical depths to AFGL 961 in the water-ice and solid CO features show no hint of 15.2 μm absorption (Whittet and Walker 1991). The spectrum of AFGL 961 is not typical of young stellar objects in the LRS database. The question of whether CO_2 is a widespread constituent of dust mantles in molecular clouds is thus currently unresolved.

Massive star formation must ultimately lead to dramatic and widespread changes in the composition and mass of the grain mantles in a molecular cloud. Absorption of energetic photons converts saturated molecules into radicals; upon warm-up, the radicals may be free to migrate through the mantle, resulting in chemical reactions with other radicals and the formation of more complex species. Radiative processing of the volatile outer mantle, composed initially of O_2, N_2 and CO, may lead to CO_2 production, as discussed above; nitrogen-bearing compounds are also likely to be formed, and tentative observational evidence for molecules rich in CN bonds is provided by the presence of absorption features at 4.62 and 4.90 μm

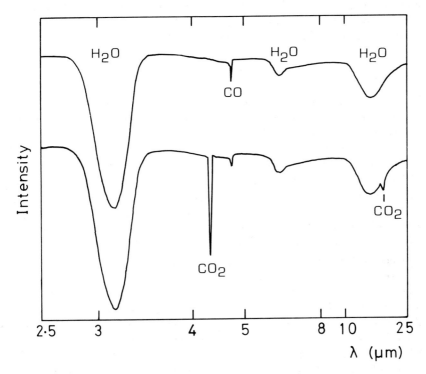

Figure 7.16   The spectral evolution of a laboratory 20:1 $H_2O:CO$ ice mixture subject to ultraviolet photolysis. Upper curve: before photolysis; lower curve: after photolysis for 2 hours. (Data from Sandford *et al* 1988.)

in the spectrum of W33A (Lacy *et al* 1984; Larson *et al* 1985). Processing which involves the H_2O-rich component of the mantle as well as the hydrogen-poor outer layer may lead to the production of complex organic molecules, some of which may be sufficiently refractory to survive the dissipation of the molecular cloud (Schutte and Greenberg 1988). As discussed elsewhere (§1.4, §2.4.4, §5.3.2), carbonaceous compounds produced in this way may account for a significant fraction of the total grain mass in low-density regions of the interstellar medium. Mantles may also be destroyed efficiently by a number of physical processes operating in regions of active star formation, including thermal evaporation, shock-induced grain–grain collisions, and explosions caused by exothermic chemical chain reactions between free radicals in UV-irradiated ices (Draine 1989b and references therein). The relative importance of destruction and reprocessing is an important issue. Schutte and Greenberg (1988) estimate on the basis of laboratory simulations that ∼ 30% of the carbon contained within the mantles will be converted to refractory organic residues during the lifetime of a typical molecular cloud.

7.2.4   The Destruction and Evolution of Refractory Grains

A number of processes may contribute to the destruction and evolution of refractory grain material in the interstellar medium. Various theoretical studies suggest that the primary destruction mechanisms are sputtering and grain–grain collisions (see Draine 1989b, McKee 1989 and Seab 1988 for reviews). Sputtering occurs when grain surfaces are eroded by impact with gas atoms or ions, and may be chemical or physical. Chemical sputtering arises when the formation of chemical bonds between the surface and impinging particles of relatively low kinetic energy leads to the desorption of molecules containing atoms which were originally part of the surface (Barlow 1978b). Physical sputtering involves the removal of surface atoms by energetic impact (Barlow 1978a): this may be thermal, due to kinetic motion of high-temperature gas, or non-thermal in cases where the dust is accelerated to high velocity with respect to the gas, as may arise, for example, when charged grains gyrate about magnetic field lines. If the grains have high differential velocities with respect to one another, the resulting grain–grain collisions may lead to shattering or evaporation of the grain material. Physical sputtering and grain–grain collisions are driven primarily by supernova-generated interstellar shock waves (e.g. Seab and Shull 1983): which of these mechanisms dominates in a particular environment depends on the shock velocity and on the pre-shock conditions. The atoms within the lattice of a refractory grain typically have binding energies $\sim 5\,\mathrm{eV}$ and thus require temperatures in excess of $10^5\,\mathrm{K}$ for efficient thermal sputtering. Calculations by various authors suggest that shocks with velocities $\sim 50\,\mathrm{km\,s^{-1}}$ or higher are required for significant destruction of refractory dust grains (Seab 1988 and references therein). In contrast, the binding energies for ices are $\leq 0.5\,\mathrm{eV}$; unless processed to more refractory materials, ice mantles formed in molecular clouds are removed easily from grains in unshielded regions of the interstellar medium by low-velocity shocks as well as by radiation fields and heating (Seab 1988; Barlow 1978b, c).

Supernova explosions generally occur in low-density phases of the interstellar medium, resulting in high-velocity shock waves which propagate through large volumes of low-density gas. Suppose a shock wave of velocity v_0 travelling in a medium of number density n_0 encounters a cloud of number density $n_c \gg n_0$. The cloud is subjected to a sudden increase in ambient pressure which drives the shock into the cloud with velocity

$$v_c = \left(\frac{n_0}{n_c}\right)^{\frac{1}{2}} v_0 \tag{7.37}$$

(Draine 1989b). This reduction in shock speed as the square root of the relative density implies that shocks capable of destroying grains in the intercloud medium are decelerated to such a degree that they do not generally destroy grains in clouds. As an illustrative example, let us assume $v_0 = 500\,\mathrm{km\,s^{-1}}$; adopting values of $n_0 = 5 \times 10^3\,\mathrm{m^{-3}}$ and $n_c = 3 \times 10^7\,\mathrm{m^{-3}}$, typical of the intercloud medium and a diffuse H I cloud, respectively (table 1.1), we deduce $v_c = 6\,\mathrm{km\,s^{-1}}$ (equation (7.37)). Similar calculations show that external shock velocities become vanishingly small in molecular clouds. Thus, destruction by shocks is efficient in the intercloud medium (where the grain number density is low) and generally inoperative in dense clouds (where the grain number density is high). On this basis, McKee (1989) argues that destruction occurs predominantly in the intermediate warm phase of the interstellar medium.

Support for the view that dust grains are being destroyed in shocks is provided by observations of element depletions (§2.4). Of particular significance is the existence of correlations between depletion and velocity. This is illustrated in figure 7.17 for the case of silicon (Cowie 1978). Recall that $D(\mathrm{Si}) \to 0$ would indicate a trend towards solar Si abundance in the gas (see §2.4.1). Whilst noting that there is considerable scatter in the data points, such a trend is, indeed, apparent towards the higher-velocity clouds. Figure 7.17 suggests that clouds with $v \simeq 100\,\mathrm{km\,s^{-1}}$ have abundances typically within a factor of 2 of solar. The trend is consistent with a model in which grains are thermally sputtered at $100\text{--}200\,\mathrm{km\,s^{-1}}$ and the shocked gas subsequently decelerated as it sweeps up matter with 'normal' depletions (Barlow and Silk 1977a; Cowie 1978). It appears that silicate grains are being destroyed with an efficiency of $\sim 50\%$ in high-velocity clouds.

The correlation of depletion with mean density illustrates the balance between destructive and constructive processes operating on grains in the interstellar medium. This is shown in figure 2.8 for titanium (see Harris *et al* 1984 and references therein for further examples). The correlation embraces a wide range of environments in which different mechanisms control the exchange of matter between the dust and the gas. Seab (1988) estimates that the mean time interval between destructive shocks in the warm phase of the ISM is $\sim 7\,\mathrm{Myr}$, compared with $\sim 3\,\mathrm{Gyr}$ for a diffuse cloud. The corresponding timescales for accretion are $\sim 400\,\mathrm{Myr}$ and $\sim 50\,\mathrm{Myr}$, respectively. Thus, in clouds, accretion dominates and very high levels of depletion are maintained for most of the condensible heavy elements. In tenuous regions, the situation is reversed and grain material is eroded more rapidly than it reaccretes, leading to reduced levels of depletion. It should be noted that evaporation of relatively small quantities of dust may lead to dramatic increases in gas phase abundances for the most highly depleted elements such as Fe, Ti, Ca and Al. Mixing of dense and diffuse phases of the ISM is estimated to occur on a timescale of $\sim 100\,\mathrm{Myr}$ (Seab 1988),

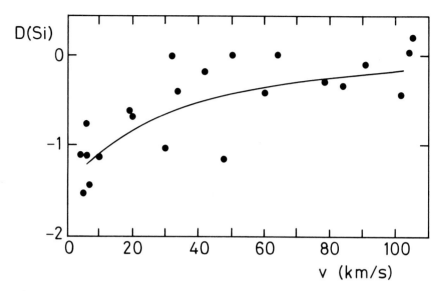

Figure 7.17   Plot of silicon depletion index against cloud velocity relative to the local standard of rest for various lines of sight (points), compared with a model (curve) for grain destruction by sputtering in shocks. (Adapted from Cowie 1978.)

which is sufficiently rapid to account for the rather high levels of depletion observed for these refractory elements even in low-density regions.

Energetic processes may lead to internal structural changes in the grains as well as to vaporization and shattering. The origin of interstellar graphite is a particularly perplexing problem (see Czyzak *et al* 1982 for a critical review). As previously noted, graphite is an essential ingredient of most grain models (§1.4; §3.4) and the most widely discussed of the proposed carriers for the λ2175 feature (§5.1.2). However, injected stardust appears to be essentially amorphous (§7.1), and carbon-rich compounds formed in the interstellar medium by accretion or by radiative processing of ice mantles appear likely to be amorphous or organic in nature. A possible solution to this dilemma is discussed by Sorrell (1990, 1991) who proposes that a sub-population of the carbon grains are graphitized by ultraviolet photons in the diffuse interstellar medium. It is important to bear in mind that amorphous carbons are porous, inhomogeneous materials (Robertson and O'Reilly 1987), containing PAH clusters which may undergo transient heating on absorption of energetic photons (§6.1.4). Temperature spikes may lead to annealing and dehydrogenation of the clusters, producing small graphitic grains capable of explaining the observed properties of the λ2175 feature (Sorrell 1990). The ratio of graphitic to polymeric carbon is sensitive to environment, and this may also account for the behaviour of the

$3.4\,\mu$m absorption feature in different lines of sight (§5.3.2; see Sorrell 1991).

Grain–grain collisions may lead to further structural evolution of carbon-rich grain material. Tielens *et al* (1987) discuss the conversion of graphitic or amorphous carbon to diamond by collisionally induced shock processing. Diamond is the thermodynamically favoured form of solid carbon at high temperature and pressure. The graphite–diamond phase change has a high activation energy ($7.5\,$eV per C atom) because complete rearrangement of the crystal lattice is required. This process has been reproduced in the laboratory. Tielens *et al* (1987) argue that diamond microcrystals ranging up to $\sim 0.01\,\mu$m in size may be formed by interstellar grain–grain collisions, and that these may contain some 5% of the available carbon. As diamonds are hard to destroy, once formed, they may survive for long periods in the interstellar medium.

Although shock-induced processes dominate current discussion of grain destruction, the possibility of significant destruction by chemical sputtering should not be overlooked. This mechanism will be more important for carbon than silicate grain material as the former naturally tends to have a much higher density of chemically active surface sites. The attachment of incident H atoms and ions to the surface of graphite may lead, for example, to the exothermic production and desorption of CH_4, although Draine (1989b) argues that this process is inefficient at low temperature and that surface attachment of H is more likely to lead to recombination and desorption of H_2. Graphite may be preferentially destroyed by chemisputtering in warm phases of the ISM, notably in compact H II regions (Barlow 1978b): this is suggested by models for the observed FIR emission from these sources (Churchwell 1990), which require a graphite/silicate abundance ratio of about half that assumed in models of diffuse-cloud extinction. Chemisputtering involving incident O atoms may also be important. Whittet *et al* (1990) suggest that such processes may account for the unexpectedly low abundance of SiC in interstellar grains.

The effect of disruptive processes on the size distribution of interstellar dust is an important issue which should be addressed by any comprehensive grain model. Thermal sputtering leads to size-independent erosion because the sputtering rate is proportional to the cross-sectional area of the grain: the situation is analogous to our discussion of *growth* (§7.2.2) by impact of gas phase particles at much lower thermal velocities. McKee (1989) estimates that shocks with $v \sim 300\,$km s^{-1} will typically remove a surface layer $\sim 0.02\,\mu$m thick, leading to annihilation of small grains whilst leaving the cores of larger grains intact. In contrast, grain–grain collisions, which may dominate in lower-velocity ($v \sim 100\,$km s^{-1}) shocks, tend to enhance the number of small grains by shattering larger ones (Seab 1988). As discussed in §7.1.3, shattering tends naturally to impose a power-law size distribution consistent with that used successfully to model the wavelength dependence of interstellar extinction (§3.4). A population of small grains in the low density ISM is required to explain the FUV rise in the extinction curve, and shattering may be an important mechanism for maintaining such a population against destruction by sputtering.

Average lifetimes for refractory grains are notoriously difficult to estimate. If it is accepted that supernova-driven shocks are the dominant destructive agent, the destruction rate will depend on the supernova rate (known to within a factor of ~ 2), the mean expansion rate, the cloudy state of the ISM, and the galactic distribution of supernova progenitors. Estimates of the timescale for destruction in shocks, t_{sh}, typically lie in the range 0.1–1 Gyr (Seab 1988; McKee 1989). Thus, $t_{sh} \leq t_{in}$ where t_{in} is the injection timescale for stardust formed in stellar atmospheres (equation (7.12)). These results suggest that stardust is not the only, or perhaps even the dominant, source of refractory interstellar dust (see §7.1.3).

7.3   DUST IN THE ENVIRONMENTS OF YOUNG STARS

"At one time, it was thought that an X-solar-mass star resulted from the collapse of an X-solar-mass cloud..."

F H Shu *et al* (1989)

Star formation is a rather inefficient process: the parent molecular clouds are much more massive than the stellar populations they spawn, and so, during each generation of starbirth, only a small percentage of associated interstellar material is converted into stars. The disruption star formation imposes upon the interstellar medium arises from the impact of resultant radiation fields, winds and shocks, as previously discussed. This section is concerned primarily with the fate of material contained within the circumstellar envelopes of young stars rather than with that of the molecular clouds as a whole. We begin with an overview of the early phases of stellar evolution (§7.3.1) and the properties of protoplanetary disks (§7.3.2), placing emphasis on objects of low and intermediate mass which seem likely to be realistic analogues of the early Solar System. Evidence that diffuse material in the *present-day* Solar System may contain significant quantities of essentially unmodified presolar interstellar dust is discussed in §7.3.3. Finally, in §7.3.4, we consider the possible relevance of organic matter in the envelopes of young stars to the origin of planetary life

7.3.1   The Early Phases of Stellar Evolution

New stars are born when part of an interstellar cloud collapses under its own gravity. The densities required for this to occur are expected to arise only in molecular clouds (e.g. Elmegreen 1985), and millimetre-wave observations of CO and other tracers of molecular gas, coupled with searches made at infrared wavelengths, demonstrate that stars in the earliest phases of their evolution are, indeed, embedded within dense regions of molecular clouds. Initially, the clouds are supported against their self-gravity by interstellar magnetic fields, but ambipolar diffusion leads to the condensation of cold, dense pockets of gas and dust (Mestel 1985). The way in which a cloud fragments into individual prestellar condensations is a central problem in the theory of star formation which will not be addressed here (see Larson 1989 for a review). In general, high-mass stars form with much lower frequency than low-mass stars, as first discussed in detail by Salpeter (1955), who determined the mass function for field stars in the solar neighbourhood. Two fundamentally different modes of star formation appear to exist (Shu *et al* 1989), with stars of high and low mass condensing in morphologically different regions. Young high-mass stars tend to occur within the dense cores of giant molecular clouds (Churchwell 1990), many of which are likely to evolve to become gravitationally bound OB associations or open clusters. In contrast, young low-mass stars are often seen scattered loosely throughout dark clouds and seem destined to become part of the field star population rather than members of bound clusters (Wilking 1989). Regions of high- and low-mass star formation are typified by the Orion (M42/OMC-1) and Taurus clouds, respectively.

The early evolution of a star is characterized by distinct 'protostellar' and 'pre-main-sequence' phases. During the protostellar phase, luminosity is derived primarily from gravitational energy released by the infall of accreting material (Wynn-Williams 1982; Beichman *et al* 1986). Net rotation of the original condensation causes the development of a disk, and both disk and nucleus are embedded within an infalling envelope of dust and gas, as illustrated schematically in figure 7.18. Protostars are not only intrinsically cool but also highly reddened, and they hence emit most of their luminosity in the mid- to far-infrared region of the spectrum. Later, during the pre-main-sequence (PMS) phase, thermonuclear reactions are ignited and the star emerges from its obscuring envelope to progress towards the zero-age main sequence (ZAMS) along predictable evolutionary tracks in the HR diagram (Bodenheimer 1989 and references therein). The entire formation process from initial condensation to arrival on the ZAMS takes approximately 40 Myr for a $1 \, M_\odot$ star.

The spectral energy distribution (SED) emerging from a young star changes dramatically during the course of its evolution towards the main sequence. Models have been constructed which predict the shape of the SED as a function of evolutionary state (Adams *et al* 1987), and the comparison of observed and predicted fluxes provides a valuable diagnostic technique for classifying infrared sources and identifying candidate protostars and PMS stars still hidden by foreground extinction (Wilking 1989; Wilking *et*

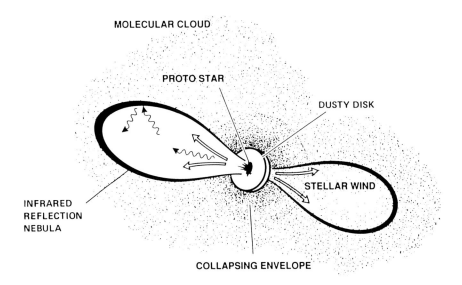

Figure 7.18 Schematic illustration of the circumstellar environment of a young stellar object with bipolar outflow. The central star is surrounded by a dusty disk and a collapsing spherical envelope, embedded within a molecular cloud. The strong stellar wind drives the outflow, which blows cavities in the envelope. Reflection nebulae may be illuminated in the outer regions of the flow. (From Pendleton *et al* 1990, Astrophysical Journal, reproduced with permission).

al 1989; Whittet *et al* 1991). A useful parameter of the SED is the spectral index α, defined as the mean value of

$$\alpha = \frac{d(\log \lambda F_\lambda)}{d(\log \lambda)} \tag{7.38}$$

in the mid-infrared (Wilking 1989). In this scheme, protostars (designated Class I) have much broader SEDs than single-temperature Planck functions and have positive slopes ($\alpha > 0$), as a result of infrared emission from cool circumstellar dust; sources with these properties rarely exhibit detectable flux in the visible and are evidently still deeply embedded. Pre-main-sequence (Class II) sources also have circumstellar infrared emission, but to a lesser degree, such that the SEDs have negative slopes ($-2 < \alpha < 0$) in the mid-infrared; they are often associated optically with T Tauri stars. Finally, sources with SEDs resembling single-temperature blackbodies and with slopes approaching the value expected in the Rayleigh–Jeans limit ($\alpha = -3$) are designated Class III; these may be associated optically with stars in the late T Tauri phase or on the ZAMS, and are often strong X-ray emitters due to the presence of active chromospheres. Observed broadband SEDs for sources of each type are illustrated in figure 7.19.

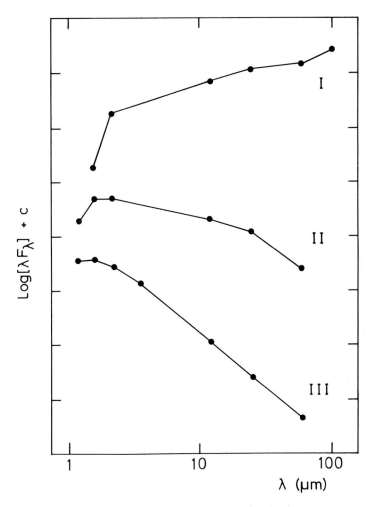

Figure 7.19   Infrared spectral energy distributions for three contrasted IRAS sources in the line of sight to the Chamaeleon I dark cloud: 11057−7706 ($\alpha = 0.8$; Class I), 11065−7701 ($\alpha = -0.7$; Class II), and 11180−7612 ($\alpha = -2.1$; Class III). Data are from Whittet, Prusti and Wesselius (1991).

During the epoch of transition between the protostellar and pre-main-sequence phases, strong winds develop which may lead to bipolar outflow (figure 7.18). The wind is rapidly decelerated when it encounters the equatorial disk, but travels more freely in other directions, blowing cavities in the polar regions of the envelope (e.g. Lada 1988). At this point, it is clear that the visibility of the central object becomes highly dependent on the viewing angle (assuming the disk to be optically thick at visible wavelengths). For most lines of sight, circumstellar extinction is greatly diminished, but sources seen through the disk edge-on will remain highly obscured at wavelengths below $\sim 1\,\mu$m. In some cases, material swept up by the bipolar wind may form Herbig–Haro objects, i.e. visible nebulae arising from shock excitation of the gas or from scattering of radiation from the central star by dust (Schwartz 1983).

The spectra of young stars frequently exhibit structure due to dust-related infrared features (§5.3.3; §7.2.3). In general, the emergent spectrum will contain contributions both from dust within the circumstellar envelope and from dust in the foreground molecular cloud. The silicate feature at $9.7\,\mu$m may occur in net absorption or net emission, and which is seen in a particular line of sight appears to depend primarily on viewing geometry (Cohen and Witteborn 1985): in cases where a thick circumstellar disk is viewed edge-on, the feature should appear strongly in absorption, whereas for more typical orientations emission from warm silicate grains may dominate the spectrum. Cohen and Witteborn (1985) detect $9.7\,\mu$m absorption in $\sim 30\%$ of their sample. It is interesting to note that stars with silicate absorption have a high incidence of associated Herbig–Haro nebulae, suggesting common constraints on viewing geometry. A number of objects with $9.7\,\mu$m silicate absorption also show ice absorption at $3.0\,\mu$m (Sato et al 1990). In most cases, the ice probably lies in foreground material, but towards HL Tau there is strong evidence to suggest that both ice and silicate dusts reside in an edge-on disk which is responsible for most of the visual extinction towards the star (Cohen 1983).

7.3.2   Protoplanetary Disks

Both theoretical models and observational results suggest that the formation process for a single star of low or intermediate mass ($0.1 < M < 5\,M_\odot$) results naturally in the development of a circumstellar disk which may subsequently become a planetary system. Evidence is found for the presence of disks around both PMS objects and more mature stars on or beyond the ZAMS. Skrutskie et al (1990) estimate that $\sim 50\%$ of T Tauri stars with ages $< 3\,$Myr have optically thick disks. Towards HL Tau, the disk is spatially resolved in both molecular line and infrared continuum emission, and has a radius of $\sim 2000\,$AU and a mass of $\sim 0.1\,M_\odot$ (Sargent and Beckwith 1987). Gas and dust temperatures in the disk are typically in the range 40–100 K, consistent with the observed detection of water-ice and the non-detection of solid CO frosts in the line of sight (§7.2.3; see figure 7.13). Sargent and Beckwith (1987) present high-velocity-resolution ^{13}CO data which demonstrate that the disk is gravitationally bound to and in Keplerian rotation about the central star. The size, mass and kinematics of the disk are similar to those proposed for the early solar nebula, prior to the formation of planets, suggesting that the HL Tau system may be in the early stages of planet formation. The incidence of optically thick disks is much lower for more evolved T Tauri stars (Skrutskie *et al* 1990). This implies a rapid thinning of the disk during the final stages of a star's evolution towards the ZAMS.

The detection of strong far-infrared emission from bright main-sequence stars, the so-called 'Vega phenomenon', was perhaps one of the most surprising results to emerge from the IRAS mission (Aumann *et al* 1984). The excess flux detected at 25–$100\,\mu$m is attributed to radiation from dust in a circumstellar disk of temperature typically ~ 80 K. The number of such objects known within 50 pc of the Sun exceeds 30 (Norman and Paresce 1989). They are predominantly single stars of age > 1 Gyr; most are of spectral classes A and F, but the sample also includes classes G and K, which are probably under-represented by observational selection (Aumann 1988). The FIR emission is generally spatially extended, and, in the case of β Pic, the presence of a resolved circumstellar disk is confirmed optically (Smith and Terrile 1984). The radii of the disks are typically up to a few hundred AU. Circumstellar molecular *gas* is not detected in the disks (Walker and Wolstencroft 1988), which suggests that they are not the product of a stellar mass-loss process, consistent with the absence of shell characteristics in the optical spectra of the stars. The disks appear to be remnants of PMS evolution which survive the outflow phase but fail to accrete into planetary systems. Their masses are typically $\sim 10^{-7}$ M$_\odot$ (Gillett 1986): for comparison, this is a factor of $\sim 10^5$–10^6 *less* than the masses of disks around PMS stars (Strom *et al* 1989), but a factor of $\sim 10^7$ *more* than the estimated total mass of interplanetary dust in the Solar System (Millman 1975). In this respect, the Sun may be unusual compared with other G stars of similar age (Aumann 1988).

Clearly, the disks around young stars tend to evolve towards a more tenuous state. Physical processes which lead to their dissipation include stellar radiation pressure on the grains leading to outflow, and the Poynting–Robertson effect, in which the absorption and re-emission of radiation leads to the decay of a particle's orbit such that it spirals into the star (Hodge 1981). Both of these processes tend selectively to remove the smaller particles. Detailed calculations by Artymowicz (1988) suggest that radiation pressure is the dominant process, and that particles of a given composition and with dimensions less than some critical size are efficiently expelled from an optically thin disk around a star of intermediate spectral type. For any realistic grain composition, this critical size is $\sim 1\,\mu$m. If the initial size distribution were typical of that deduced for interstellar grains, radiation pressure would thus lead to complete dissipation of the disk in the absence of a competing process. However, bearing in mind that the growth of solids by coagulation and accretion is fundamental to current models for the formation of planetesimals (e.g. Wetherill 1989), it is natural to assume that a protoplanetary disk develops a population of particles larger than the critical size before it becomes optically thin. The trend towards larger particles affects the optical properties of the disk. For a given volume of dust, coagulation diminishes the total surface area, and both the optical depth and the infrared luminosity are thus reduced.

An observational constraint on particle size arises from the equation describing the balance of energy absorbed and re-emitted in a circumstellar disk. Considering a grain of radius a at a distance r from a star of radius R_s and surface temperature T_s, we have

$$\left(\frac{\pi a^2}{4\pi r^2} \right) 4\pi R_s^2 \sigma T_s^4 \langle Q_V \rangle = 4\pi a^2 \sigma T_d^4 \langle Q_{FIR} \rangle \qquad (7.39)$$

which reduces to

$$\frac{r}{R_s} = 0.5 \left(\frac{T_s}{T_d} \right)^2 \left\langle \frac{Q_V}{Q_{FIR}} \right\rangle^{\frac{1}{2}} \qquad (7.40)$$

(Walker and Wolstencroft 1988), where $\langle Q_V \rangle$ and $\langle Q_{FIR} \rangle$ are the mean absorptivity and emissivity of the dust grains evaluated at the appropriate wavelengths (see §6.1.1). Aumann $et\ al$ (1984) and Gillett (1986) argue that a self-consistent model for the disks of Vega-like stars which accounts simultaneously for the dust temperature (estimated from the spectral energy distribution) and the disk radius (estimated from the angular size) requires $\langle Q_V/Q_{FIR} \rangle \sim 1$. As $Q_V \sim 1$ for most grain materials, Q_{FIR} must also be approximately unity, i.e. the particles behave as blackbodies†. This may be understood if they are large compared with the wavelength emitted: quantitatively, we require $2\pi a > \lambda_{peak}$, where λ_{peak} is the wavelength of peak emission, and for $\lambda_{peak} \sim 60\,\mu m$, $a > 10\,\mu m$. In the case of Vega itself, a recent revision of the angular extent of the FIR emission detected by IRAS has somewhat relaxed this constraint (Prusti, personal communication). Nevertheless, it appears that the grains surrounding Vega-like stars are, indeed, larger than the critical size for retention against radiation pressure.

The various factors which govern the evolution of a circumstellar disk and determine the distribution of material between planetary bodies and remnant diffuse matter around a mature main-sequence star are still poorly understood (e.g. Wetherill 1989; Weidenschilling $et\ al$ 1989). Of course, we do not know whether the Vega-like stars possess planetary systems in addition to diffuse matter as we lack techniques of sufficient sensitivity to detect planets beyond the Solar System. Disks are much more easily detectable because of their vastly greater surface areas, and the study of

---

† For comparison, classical silicate grains with $a \sim 0.1\,\mu m$ have strongly wavelength-dependent FIR emissivity (§6.1.1; equation (6.6)) with a value of $Q_{FIR} \sim 0.007$ at $40\,\mu m$ (Jones and Merrill 1976).

disks around pre-main-sequence and main-sequence stars may thus offer the most promising means of constraining models for the formation of planetary systems (Norman and Paresce 1989).

7.3.3   Diffuse Matter in the Solar System—the Fossil Record

As noted in Chapter 1, the Sun lies in a low-density phase of the interstellar medium, consistent with available evidence to suggest that the influx of interstellar dust into the Solar System is currently negligible. Although the Sun will encounter an interstellar cloud on a timescale of 10–100 Myr, solar radiation pressure strongly inhibits the approach of submicron-sized dust particles. The collection and laboratory analysis of 'ambient' interstellar dust is not therefore a realistic proposition. The basis of the discussion below is that samples of *presolar* interstellar grains are preserved in relatively unmodified form in primitive bodies in the Solar System (asteroids and comets), and hence in diffuse matter derived from them (i.e. meteorites and interplanetary dust). At one time, it was thought that all solid material in the Solar System condensed from a homogeneous hot gas which erased the chemical and mineralogical fingerprints of presolar material. The discovery of isotopic anomalies in collected samples of diffuse matter has completely altered this picture and, in some sense, opened up a new branch of astrophysics (Clayton 1982, 1988; Nuth 1990). It is now accepted that a significant fraction of the dust grains present in the solar nebula did not undergo vaporization and recondensation, but accreted directly into macroscopic bodies during the early stages of planet formation. Undifferentiated bodies of subplanetary mass may thus contain aggregates of presolar interstellar grains which retain their physical structure and chemical identity. The aim of this section is to assess the significance of this 'fossil record'.

No permanent population of diffuse matter exists in the inner Solar System. Much of the debris in Earth-crossing orbits is likely to have been dispersed within the past 10,000 years by collisions between asteroids or by ablation of comets at perihelion. The fate of such debris upon entering the Earth's atmosphere is a function of size: the smallest (micron-sized) grains are collisionally decelerated without significant melting, and are thus available for collection, whereas millimetre-sized grains are completely ablated. A meteorite fall arises from the arrival of a large fragment, some fraction of which survives passage through the atmosphere. Spectroscopic evidence suggests that asteroids are the parent bodies of most classes of meteorite, whereas the dust grains originate primarily in comets.

For reasons noted in §2.2.1, we limit our discussion of meteorites to a single class, the carbonaceous chondrites. Their properties are reviewed by Wood *et al* (1986). They contain millimetre-sized spherical chondrules and Ca, Al-rich inclusions, embedded in a fine-grained aggregate or 'matrix'. The chondrules are glassy and evidently condensed from liquid silicate droplets in the solar nebula. The Ca, Al-rich inclusions must have formed at higher temperature than most other minerals present, and have received attention because they often contain anomalous isotopic abundances for oxygen (see below). The matrix is constructed of particles typically $\sim 1\,\mu m$ in size, composed predominantly of magnesium silicates with varying degrees of hydration. In some cases, hydration appears to have occurred subsequent to accretion. Other mineral ingredients include magnesium sulphate ($MgSO_4$), magnetite (Fe_3O_4) and troilite (FeS). The carbonaceous fraction, which is predominantly organic, contributes no more than a few per cent by mass. Most of the minerals represented in these meteorites evidently condensed in the solar nebula (Barshay and Lewis 1976), as a result of processes analogous to those occurring in the envelopes of evolved stars (§7.1.1; see figure 7.2).

The primary evidence for a connection between interstellar grains and meteorites arises from the detection of non-solar abundance ratios for various isotopes. Deuterium enrichment in the organic fraction suggests synthesis by interstellar processes (Kerridge 1989), a topic discussed further in §7.3.4 below. Anomalies involving a number of heavier elements are expected to occur in dust that condensed in, or was contaminated by, stellar outflows modified by nucleosynthesis (e.g. Clayton 1982, 1988). Likely scenarios include nova and supernova ejecta and red giant winds. Consider the case of oxygen. On Earth, natural processes lead to fractionation of the three stable isotopes due to the differences in their masses, resulting in a well-understood relationship between the $^{17}O/^{16}O$ and $^{18}O/^{16}O$ ratios. The isoptopic ratios measured in meteorites do not conform to this pattern, and the simplest explanation is a 5% enrichment of ^{16}O in the meteoritic material. Clayton (1982) argues that this arises due to the condensation of Al_2O_3 from supernova ejecta containing pure ^{16}O; the Al_2O_3 is locked up in interstellar grains which are eventually incorporated into the meteorite. If the grains had been vaporized in the solar nebula, the isotopically anomalous fraction would have been mixed with a much greater mass of isotopically 'normal' oxygen derived from other sources, and subsequent condensates would have lost their distinctive chemical identity.

Noble gases are retained in measurable quantities in carbonaceous chondrites despite their chemical inertia, and these have proved particularly valuable as chemical fingerprints of interstellar grains. Two examples are mentioned here. The isotope ^{22}Ne is produced by decay of ^{22}Na, and it is widely accepted that an excess of ^{22}Ne observed in meteorites results from the presence of grains containing sodium-rich minerals which originally condensed in stellar ejecta. As the half-life for ^{22}Na decay is only 2.6 years, it must have been incorporated into carrier grains quite rapidly after the nucleosynthesis event in which it was formed, in order to preserve its isotopic identity. The product ^{22}Ne is rapidly desorbed from meteoritic samples at temperatures $> 800\,K$ in the laboratory, which implies that the grains cannot have been heated to such a degree in the solar nebula. Xenon, another widely studied meteoritic noble gas, has no less than nine stable isotopes. A component of anomalous Xe, cryptically named Xe–HL, is enriched in isotopes produced by r-process reactions, and meteoritic Xe–HL is thus thought to be a signature of presolar interstellar grains implanted with supernova ejecta. Other configurations identify the products of s-process reactions.

Isotopic analyses have greatest influence on our understanding of the carbon-rich component of the chondrites†. The abundances of the various carriers of meteoritic carbon are illustrated in figure 7.20, normalized to the solar C abundance. The organic component seems likely to be partly presolar in origin (§7.3.4), whereas the carbonate component originated entirely within the Solar System by aqueous processes on primitive parent bodies. The diamond, graphite and silicon carbide components are thought to be of essentially presolar origin.

Meteoritic diamonds exist in the form of microcrystals of median diameter ~ 3 nm (Lewis *et al* 1987, 1989), i.e. they are very small compared with grains commonly used to model interstellar extinction. Although accepted to be interstellar on the basis of Xe–HL content, their origin remains controversial. The size distribution is remarkable, exhibiting a log–normal dependence characteristic of formation by growth rather than a power-law dependence characteristic of shattering. This appears to be inconsistent with an origin in interstellar shock-induced grain–grain collisions (§7.2.4). As the efficiency of shock production is only $\sim 5\%$, Lewis *et al* (1989) argue that there should be a ~ 20-fold excess of matching interstellar graphite or amorphous carbon in meteorites, which is not seen: the graphitic component is not only less abundant (figure 7.20) but also isotopically distinct from the diamond, lacking Xe–HL, which suggests a different origin. Lewis *et al* propose that the diamonds formed by metastable condensation in a C-rich stellar atmosphere, and that Xe–HL was subsequently implanted by a supernova explosion.

Large ($a \sim 1\,\mu$m) meteoritic graphite grains have been isolated by Amari *et al* (1990). A presolar origin is established on the basis of anomalous $^{12}C/^{13}C$ ratios and a large ^{22}Ne excess. In the context of models for interstellar grains, meteoritic graphite is remarkable for its *rarity*. Amorphous carbon is more difficult to isolate and has not been detected directly: Amari *et al* deduce an upper limit on its abundance of between 5 and 50 times the graphite abundance. Thus, the total abundance of presolar amorphous, graphitic and diamond-like carbon in meteorites appears to be no more than about 1% of the solar abundance, in contrast to the estimated 50% depletion of interstellar C into grains (§2.4). It may be the case that these

---

† Reliable isotopic signatures of presolar silicate grains have yet to be identified (see Nuth 1990).

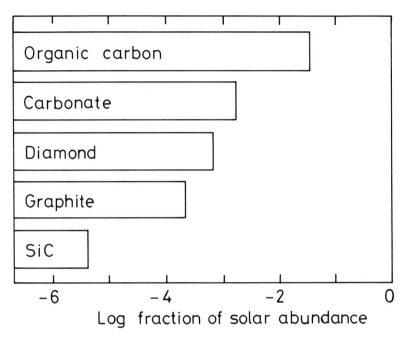

Figure 7.20   The chemical state of carbon in primitive meteorites. Studies of isotopic abundances indicate that the diamond, graphite and SiC components are probably of interstellar origin, whereas the carbonates are synthesized in the Solar System. The dominant organic component may be partially interstellar. (Adapted from Anders *et al* 1989.)

forms of solid carbon are converted efficiently into organic molecules in the solar nebula (Barlow and Silk 1977b).

   The detection of meteoritic SiC grains typically $0.1–1\,\mu$m in size was reported by Bernatowicz *et al* (1987). Like graphite, they have non-solar $^{12}C/^{13}C$ ratios and dramatic excesses of ^{22}Ne, suggesting an origin in red giants (Tang *et al* 1989), in qualitative agreement with the spectroscopic identification of SiC dust in the envelopes of carbon stars (§7.1.2). However, SiC is a very minor constituent of carbonaceous chondrites: the fraction of all Si atoms tied up in SiC is a mere 0.004% (Tang *et al* 1989), the vast majority being in silicates. It has been argued (e.g. Anders *et al* 1989) that this result implies a deficiency of carbon star material in the solar nebula, which could explain the rarity of other forms of presolar solid carbon as well. However, the non-detection of SiC dust in the ISM (§5.3.2) also requires an explanation: although the limit of < 5% interstellar Si in SiC is three orders of magnitude above the value for meteorites, it nevertheless suggests that selective destruction must occur if carbon stars are significant sources of interstellar SiC. One such mechanism is chemisputtering by atomic oxygen, which may operate on a timescale of 50 Myr in diffuse clouds (Whittet *et al* 1990). This possibility is consistent with the results of Tang and Anders (1988), who deduce that meteoritic SiC has a probable presolar cosmic-ray exposure age of ~ 40 Myr. These timescales are an order of magnitude shorter than the usual estimates of lifetimes for refractory interstellar grains (§7.2.4). It is thus consistent to argue that oxidation causes selective destruction of solid SiC in the ISM within typical cloud lifetimes, and that this process is at least partially responsible for the low SiC abundance in meteorites.

The collection and study of interplanetary dust grains was pioneered by Brownlee (1978). Grains with diameters ranging from 1μm up to about 100μm are sufficiently decelerated by the Earth's atmosphere to allow non-destructive collection in the stratosphere. The extraterrestrial origin of the samples is confirmed by evidence for cosmic-ray and solar-wind exposure, and by the presence of isotopic anomalies, notably an excess of deuterium. Mineralogical studies show that individual grains may be grouped into two general classes (Sandford 1989). The first consists of single minerals composed of hydrous layer-lattice silicates such as smectites. Of greater interest is the second group, which consists of porous aggregates of much smaller particles, typically $\sim 0.1 \mu$m in size, some of which are single crystals; others are themselves aggregates of smaller particles 0.01μm or less in size. They are composed predominantly of anhydrous silicates such as olivines and pyroxenes, held together by a matrix of amorphous carbonaceous material. Other minerals present include magnetite, nickel–iron sulphides, carbides and carbonates. The mass fraction of carbonaceous material is typically 5–10%, somewhat higher than that in chondrites. An important point to note is that a variety of chemically and structurally distinct mineral grains generally coexist in the same aggregate, implying diverse origins. The macroscopic particles evidently formed by an accretion process from a heterogeneous mixture of grains with sizes characteristic of interstellar dust. This does not, of course, *prove* that the constituent grains are all, or even partly, of interstellar origin. Studies of isotopic abundances, which could, in principle, provide strong support for this view, are not as well developed for interplanetary dust as they are for meteorites, and it is not possible to assign specific anomalies to specific mineral components within the aggregates.

Interplanetary dust grains are dark in appearance: their optical properties at visible wavelengths are dominated by absorption in the minor carbonaceous fraction, which tends to mask the more abundant silicates. The spectral signatures of silicates become dominant in the infrared (Sandford and Walker 1985). The 9.7μm profiles of the fine-grained anhydrous particles show significant structure (see figure 7.21) which is characteristic of crystalline rather than amorphous mineralogy. No corresponding structure is seen in spectra of interstellar dust (e.g. figures 5.10, 5.14 and

6.5), suggesting that any presolar silicates represented in the interplanetary aggregates must have undergone significant annealing in the Solar System. The layer-lattice silicate particles exhibit smoother $9.7\,\mu m$ profiles, but their spectra also contain hydrate $(2.7, 6.0\,\mu m)$ and carbonate $(6.9, 11.4\,\mu m)$ features which are unrepresented in interstellar spectra of refractory (diffuse-cloud) dust to current limits of sensitivity. The interstellar 6.0 and $6.85\,\mu m$ features seen in molecular clouds (§5.3.3) are unlikely to originate in mineral grains.

Comets display spectral signatures of both silicate and carbonaceous dust in the infrared. Extensive data were obtained during the 1985/6 apparition of Comet Halley. The 6–$13\,\mu m$ spectrum is illustrated in figure 7.21 and compared with a model based on laboratory data for interplanetary particles. The model assumes a mixture of olivine-rich, pyroxene-rich and layer-lattice silicate classes in the ratio 11:7:2, which gives excellent agreement with the observations. Structure near $11\,\mu m$, characteristic of crystallinity, is common to both spectra. This comparison lends support to the view that the collected interplanetary particles are of cometary origin. It also provides spectral evidence for the presence of crystalline silicates in the comet.

A 3.1–$3.6\,\mu m$ spectrum of Comet Halley is illustrated in figure 7.22. This shows a broad emission feature centred near $3.35\,\mu m$, with significant subpeaks at 3.28 and $3.52\,\mu m$. The profile bears a superficial resemblance to an inversion of that seen in absorption in the interstellar medium (figure 5.13). The position and shape of the emission is suggestive of C–H resonances in hydrocarbons. Baas *et al* (1986) propose that it arises in small grains or large molecules pumped by solar ultraviolet radiation, although the observed spectrum does not resemble the characteristic interstellar 'PAH' spectrum at 3–$4\,\mu m$ (§6.3.2; e.g. figure 6.10) or, indeed, at other wavelengths. Chyba and Sagan (1987) demonstrate the presence of a 3.3–$3.4\,\mu m$ absorption feature in the spectrum of a synthetic organic residue produced in the laboratory by irradiation of a $H_2O:CH_4$ ice mixture. These authors propose a simple emission model for the Halley spectrum, based on the measured transmission curve of the residue and the spacecraft-determined dust distribution in the coma of the comet: this gives a reasonable fit to the 3.3–$3.4\,\mu m$ feature, whilst simultaneously accounting for the absence of other features at longer wavelengths which would be expected (table 6.1) for emission in free PAHs.

The mass spectrometers carried by the Giotto and Vega missions to Halley produced the first direct analyses of the composition of cometary dust (see Brownlee and Kissel 1990 for a review). Abundances were determined for individual particles with diameters typically in the range 0.1–$1\,\mu m$. Averaged over many particles, the results for elements commonly found in refractory minerals (O, Mg, Al, Si, Ca, Fe) match data for carbonaceous chondrites to within a factor of 2. However, C and N show a

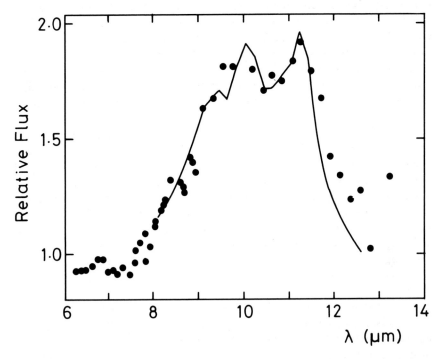

Figure 7.21   Silicate emissivity curve $F_\lambda/B_\lambda(T_d)$ from 6 to 13 μm for Comet Halley (Bregman *et al* 1987, points), derived with respect to a blackbody continuum of temperature 320 K. This is compared with a model based on the optical properties of collected interplanetary dust particles (Sandford and Walker 1985, curve). Structure in the profile indicates the presence of crystalline silicates.

significant excess compared with chondrites, their abundances more closely matching those in the solar atmosphere (see figure 2.2). Halley dust may contain as much as 20% by mass of carbon. Mass spectra for individual impacts indicate a range in compositions from grain to grain. Some grains (called 'CHON' particles) are composed almost exclusively of the abundant lighter elements and are probably organic in composition, others contain only heavier elements and oxygen and are presumably pure minerals, whilst members of a third group contain comparable quantities of both organic and mineral-forming elements. The three types of particle were detected in roughly equal numbers (Langevin *et al* 1987). Statistical investigations of ionic molecular lines in the mass spectra allow some characterization of the organic fraction of the dust, and results suggest that it is composed primarily of unsaturated hydrocarbons of low oxygen content (Kissel and Krueger 1987).

Greenberg (1982) discusses an interstellar model for comets, in which

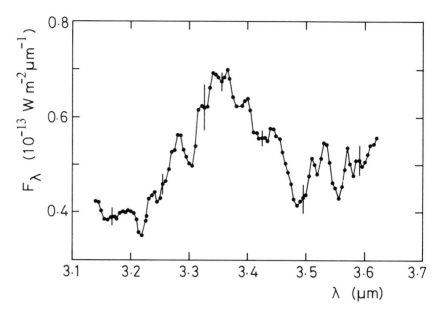

Figure 7.22   Spectrum of Comet Halley from 3.1 to 3.6 μm, illustrating the presence of a broad emission peak centred near 3.35 μm (Baas *et al* 1986). Representative 1 σ error bars are shown for selected points.

it is proposed that they originate as accumulations of presolar interstellar grains composed of silicates, ices and organic carbon. This hypothesis is successful in predicting a number of the observed properties of comets described above and elsewhere (e.g. Whipple 1987), including mean abundances (notably, the roughly fivefold enrichment of carbon over chondritic values), and the presence of spectral signatures of silicates and hydrocarbons. The profile of the silicate feature is sensitive to the size distribution and porosity of the particles. Greenberg and Hage (1990) show that available data on the size distribution from the Vega 2 and Giotto instruments lead to acceptable fits if it is assumed that the grains are extremely porous, with bulk densities as low as $300 \, \text{kg} \, \text{m}^{-3}$, consistent with the properties of the anhydrous interplanetary particles. One problem which needs to be addressed in more detail is the question of the degree of crystallinity of the silicates. Greenberg and Hage (1990) argue that crystalline forms account for only ~ 5% by mass of the silicates in Comet Halley, but the spectra they fitted were of considerably lower resolution than those obtained by Bregman *et al* (1987) which show evidence for annealing (figure 7.21). One possibility is that annealing occurs during release at perihelion and is not characteristic of the grains which were originally included in the comet.

7.3.4   Organic Matter and the Origin of Life

Interstellar and cometary dust grains are carriers of carbon and other 'biogenic' elements which appear to have undergone at least partial synthesis into complex organic molecules. If such molecules were present in the atmosphere and oceans of the primitive Earth, they may have played a significant part in the early stages of prebiotic evolution which led to the emergence of life†. The potential importance of extraterrestrial organic matter is emphasized by developments in our understanding of the origin and evolution of planetary atmospheres. For many years, it was assumed that the Earth's atmosphere was originally highly reducing, containing hydrogen-rich gases captured directly from the solar nebula. The classic experiment of Miller and Urey (1959) showed that amino acids are produced when an electric discharge occurs in an atmosphere composed of H_2, H_2O, NH_3 and CH_4. However, it is now widely accepted that the Earth's atmosphere originated primarily through the release of volatiles from the surface (e.g. Henderson-Sellars *et al* 1980), and was thus much less reducing in nature, being composed predominantly of CO_2, N_2 and H_2O. This model is strongly supported by the present rarity of the noble gases, most notably neon, which should have an abundance comparable with that of nitrogen if the atmosphere were derived from gases of solar composition. Significantly, Miller–Urey-type experiments in CO_2-rich atmospheres give much lower yields of amino acids and other biologically significant molecules (Schlessinger and Miller 1983). Of course, this finding does not necessarily imply that the amino acids and other prebiotic molecules implicated in the origin of terrestrial life were of cosmic origin, but it illustrates the importance of exploring this possibility.

Carbonaceous chondrites contain small but significant quantities of amino acids (Kvenvolden *et al* 1970), the extraterrestrial nature of which is established on the basis of molecular structure and isotopic composition. The atoms of an individual amino acid molecule may be arranged in one of two ways which are mirror images (isomers) of each other (e.g. Chyba 1990). Left-handed (L) and right-handed (D) isomers rotate the plane of polarization of light in opposite directions, a property which provides a straightforward laboratory technique for investigation. Samples containing equal numbers of L and D isomers are said to be racemic, whereas life on Earth is based entirely on L isomers. The origin of this terrestrial bias is unknown. Meteoritic amino acids are generally racemic, a result which supports an extraterrestral origin and suggests abiotic synthesis. Occasional claims have been made for a predominance of L isomers in chondrites, but these are probably due largely to terrestrial contamination (e.g. Bada *et al*

---

† Of necessity, the present discussion is focussed on the origin of life *on Earth*; we may speculate that similar events are likely to have occurred on hospitable planets elsewhere.

1983). The importance of these results is twofold: firstly, they demonstrate that molecules as complex as amino acids may, indeed, be synthesized in interstellar or interplanetary environments; and secondly, they illustrate one viable method for their delivery to Earth.

Extensive observational evidence exists for the presence of organic gas phase molecules in interstellar clouds. Although there is no direct support for the presence of amino acids, significant prebiotic molecules such as hydrogen cyanide (HCN) and formaldehyde (H_2CO) are quite widespread, and more complex precursors such as methanimine (CH_2NH) and formic acid (HCOOH) have been detected in the densest regions (e.g. Mann and Williams 1980; van Dishoeck *et al* 1991). Stochastic reactions leading to the synthesis of such molecules at interstellar temperatures tend to favour deuterated species over normal isotopic forms due to differences in binding energy (Watson 1976). Of particular significance is the ion H_3^+, which initiates a major chain of reactions in models for gas phase chemistry and is thus a source of hydrogen for a number of polyatomic molecules. The exothermic reaction

$$H_3^+ + HD \rightarrow H_2D^+ + H_2 \qquad (7.41)$$

can therefore influence the deuterium content in many interstellar molecules. The abundance of DCO^+, formed by the reaction

$$H_2D^+ + CO \rightarrow DCO^+ + H_2 \qquad (7.42)$$

provides observational confirmation of such fractionation. The chemistry of deuterated molecules on grain surfaces is discussed by Tielens (1983), who predicts enhancements over the cosmic D/H ratio resulting from both adsorption of deuterated species from the gas and the occurrence of grain–surface reactions which lead to further fractionation. The detection of doubly deuterated molecules such as D_2CO suggests active surface chemistry (Turner 1990). The molecules discussed are relatively simple, but it is clear that any complex species which arise will inevitably tend to reflect the deuterium content of their precursors. The presence of deuterium enrichment in carbonaceous chondrites and interplanetary dust particles (e.g. Kerridge 1989; Sandford 1989) is therefore suggestive: the balance of evidence supports the view that this is at least partially due to inclusion of organic molecules formed by interstellar processes. The chemical evolution of the solar nebula is likely to have been complex (see van Dishoeck *et al* 1991 for a review), and it is not known whether these molecules have survived in essentially their interstellar form or whether they have undergone further evolution in the Solar System.

A number of processes may have led to the synthesis of organic molecules in the early Solar System. Active surface reactions on presolar carbon grains may form hydrocarbons (Barlow and Silk 1977b), but a more abundant potential source of organic carbon is locked up in gas phase CO. We noted in §7.1.1 that the stability of CO normally prevents the condensation of carbon-rich solids in a stellar atmosphere of solar composition. However, at temperatures as low as 300–400 K, this reservoir may be tapped by Fischer–Tropsch reactions, which involve the catalytic hydrogenation of CO on a warm surface (e.g. Anders 1971). The parent bodies of meteorites are likely sites for such reactions in the solar nebula. The products include both linear and aromatic hydrocarbons of the general formula $C_n H_{2n+2}$, synthesized by reactions such as

$$10\,CO + 21\,H_2 \rightarrow C_{10}H_{22} + 10\,H_2O. \qquad (7.43)$$

Presolar or newly formed hydrocarbons may, in turn, be converted to amino acids by aqueous alteration on the parent bodies (e.g. Shock and Schulte 1990), and this provides a basis for understanding their presence in meteorites.

The delivery of cometary and asteroidal organic matter to the primitive Earth is discussed in detail by Anders (1989) and Chyba $et\ al$ (1990). Organics will not survive impacts at velocities above about $10\,km\,s^{-1}$, and this precludes delivery by bodies greater than about 100 m in radius through the atmosphere in its present form. The current infall rate of unablated organic matter is estimated to be $\sim 3 \times 10^5\,kg$ per year (Anders 1989), and the dominant contribution comes not from meteorites but from interplanetary dust particles with diameters in the range 1–$100\,\mu m$. If this rate were to remain constant, the timescale for accretion of a mass equal to the current total biomass ($\sim 6 \times 10^{14}\,kg$; Chyba $et\ al$ 1990) would be $\sim 2\,Gyr$. However, during the period of heavy bombardment between 4.5 and 3.8 Gyr ago, the infall rate must have been much greater than it is today, and the atmosphere may have been denser, broadening the size range of particles which would be sufficiently decelerated to preserve their organic content. On the basis of reasonable quantitative assumptions, Chyba $et\ al$ estimate that the organic infall rate may have been as high as $10^7\,kg$ per year at an early epoch, reducing the timescale for accretion of a biomass to $\sim 60\,Myr$. It is thus plausible to suppose that a large endowment of extraterrestrial organic matter was present on the surface of the Earth at the time of the emergence of life some 3.5 Gyr ago.

In the light of the above discussion, it is natural to question whether the entire process of the origin of life may have preceded its appearance on our planet. The concept of panspermia, i.e. the seeding of the Earth with life from space, was introduced many years ago by Arrhenius (1903). There is no $a\ priori$ objection to the concept of interstellar organisms (Sagan 1973) but the densities prevailing in molecular clouds are insufficent to allow appreciable evolution within cloud lifetimes, and organisms exposed to the interstellar radiation field outside of molecular clouds are rapidly destroyed by ultraviolet irradiation (Greenberg 1984). Their lack of viability in extreme environments appears to exclude bacteria as carriers of the general interstellar extinction (see §1.4), a proposal which is also incompatible with abundance constraints (§2.4.4) and spectroscopic evidence, notably the absence of strong absorption features in the 2500–3000 Å region of the extinction curve (§5.1.1). If biological organisms do exist outside of planetary environments, the most viable site would appear to be comets. However, to invoke an extraterrestrial source does not solve the problem of the origin of life on Earth, but merely displaces it. Philosophically, Occam's razor may provide suitable guidance (Chyba 1990): *"pluralitas non est ponenda sine necessitate"* (plurality is not to be posited without necessity).

Towards a Unified Model for Interstellar Dust

"This phenomenon is so puzzling... one is almost tempted to give up a satisfactory explanation. However, the man of science ought not to recoil, either before the obscurity of a phenomenon or before the difficulty of a research. Whether he is in possession of earlier work, or whether he tries to increase the knowledge of a phenomenon by new, precise observations, he can be sure of a certain success in his studies, if he employs calm speculation, without abandoning himself to an excited and preoccupied imagination. However little ground he gains, he will always enlarge it by returning to his problem with that persistence which is the indispensible condition of study. It is thus, guided by analysis and calculation, that he can even derive unexpected results which, however, enjoy considerable certainty."

F G W Struve (1847)

A wide variety of observational constraints on the properties of interstellar grains have been described in previous chapters. The key results are the abundances of the condensible elements and their observed depletions (§2.2–§2.4), the wavelength dependence of the extinction, albedo and phase function (§3.2–§3.3), the polarization and alignment properties of the grains (§4.2–§4.4), the spectral absorption features (§5.1–§5.3), and the continuum and spectral line emission (§6.2–§6.3). Variations in all of these phenomena as a function of interstellar environment are of interest. Additional relevant information is provided by the properties of dust in the circumstellar shells of dust-ejecting stars (§7.1.2) and by identification and study of presolar interstellar grains in the Solar System (§7.3.3). Finally, estimates of the formation and destruction rates for interstellar grains must also be explained (§7.1.3, §7.2.4). We noted in Chapter 1 (§1.4) that the observations imply a need for multi-component models in which different grain populations account for different aspects of the data. A unified model for interstellar dust should be capable of matching all known contraints unambiguously and in a self-consistent way, invoking substances and size distributions which are physically reasonable and have an obvious source in terms of stellar mass-loss processes or production in the interstellar medium itself. Models which are selective in the data they attempt to explain generally suffer from lack of uniqueness, a problem which has arisen historically in proposals based primarily on fits to the extinction curve (§3.4).

The importance of the unified approach was emphasized by Hong and Greenberg (1980), who attempted simultaneous fits to extinction, polarization and abundance data. More recent theoretical work has addressed the additional problem of accounting for the emissive properties of the grains (Draine and Lee 1984; Rowan-Robinson 1986; Chlewicki and Laureijs 1988; Désert *et al* 1990). No model so far discussed can claim to be entirely successful in uniquely explaining all the observed properties of interstellar dust (see §1.4 for additional comments and references). However, it is notable that a number of diverse observational results place common or similar constraints on grain properties, and consequently a number of models have common factors, suggesting a degree of convergence. The purpose of the ensuing discussion is not to enter into a detailed assessment of the merits and failures of individual proposal, or to present a rival theory, but rather to attempt an overview which focuses on common factors and highlights the main areas of uncertainty, with a view to future progress towards a truly unified and rigorous dust model.

There is broad agreement on the need for both O-rich and C-rich grain materials in any successful model. Silicates are identified spectroscopically in interstellar clouds and in the envelopes of O-rich dust-forming stars. Although the chemical composition of interstellar silicates is not closely constrained, minerals such as $MgSiO_3$ and $(Mg, Fe)SiO_4$ provide a natural explanation for the depletions of these abundant heavy elements. The observed strength of the $9.7\,\mu m$ feature *requires* that most of the available Si atoms are bonded to O in silicates (§5.3.2), and these may account for up to about 60% of the total mass of refractory interstellar dust (§2.4.4). This result is consistent with the need for a dielectric (non-absorbing) dust component to account for observations of scattering (§3.2.2) and polarization (§4.3). Studies of the $9.7\,\mu m$ feature demonstrate that aligned silicate grains are, indeed, a source of polarization (§5.3.4). However, they cannot be responsible for all of the visible extinction as the ratio $A_V/\tau_{9.7}$ is too low in pure silicates of appropriate size. Metallic grains are unstable to oxidation (Jones 1990), and abundance criteria thus seem to force the conclusion that most of the additional extinction arises in some form of solid carbon. The nature and origin of the C-rich component has been one of the main distinguishing factors between the dust models of different authors.

There is also broad agreement on the need for a wide spectrum of grain sizes. In order to explain the λ^{-1} extinction law in the visible, we require large ('classical') grains with dimensions approaching the wavelength of visible light ($a \sim \lambda/\pi \sim 0.2\,\mu m$ for silicates). Evidence for a distinct population of much smaller ('subclassical') grains ($a < 0.02\,\mu m$) arises from the shape of the extinction curve (§3.4) and the behaviour of the phase function (§3.2.2) in the ultraviolet, and from the detection of near- and mid-infrared emission greatly in excess of that expected from classical-sized grains exposed to typical radiation fields. The latter effect is characteristic of particles of low heat capacity which undergo transient heating (§6.1.4).

Table 8.1 Overview of grain components in a general model of interstellar dust, consistent with observations of diffuse clouds, H II regions and reflection nebulae.

	Silicate cores	C-rich mantles	C-rich small grains	PAHs
Origin	Stars (O>C)	ISM	Stars (C>O)	Stars (C>O)
Size	Classical and subclassical	(On silicate cores)	Subclassical	Macromolecular
Composition	Amorphous $MgSiO_3$, etc. with Fe-rich inclusions	Organic or amorphous carbon	Amorphous carbon (partially graphitized)	PAHs
Elements depleted	O, Mg, Al, Si, Fe, etc.	C, N, O	C	C
Extinction	— Contribution at all λ —		Mid-UV	Far-UV
Alignment	Yes	Yes	No	No
Polarization	— Visible–infrared —		None	None
Absorption features	9.7, 18.5 μm	3.4 μm	2175 Å	Not seen
Emission	— Far-infrared continuum —		— Mid-infrared continuum —	
Emission features	9.7, 18.5 μm	ERE	ERE	3.3, 6.2, 7.7, 8.6, 11.3 μm

As a basis for discussion, I summarize in table 8.1 the essential properties of a general grain model. Four components are needed to explain the observed properties of dust in diffuse clouds, H II regions, and reflection nebulae: (i) silicate grains, (ii) carbon-rich mantles, (iii) carbon-rich small grains, and (iv) free PAH molecules.

The silicates are assumed to originate in O-rich stellar atmospheres

(§7.1), and to have a power-law size distribution imposed by grain–grain collisions in the parent stellar winds and interstellar clouds. Icy mantles accumulate on silicate cores in molecular clouds and are subsequently processed to organic residues, and, ultimately, amorphous carbon by prolonged exposure to ultraviolet radiation. The relative abundance of organics and amorphous carbon in the mantles depends on the degree of photoprocessing. Hydrocarbon molecules in the mantles are predominantly aliphatic (and therefore absorb at $3.4\,\mu$m), in contrast to those produced in stellar winds, which are predominantly aromatic. The use of idealized core–mantle particles in theoretical calculations is, of course, a mathematical convenience, and one can envisage that the larger grains are, in reality, complex in structure; they may include a population of 'clean' silicates recently injected from stellar atmospheres as well as coated grains which have been cycled through molecular clouds and regions of recent star formation. The latter may be similar in structure to interplanetary particles, containing silicate clusters formed by accretion as well as isolated silicate grains, all embedded in a carbonaceous matrix (§7.3.3).

The formation of PAHs and amorphous carbon in C-rich stellar atmospheres is described in §7.1. Free PAHs are two-dimensional, with radii typically in the range 0.4–1.2 nm (Désert *et al* 1990), whereas amorphous carbon grains are three-dimensional and composed of loosely connected PAH clusters. The size distribution is governed by the efficiency of the growth and injection processes and by destructive events in the interstellar medium. Theoretical arguments and observational results suggest that highly absorbing particles are expelled from red giants by radiation pressure before they grow to classical dimensions, and this provides a physical basis for the exclusion of large pure carbon grains from the model. C-rich grains of stellar origin are assumed to be in the small-particle limit at all wavelengths of interest: in this respect, table 8.1 differs from the widely discussed 'MRN' model (Mathis *et al* 1977; Draine and Lee 1984). Exposure to the interstellar radiation field may lead to dehydrogenation and partial graphitization (§5.1.2; §7.2.4) of the C-rich small grains. Conversion to diamond may occur in those subjected to shock-induced collisions (§7.2.4).

The observational phenomena attributed to each component of the general model are listed in table 8.1. The silicates contain essentially all the depleted metals and most of the depleted oxygen. Approximately half of all interstellar C is assumed to be in dust in some form, with $\sim 5\%$ in free PAHs and the remainder roughly equally divided between mantles and small carbon grains. Individual contributions to the extinction curve resemble the lower frame of figure 3.11, with core–mantle grains, small carbon grains and PAHs accounting for the visual extinction, mid-ultraviolet peak and far-ultraviolet rise, respectively (see Désert *et al* 1990 for a similar formulation). The $\lambda2175$ feature is tentatively assigned to graphite: the arguments for and against this identification are discussed in detail in §5.1.2. PAHs are assumed to be insufficiently abundant to produce measurable absorption features, but their excitation by soft ultraviolet radiation accounts for characteristic infrared emission lines and continuum emission in the 3–12 μm region of the spectrum. Extended red emission (§6.3.3) is attributed to luminescence in either C-rich small grains or amorphous carbon mantles. The large grains are aligned, most probably by the Davis–Greenstein mechanism operating on silicates with superparamagnetic properties due to the presence of Fe-rich inclusions (§4.4.3). The alignment mechanism becomes inefficient for progressively smaller grains and is probably inoperative for small C-rich grains and PAHs.

The scenario outlined in table 8.1 does not constitute a unique or complete dust model for a number of reasons. The structure of the large grains and the mechanism for their alignment are open to alternative interpretations. Lingering doubts concerning the identity of the λ2175 feature have already been noted, and no solution to the problem of identifying the optical diffuse interstellar bands (§5.2) has been proposed. The FUV extinction rise is attributed to PAHs on the basis of observational evidence and theoretical calculations (see Désert et al 1990), but other assignments, such as an EUV feature which extends longward of the Lyman limit (§5.1.2), are not excluded.

Comparisons between the observed properties of interstellar grains in the Milky Way and other galaxies with significantly different chemical compositions and dust to gas ratios are likely to be revealing. The Small Magellanic Cloud (SMC) is currently the best example. It has low metallicity (see Roche et al 1987 for discussion and references) leading to a large number density of carbon stars relative to that for normal red giants (§2.3.2). The extinction curve of SMC stars displays a steep FUV rise and an apparent absence of λ2175 absorption (see figure 3.9). If the scenario in table 8.1 is broadly correct and applicable to the SMC as well as the Galaxy, it follows that dust formation and modification processes in the SMC lead to the production of PAHs but not to (graphitized) small carbon grains. It may be significant that galactic post-AGB stars exhibiting PAH features (e.g. HR 4049 and HD 44179; see §6.3) also have steep FUV extinction and an absence of λ2175 absorption (§7.1.2), and are also metal-poor, suggesting that metallicity influences the quality of dust produced in carbon stars. One possibility is that small SiC grains play a similar rôle in C-rich atmospheres to that suggested for Al_2O_3 in O-rich atmospheres (§7.1.1), providing nucleation centres for the condensation of more abundant grain materials (see Frenklach et al 1989). In a metal-poor, carbon-rich environment, growth of amorphous carbon grains would then be inhibited.

An important aspect of the large, mantled grains proposed in table 8.1 is that they are rendered absorbing in nature by their C-rich coatings. Again, the analogy with interplanetary dust, blackened by a relatively low mass fraction of carbon, is appropriate. The primary motivation for this constraint on the interstellar particles is the fact that simultaneous fits to the extinction curve and the diffuse FIR emission on a 'per grain' basis are only possible if the albedo is assumed to be rather low. The local minimum in the observed albedo at $0.22\,\mu m$ (figure 3.4) is easily explained by absorption in the particles carrying the $\lambda2175$ feature. However, the infrared luminosity of the galactic disk can only be explained if the albedo of the particles responsible for the continuum extinction in the visible and ultraviolet is also quite low ($\gamma \leq 0.5$ at wavelengths of interest) compared with current observational constraints, unless the intensity of the interstellar radiation field has been underestimated by at least 50% (see Désert et al 1990). The difficulty in obtaining precise data on the albedo has already been noted (§1.1.3, §3.2.2). The extent to which the imaginary (absorptive) component of the grain refractive index can be constrained by observations of circular polarization (§4.3.2) has also been questioned (Chlewicki and Greenberg 1990), and it should be recalled that the available data on circular polarization are, in any case, limited to very few lines of sight. It is particularly important to distinguish between absorption-dominated and scattering-dominated extinction in the near infrared where the extinction law converges to a common form (§3.3.2). For extinction due to pure absorption in a semiconductor material such as amorphous carbon, the form of the extinction curve is determined by the intrinsic properties of the solid (specifically, the bandgap energy) rather than by the size distribution of the particles (see Duley 1988; Duley and Whittet 1992), and this might explain its insensitivity to environmental variations.

Another problem which requires further comment is the rather stringent constraint that essentially all interstellar Si should be in silicates to explain the observed strength of the $9.7\,\mu m$ absorption feature. There are two issues to consider. Firstly, carbon stars produce a non-trivial mass fraction of refractory interstellar dust (perhaps 20%, but as high as 50% according to some estimates), in which silicon is expected to be tied up in SiC. It is not simply a case of destroying SiC to explain the observational limit on its interstellar abundance (§5.3.2), but also of somehow reaccreting the Si into silicates. The problem is compounded by the fact that hot stars with purely gaseous winds inject gas phase Si into the interstellar medium. The second issue concerns destruction rates for the silicates themselves, which appear to be uncomfortably high compared to injection rates from evolved stars (§7.2.4). It is possible that the presence of mantles upon the silicate grains provides some degree of protection from sputtering, but the correlation of Si depletion with cloud velocity (figure 7.17) argues that silicates are, indeed, being destroyed to a significant degree in shocks. It is physically reasonable to suppose that Si atoms rapidly reaccrete onto existing grains in dense clouds, but it is less obvious that they will be incorporated into appropriate silicate mineral structures which contribute to the observed

9.7 μm feature.

The possibility that *young* stars are significant sources of interstellar dust was suggested some years ago by Herbig (1970). In view of the high pre-existing depletions of the heavy elements in dense clouds, it is, perhaps, unlikely that star formation can produce more dust than it consumes, but it may provide a means of boosting the silicate fraction. Suppose that a certain mass of interstellar matter in a protostellar nebula is completely vaporized, and subsequently recondenses to form grains which are ejected into the interstellar medium during the outflow phase of pre-main-sequence evolution (Burke and Silk 1976). Because the interstellar (and hence the protostellar) C/O ratio is less than unity, the solids that condense will be oxygen-rich, and Si atoms present in the gas (including those resulting from vaporization of any SiC dust included in the nebula) will recondense as silicates (figure 7.2). This proposal is qualitatively reasonable, but unfortunately almost impossible to quantify, as we have no way of estimating what fraction of a protostellar envelope is processed in this way (Salpeter 1977).

The general model discussed above concerns only dust in diffuse regions of the ISM. Existing observations of dust in molecular clouds (§5.3.3) broadly confirm theoretical predictions concerning the growth of ice mantles (§7.2.3), although only two molecules (H_2O and CO) have been directly identified in the solid phase in many lines of sight. Some additional molecules (e.g. CO_2, CH_3OH and H_2S) are detected only tentatively or in a very few sources; a number of others, notably NH_3, O_2 and N_2, are also predicted to be present, but observational confirmation is either lacking (as in the case of NH_3) or excluded by an absence of appropriate spectroscopic signatures. Further opportunities for identification of molecules in ices are likely to arise with the launch of the Infrared Space Observatory (currently scheduled for 1993). This mission will have spectroscopic capability in the 4.1–4.5 μm region, one of the 'final frontiers' of near-infrared astronomical spectroscopy, and will thus allow direct study of the principal vibrational mode in solid CO_2. Spectroscopic data at wavelengths beyond 30 μm will also be of great interest for the study of molecular lattice modes in interstellar and circumstellar ices (see Omont *et al* 1990 for an example). Corresponding studies of gas phase abundances in the same lines of sight will also be possible. This is an area where great advances are anticipated over the next few years.

References

Aannestad P A and Greenberg J M 1983 *Astrophys. J.* **272** 551

Aannestad P A and Kenyon S J 1979 *Astrophys. Space Sci.* **65** 155

Aannestad P A and Purcell E M 1973 *Ann. Rev. Astron. Astrophys.* **11** 309

Adams F C Lada C J and Shu F H 1987 *Astrophys. J.* **312** 788

Adamson A J and Whittet D C B 1990 *Astron. Astrophys.* **232** 27

Adamson A J Whittet D C B and Duley W W 1990 *Mon. Not. R. Astron. Soc.* **243** 400

Aitken D K 1989 in *22nd ESLAB Symposium, Infrared Spectroscopy in Astronomy*, ed. B H Kaldeich (ESA publication SP-290) p 99

Aitken D K and Roche P F 1985 *Mon. Not. R. Astron. Soc.* **213** 777

Aitken D K Roche P F Spenser P M and Jones B 1979 *Astrophys. J.* **233** 925

Aitken D K Bailey J A Roche P F and Hough J H 1985 *Mon. Not. R. Astron. Soc.* **215** 815

Aitken D K Roche P F Bailey J A Briggs G P Hough J H and Thomas J A 1986 *Mon. Not. R. Astron. Soc.* **218** 363

Aitken D K Roche P F Smith C H James S D and Hough J H 1988 *Mon. Not. R. Astron. Soc.* **230** 629

Aitken D K Smith C H and Roche P F 1989 *Mon. Not. R. Astron. Soc.* **236** 919

Allamandola L J and Sandford S A 1988 in *Dust in the Universe*, eds M E Bailey and D A Williams (Cambridge University Press) p 229

Allamandola L J and Tielens A G G M (eds) 1989 *IAU Symp. no. 135, Interstellar Dust* (Kluwer, Dordrecht)

Allamandola L J Tielens A G G M and Barker J R 1985 *Astrophys. J. Lett.* **290** L25

Allamandola L J Tielens A G G M and Barker J R 1989 *Astrophys. J. Suppl.* **71** 733

Allen C W 1973 *Astrophysical Quantities*, 3rd edn (Athlone Press, London) p 268

Allen D A Baines D W T Blades J C and Whittet D C B 1982 *Mon. Not. R. Astron. Soc.* **199** 1017

Amari S Anders E Virag A and Zinner E 1990 *Nature* **345** 238

Anders E 1971 *Ann. Rev. Astron. Astrophys.* **9** 1

Anders E 1989 *Nature* **342** 255

Anders E and Grevesse N 1989 *Geochim. Cosmochim. Acta* **53** 197

Anders E Lewis R S Tang M and Zinner E 1989 in *IAU Symp. no. 135, Interstellar Dust*, eds L J Allamandola and A G G M Tielens (Kluwer, Dordrecht) p 389

Arrhenius S 1903 *Lehrbuch der Kosmischen Physik* (Leipzig)

Artymowicz P 1988 *Astrophys. J. Lett.* **335** L79

Aumann H H 1988 *Astron. J.* **96** 1415

Aumann H H Gillett F C Beichman C A de Jong T Houk J R Low F J Neugebauer G Walker R G and Wesselius P R 1984 *Astrophys. J. Lett.* **278** L23

Avery R W Stokes R A Michalsky J J and Ekstrom P A 1975 *Astron. J.* **80** 1026

Axon D J and Ellis R S 1976 *Mon. Not. R. Astron. Soc.* **177** 499

Baade W and Minkowski R 1937 *Astrophys. J.* **86** 123

Baas F Allamandola L J Geballe T R Persson S E and Lacy J H 1983 *Astrophys. J.* **265** 290

Baas F Geballe T R and Walther D M 1986 *Astrophys. J. Lett.* **311** L97

Baas F Grim R J A Geballe T R Schutte W and Greenberg J M 1988 in *Dust in the Universe*, eds M E Bailey and D A Williams (Cambridge University Press) p 55

Bada J L Cronin J R Ho M S Kvenvolden K A Lawless J G Miller S L Oro J and Steinberg S 1983 *Nature* **301** 494

Bailey M E and Williams D A (eds) 1988 *Dust in the Universe* (Cambridge University Press)

Baines D W T and Whittet D C B 1983 *Mon. Not. R. Astron. Soc.* **203** 419

Banwell C N 1983 *Fundamentals of Molecular Spectroscopy*, 3rd edn (McGraw-Hill, London)

Barker J R Allamandola L J and Tielens A G G M 1987 *Astrophys. J. Lett.* **315** L61

Barlow M J 1978a *Mon. Not. R. Astron. Soc.* **183** 367

Barlow M J 1978b *Mon. Not. R. Astron. Soc.* **183** 397

Barlow M J 1978c *Mon. Not. R. Astron. Soc.* **183** 417

Barlow M J and Silk J 1977a *Astrophys. J. Lett.* **211** L83

Barlow M J and Silk J 1977b *Astrophys. J.* **215** 800

Barnard E E 1910 *Astrophys. J.* **31** 8

Barnard E E 1913 *Astrophys. J.* **38** 496

Barnard E E 1919 *Astrophys. J.* **49** 1

Barnard E E 1927 *Atlas of Selected Regions of the Milky Way*, eds E B Frost and M R Calvert (Carnegie Institute, Washington)

Barshay S S and Lewis J S 1976 *Ann. Rev. Astron. Astrophys.* **14** 81

Batten A H 1988 *The Lives of Wilhelm and Otto Struve* (Reidel, Dordrecht) p 149

Becklin E E and Neugebauer G 1975 *Astrophys. J. Lett.* **200** L71

Beichman C A 1987 *Ann. Rev. Astron. Astrophys.* **25** 521

Beichman C A Myers P C Emerson J P Harris S Mathieu R Benson P J Jennings R E 1986 *Astrophys. J.* **307** 337

Benvenuti P and Porceddu I 1989 *Astron. Astrophys.* **223** 329

Bernatowicz T Fraundorf G Tang M Anders E Wopenka B Zinner E and Fraundorf P 1987 *Nature* **330** 728

Bertaux J L and Blamont J E 1976 *Nature* **262** 263

Biermann P and Harwit M 1980 *Astrophys. J. Lett.* **241** L105

Black J H and Willner S P 1984 *Astrophys. J.* **279** 673

Blades J C and Somerville W B 1977 *Mon. Not. R. Astron. Soc.* **181** 769

Blades J C and Somerville W B 1981 *Mon. Not. R. Astron. Soc.* **197** 543

Blades J C and Whittet D C B 1980 *Mon. Not. R. Astron. Soc.* **191** 701

Bless R C and Savage B D 1972 *Astrophys. J.* **171** 293

Bode M F 1988 in *Dust in the Universe*, eds M E Bailey and D A Williams (Cambridge University Press) p 73

Bode M F 1989 in *22nd ESLAB Symposium, Infrared Spectroscopy in Astronomy*, ed. B H Kaldeich (ESA publication SP-290) p 317

Bode M F and Evans A 1983 *Mon. Not. R. Astron. Soc.* **203** 285

Bode M F Evans A Whittet D C B Aitken D K Roche P F and Whitmore B 1984 *Mon. Not. R. Astron. Soc.* **207** 897

Bodenheimer P 1989 in *The Formation and Evolution of Planetary Systems*, eds H A Weaver and L Danly (Cambridge University Press) p 243

Boesgaard A M and Steigman G 1985 *Ann. Rev. Astron. Astrophys.* **23** 319

Bohlin R C and Savage B D 1981 *Astrophys. J.* **249** 109

Bohlin R C Savage B D and Drake J F 1978 *Astrophys. J.* **224** 132

Bohren C F and Huffman D R 1983 *Absorption and Scattering of Light by Small Particles* (John Wiley & Sons, New York)

Bok B J 1956 *Astron. J.* **61** 309

Boulanger F and Pérault M 1988 *Astrophys. J.* **330** 964

Breger M Gehrz R D and Hackwell J A 1981 *Astrophys. J.* **248** 963

Bregman J 1989 in *IAU Symp. no. 135, Interstellar Dust*, eds L J Allamandola and A G G M Tielens (Kluwer, Dordrecht) p 109

Bregman J D Campins H Witteborn F C Wooden D H Rank D M Allamandola L J Cohen M and Tielens A G G M 1987 *Astron. Astrophys.* **187** 616

Bromage G E 1987 *Quart. J. R. Astron. Soc.* **28** 301

Bromage G E and Nandy K 1973 *Astron. Astrophys.* **26** 17

Brownlee D E 1978 in *Protostars & Planets*, ed. T Gehrels (University of Arizona Press, Tucson) p 134

Brownlee D E and Kissel J 1990 in *Comet Halley: Investigations, Results and Interpretations, vol. 2: Dust, Nucleus, Evolution*, ed. J W Mason (Ellis Horwood, Chichester) p 89

Burke J R and Silk J 1976 *Astrophys. J.* **210** 341

Burstein D and Heiles C 1982 *Astron. J.* **87** 1165

Buss R H and Snow T P 1988 *Astrophys. J.* **335** 331

Buss R H Lamers H J G L M and Snow T P 1989 *Astrophys. J.* **347** 977

Butchart I and Whittet D C B 1983 *Mon. Not. R. Astron. Soc.* **202** 971

Butchart I McFadzean A D Whittet D C B Geballe T R and Greenberg J M 1986 *Astron. Astrophys.* **154** L5

Campbell M F Elias J H Gezari D Y Harvey P M Hoffmann W F Hudson H S Neugebauer G Soifer B T Werner M W and Westbrook W E 1976 *Astrophys. J.* **208** 396

Capps R W and Knacke R F 1976 *Astrophys. J.* **210** 76

Cardelli J A and Savage B D 1988 *Astrophys. J.* **325** 864

Cardelli J A Clayton G C and Mathis J S 1989 *Astrophys. J.* **345** 245

Carnochan D J 1989 in *Interstellar Dust*, eds L J Allamandola and A G G M Tielens (NASA Conference Publication 3036) p 11

Charnley S B Dyson J E Hartquist T W and Williams D A 1988 *Mon. Not. R. Astron. Soc.* **231** 269

Chiosi C and Maeder A 1986 *Ann. Rev. Astron. Astrophys.* **24** 329

Chlewicki G and Greenberg J M 1990 *Astrophys. J.* **365** 230

Chlewicki G and Laureijs R J 1988 *Astron. Astrophys.* **207** L11

Churchwell E 1990 *Astron. Astrophys. Rev.* **2** 79

Chyba C F 1990 *Nature* **348** 113

Chyba C F and Sagan C 1987 *Nature* **330** 350

Chyba C F Thomas P J Brookshaw L and Sagan C 1990 *Science* **249** 366

Clark D H and Stephenson F R 1977 *The Historical Supernovae* (Pergamon Press, Oxford)

Clarke D 1984 *Mon. Not. R. Astron. Soc.* **206** 739

Clarke D and Al-Roubaie A 1983 *Mon. Not. R. Astron. Soc.* **202** 173

Clayton D D 1982 *Quart. J. R. Astron. Soc.* **23** 174

Clayton D D 1988 in *Dust in the Universe*, eds M E Bailey and D A Williams (Cambridge University Press) p 145

Clayton G C and Mathis J S 1988 *Astrophys. J.* **327** 911

Clayton G C Martin P G and Thompson I 1983 *Astrophys. J.* **265** 194

Clayton G C Anderson C M Magalhães A M Code A D Nordsieck K H Meade M R Wolff M J Babler B Bjorkman K S Schulte-Ladbeck R Taylor M J and Whitney B A 1992 *Astrophys. J.* in press

Code A D 1958 *Publ. Astron. Soc. Pacific* **70** 407

Codina-Landaberry S and Magalhães A M 1976 *Astron. Astrophys.* **49** 407

Cohen M 1983 *Astrophys. J.* **270** L69

Cohen M 1984 *Mon. Not. R. Astron. Soc.* **206** 137

Cohen M and Witteborn F C 1985 *Astrophys. J.* **294** 345

Cohen M Allamondola L J Tielens A G G M Bregman J Simpson J P Witteborn F C Wooden D and Rank D M 1986 *Astrophys. J.* **302** 737

Cohen M Tielens A G G M Bregman J Witteborn F C Rank D M Allamondola L J Wooden D H and de Muizon M 1989 *Astrophys. J.* **341** 246

Cowie L L 1978 *Astrophys. J.* **225** 887

Cowie L L and Songaila A 1986 *Ann. Rev. Astron. Astrophys.* **24** 499

Cox D P and Reynolds R J 1987 *Ann. Rev. Astron. Astrophys.* **25** 303

Cox P 1989 *Astron. Astrophys.* **225** L1

Cox P and Mezger P G 1987 in *Star Formation in Galaxies*, ed. Carol J Lonsdale Persson (NASA Conference Publication 2466) p 23

Cox P and Mezger P G 1989 *Astron. Astrophys. Rev.* **1** 49

Cox P Krugel E and Mezger P G 1986 *Astron. Astrophys.* **155** 380

Coyne G V Gehrels T and Serkowski K 1974 *Astron. J.* **79** 581

Coyne G V Tapia S and Vrba F J 1979 *Astron. J.* **84** 356

Crutcher R M 1985 *Astrophys. J.* **288** 604

Czyzak S J Hirth J P and Tabak R G 1982 *Vistas Astron.* **25** 337

Dame T M and Thaddeus P 1985 *Astrophys. J.* **297** 751

Dame T M Ungerechts H Cohen R S de Geus E J Grenier I A May J Murphy D C Nyman L A and Thaddeus P 1987 *Astrophys. J.* **322** 706

Danks A C 1980 *Publ. Astron. Soc. Pacific* **92** 52

Danks A C and Lambert D L 1976 *Mon. Not. R. Astron. Soc.* **174** 571

Davies R E Delluva A M and Kock R H 1984 *Nature* **311** 748

Davis L and Greenstein J L 1951 *Astrophys. J.* **114** 206

Day K L 1979 *Astrophys. J.* **234** 158

Day K L 1981 *Astrophys. J.* **246** 110

Day K L and Huffman D R 1973 *Nature Phys. Sci.* **243** 50

Debye P 1909 *Ann. Phys., NY* **30** 59

Debye P 1912 *Ann. Phys., NY* **39** 789

de Muizon M Geballe T R and d'Hendecourt L B 1986 *Astrophys. J. Lett.* **306** L105

de Muizon M d'Hendecourt L B and Geballe T R 1990 *Astron. Astrophys.* **235** 367

Désert F X and Dennefeld M 1988 *Astron. Astrophys.* **206** 227

Désert F X Boulanger F Léger A Puget J L and Sellgren K 1986 *Astron. Astrophys.* **159** 328

Désert F X Boulanger F and Puget J L 1990 *Astron. Astrophys.* **237** 215

de Vaucouleurs G and Buta R 1983 *Astron. J.* **88** 939

d'Hendecourt L B and Jourdain de Muizon M 1989 *Astron. Astrophys.* **223** L5

d'Hendecourt L B Allamandola L J and Greenberg J M 1985 *Astron. Astrophys.* **152** 130

d'Hendecourt L B Allamandola L J Grim R J A and Greenberg J M 1986 *Astron. Astrophys.* **158** 119

Disney M 1990 *Nature* **346** 105

Donn B 1968 *Astrophys. J. Lett.* **152** L129

Donn B and Nuth J A 1985 *Astrophys. J.* **288** 187

Dorschner J and Henning T 1986 *Astrophys. Space Sci.* **128** 47

Dorschner J Friedemann C and Gürtler J 1977 *Astron. Astrophys.* **58** 201

Dorschner J Friedemann C Gürtler J and Henning T 1988 *Astron. Astrophys.* **198** 223

Douglas A E 1977 *Nature* **269** 130

Dragovan M 1986 *Astrophys. J.* **308** 270

Draine B T 1984 *Astrophys. J. Lett.* **277** L71

Draine B T 1989a in *IAU Symp. no. 135, Interstellar Dust*, eds L J Allamandola and A G G M Tielens (Kluwer, Dordrecht) p 313

Draine B T 1989b in *Evolution of Interstellar Dust and Related Topics*, eds A Bonetti J M Greenberg and S Aiello (North-Holland Publishing Company, Amsterdam) p 103

Draine B T 1990 in *The Interstellar Medium in Galaxies*, eds H A Thronson and J M Shull (Kluwer, Dordrecht) p 483

Draine B T and Anderson N 1985 *Astrophys. J.* **292** 494

Draine B T and Lee H M 1984 *Astrophys. J.* **285** 89

Drilling J S and Schönberner D 1989 *Astrophys. J. Lett.* **343** L45

Duley W W 1973 *Astrophys. Space Sci.* **23** 43

Duley W W 1974 *Astrophys. Space Sci.* **26** 199

Duley W W 1975 *Astrophys. Space Sci.* **36** 345

Duley W W 1984 *Quart. J. R. Astron. Soc.* **25** 109

Duley W W 1987 *Mon. Not. R. Astron. Soc.* **229** 203

Duley W W 1988 in *Dust in the Universe*, eds M E Bailey and D A Williams (Cambridge University Press) p 209

Duley W W and Whittet D C B 1990 *Mon. Not. R. Astron. Soc.* **242** 40P

Duley W W and Whittet D C B 1992 *Mon. Not. R. Astron. Soc.* in press

Duley W W and Williams D A 1981 *Mon. Not. R. Astron. Soc.* **196** 269

Duley W W and Williams D A 1983 *Mon. Not. R. Astron. Soc.* **205** 67P

Duley W W and Williams D A 1984 *Interstellar Chemistry* (Academic Press, London)

Duley W W and Williams D A 1988 *Mon. Not. R. Astron. Soc.* **231** 969

Duley W W Jones A P and Williams D A 1989a *Mon. Not. R. Astron. Soc.* **236** 709

Duley W W Jones A P Whittet D C B and Williams D A 1989b *Mon. Not. R. Astron. Soc.* **241** 697

Dupree A K 1986 *Ann. Rev. Astron. Astrophys.* **24** 377

Dwek E 1989 in *IAU Symp. no. 135, Interstellar Dust*, eds L J Allamandola and A G G M Tielens (Kluwer, Dordrecht) p 479

Dyck H M and Jones T J 1978 *Astron. J.* **83** 594

Dyck H M and Lonsdale C J 1981 in *IAU Symp. no. 96, Infrared Astronomy*, eds C G Wynn-Williams and D P Cruikshank (Reidel, Dordrecht) p 223

Dyson J E and Williams D A 1980 *The Physics of the Interstellar Medium* (Manchester University Press) p 63

Eales S A Wynn-Williams C G and Duncan W D 1989 *Astrophys. J.* **339** 859

Eddington A S 1926 *Proc. R. Soc.* A **111** 423

Eiroa C Hefele H and Zhong-yu Q 1983 *Astron. Astrophys. Suppl.* **54** 309

Elias J H 1978a *Astrophys. J.* **224** 453

Elias J H 1978b *Astrophys. J.* **224** 857

Elmegreen B G 1985 in *Protostars & Planets II*, eds D C Black and M S Matthews (University of Arizona Press, Tucson) p 33

Evans A Whittet D C B Davies J K Kilkenny D and Bode M F 1985 *Mon. Not. R. Astron. Soc.* **217** 767

Fahlman G G and Walker G A H 1975 *Astrophys. J.* **200** 22

Fano U 1961 *Phys. Rev.* **124** 1866

Federman S R Kumar C K and Vanden Bout P A 1984 *Astrophys. J.* **282** 485

Fesen R A and Blair W P 1990 *Astrophys. J. Lett.* **351** L45

Field G B 1974 *Astrophys. J.* **187** 453

Field G B Partridge R B and Sobel H 1967 in *Interstellar Grains*, eds J M Greenberg and T P Roark (NASA, Washington) p 207

FitzGerald M P 1968 *Astron. J.* **73** 983

Fitzpatrick E L 1986 *Astron. J.* **92** 1068

Fitzpatrick E L 1989 in *IAU Symp. no. 135, Interstellar Dust*, eds L J Allamandola and A G G M Tielens (Kluwer, Dordrecht) p 37

Fitzpatrick E L and Massa D 1986 *Astrophys. J.* **307** 286

Fitzpatrick E L and Massa D 1988 *Astrophys. J.* **328** 734

Forrest W J McCarthy J F and Houck J R 1979 *Astrophys. J.* **233** 611

Frenklach M and Feigelson E D 1989 *Astrophys. J.* **341** 372

Frenklach M Carmer C S and Feigelson E D 1989 *Nature* **339** 196

Frerking M A Langer W D and Wilson R W 1982 *Astrophys. J.* **262** 590

Gail H P and Sedlmayr E 1984 *Astron. Astrophys.* **132** 163

Gail H P and Sedlmayr E 1986 *Astron. Astrophys.* **166** 225

Gail H P and Sedlmayr E 1987 in *Physical Processes in Interstellar Clouds*, eds G E Morfill and M Scholer (Reidel, Dordrecht) p 275

Gatley I 1984 in *Laboratory and Observational Infrared Spectra of Interstellar Dust*, eds R D Wolstencroft and J M Greenberg (Royal Observatory Edinburgh) p 118

Gaustad J E 1971 in *Dark Nebulae, Globules and Protostars*, ed. B T Lynds (University of Arizona Press, Tucson) p 91

Geballe T R 1986 *Astron. Astrophys.* **162** 248

Geballe T R Kim Y H Knacke R F and Noll K S 1988 *Astrophys. J. Lett.* **326** L65

Geballe T R Noll K S Whittet D C B and Waters L B F M 1989a *Astrophys. J. Lett.* **340** L29

Geballe T R Tielens A G G M Allamandola L J Moorhouse A and Brand P W J L 1989b *Astrophys. J.* **341** 278

Gehrels T 1974 *Astron. J.* **79** 590

Gehrz R D 1988 *Ann. Rev. Astron. Astrophys.* **26** 377

Gehrz R D 1989 in *IAU Symp. no. 135, Interstellar Dust*, eds L J Allamandola and A G G M Tielens (Kluwer, Dordrecht) p 445

Gehrz R D and Ney E P 1990 *Proc. Natl. Acad. Sci. USA* **87** 4354

Gehrz R D Ney E P Grasdalen G L Hackwell J A and Thronson H A 1984 *Astrophys. J.* **281** 303

Gehrz R D Grasdalen G L Greenhouse M Hackwell J A Heyward T and Bentley A F 1986 *Astrophys. J. Lett.* **308** L63

Gillett F C 1986 in *Light on Dark Matter*, ed. F P Israel (Reidel, Dordrecht) p 61

Gillett F C Forrest W J and Merrill K M 1973 *Astrophys. J.* **183** 87

Gillett F C Forrest W J Merrill K M Capps R W and Soifer B T 1975a *Astrophys. J.* **200** 609

Gillett F C Jones T W Merrill K M and Stein W A 1975b *Astron. Astrophys.* **45** 77

Gilman R C 1969 *Astrophys. J. Lett.* **155** L185

Glasse A C H Towlson W A Aitken D K and Roche P F 1986 *Mon. Not. R. Astron. Soc.* **220** 185

Goebel J H and Moseley S H 1985 *Astrophys. J. Lett.* **290** L35

Gold T 1952 *Mon. Not. R. Astron. Soc.* **112** 215

Gottlieb D M 1978 *Astrophys. J. Suppl.* **38** 287

Greenberg J M 1968 in *Nebulae and Interstellar Matter*, eds B M Middlehurst and L H Aller (University of Chicago Press) p 221

Greenberg J M 1971 *Astron. Astrophys.* **12** 240

Greenberg J M 1973 in *IAU Symp. no. 52, Interstellar Dust and Related Topics*, eds J M Greenberg and H C van de Hulst (Reidel, Dordrecht) p 3

Greenberg J M 1978 in *Cosmic Dust*, ed. J A M McDonnell (John Wiley & Sons, New York) p 187

Greenberg J M 1982 in *Comets*, ed. L L Wilkening (University of Arizona Press, Tucson) p 131

Greenberg J M 1984 *The Observatory* **104** 134

Greenberg J M 1989 in *IAU Symp. no. 135, Interstellar Dust*, eds L J Allamandola and A G G M Tielens (Kluwer, Dordrecht) p 345

Greenberg J M and Chlewicki G 1983 *Astrophys. J.* **272** 563

Greenberg J M and Hage J I 1990 *Astrophys. J.* **361** 260

Greenberg J M and Hong S S 1974 in *IAU Symp. no. 60, Galactic Radio Astronomy*, eds F J Kerr and S C Simonson (Reidel, Dordrecht) p 155

Greenberg J M and Hong S S 1975 in *The Dusty Universe*, eds G B Field and A G W Cameron (Neale Watson Academic Publications, New York) p 131

Greenberg J M and Meltzer A S 1960 *Mon. Not. R. Astron. Soc.* **132** 667

Greenberg J M and Shah G A 1971 *Astron. Astrophys.* **12** 250

Greenstein J L 1981 *Astrophys. J.* **245** 24

Grim R J A and d'Hendecourt L B 1986 *Astron. Astrophys.* **167** 161

Hall J S 1949 *Science* **109** 166

Hall J S and Serkowski K 1963 in *Basic Astronomical Data*, ed. K Aa Strand (University of Chicago Press) p 293

Harris D H 1973 in *IAU Symp. no. 52, Interstellar Dust and Related Topics*, eds J M Greenberg and H C van de Hulst (Reidel, Dordrecht) p 31

Harris A W Gry C and Bromage G E 1984 *Astrophys. J.* **284** 157

Harwit M 1970 *Nature* **226** 61

Hauser M G Silverberg R F Stier M T Kelsall T Gezari D Y Dwek E Walser D Mather J C and Cheung L H 1984 *Astrophys. J.* **285** 74

Hecht J H 1986 *Astrophys. J.* **305** 817

Hecht J H Helfer H L Wolf J Donn B and Pipher J L 1982 *Astrophys. J. Lett.* **263** L39

Hecht J H Holm A V Donn B and Wu C C 1984 *Astrophys. J.* **280** 228

Heger M L 1922 *Lick Obs. Bull.* **10** 146

Heiles C 1987 in *Interstellar Processes*, eds D J Hollenbach and H A Thronson (Reidel, Dordrecht) p 171

Helou G 1989 in *IAU Symp. no. 135, Interstellar Dust*, eds L J Allamandola and A G G M Tielens (Kluwer, Dordrecht) p 285

Henderson-Sellars A Benlow A and Meadows A J 1980 *Quart. J. R. Astron. Soc.* **21** 74

Herbig G H 1970 *Mém. Soc. R. Sci. Liège* **19** 13

Herbig G H 1975 *Astrophys. J.* **196** 129

Herbig G H 1988 *Astrophys. J.* **331** 999

Herbig G H and Soderblom D R 1982 *Astrophys. J.* **252** 610

Herbst E 1987 in *Interstellar Processes*, eds D J Hollenbach and H A Thronson (Reidel, Dordrecht) p 611

Herbst E and Leung C M 1986 *Mon. Not. R. Astron. Soc.* **222** 689

Herzberg G 1955 *Mém. Soc. R. Sci. Liège* **15** 291

Herzberg G 1967 in *IAU Symp. no. 31, Radio Astronomy and the Galactic System*, ed. H van Woerden (Academic Press, New York) p 91

Hildebrand R H 1983 *Quart. J. R. Astron. Soc.* **24** 267

Hildebrand R H 1988 *Quart. J. R. Astron. Soc.* **29** 327

Hildebrand R H Dragovan M and Novak G 1984 *Astrophys. J. Lett.* **284** L51

Hilditch R W Hill G and Barnes J V 1983 *Mon. Not. R. Astron. Soc.* **204** 241

Hiltner W A 1949 *Science* **109** 165

Hobbs L M York D G and Oegerle W 1982 *Astrophys. J. Lett.* **252** L21

Hobbs L M Blitz L and Magnani L 1986 *Astrophys. J. Lett.* **306** L109

Hodge P W 1981 *Interplanetary Dust* (Gordon and Breach, New York) p 201

Hollenbach D and Salpeter E E 1971 *Astrophys. J.* **163** 155

Hollenbach D and Thronson H A (eds) 1987 *Interstellar Processes* (Reidel, Dordrecht)

Hong S S and Greenberg J M 1980 *Astron. Astrophys.* **88** 194

Hough J H Bailey J A Rouse M F and Whittet D C B 1987 *Mon. Not. R. Astron. Soc.* **227** 1P

Hough J H Sato S Tamura M Yamashita T McFadzean A D Rouse M F Whittet D C B Kaifu N Suzuki H Nagata T Gatley I and Bailey J A 1988 *Mon. Not. R. Astron. Soc.* **230** 107

Hough J H Whittet D C B Sato S Yamashita T Tamura M Nagata T Aitken D K and Roche P F 1989 *Mon. Not. R. Astron. Soc.* **241** 71

Hough J H Peacock T and Bailey J A 1991 *Mon. Not. R. Astron. Soc.* **248** 74

Hoyle F and Wickramasinghe N C 1962 *Mon. Not. R. Astron. Soc.* **124** 417

Hoyle F and Wickramasinghe N C 1986 *Quart. J. R. Astron. Soc.* **27** 21

Huffman D R 1970 *Astrophys. J.* **161** 1157

Huffman D R 1977 *Adv. Phys.* **26** 129

Huffman D R 1989 in *IAU Symp. no. 135, Interstellar Dust*, eds L J Allamandola and A G G M Tielens (Kluwer, Dordrecht) p 329

Huffman D R and Stapp J L 1971 *Nature Phys. Sci.* **229** 45

Issa M R MacLaren I and Wolfendale A W 1990 *Astron. Astrophys.* **236** 237

Jabbir N L Jabbar S R Salih S A H and Majeed Q S 1986 *Astrophys. Space Sci.* **123** 351

Jenkins E B 1989 in *IAU Symp. no. 135, Interstellar Dust*, eds L J Allamandola and A G G M Tielens (Kluwer, Dordrecht) p 23

Jenkins E B Jura M and Loewenstein M 1983 *Astrophys. J.* **270** 88

Johnson H L 1963 in *Basic Astronomical Data*, ed. K Aa Strand (University of Chicago Press) p 204

Johnson H L 1966 *Ann. Rev. Astron. Astrophys.* **4** 193

Johnson H L 1968 in *Nebulae and Interstellar Matter*, eds B M Middlehurst and L H Aller (University of Chicago Press) p 167

Johnson P E 1982 *Nature* **295** 371

Jones A P 1990 *Mon. Not. R. Astron. Soc.* **245** 331

Jones A P and Williams D A 1984 *Mon. Not. R. Astron. Soc.* **209** 955

Jones R V and Spitzer L 1967 *Astrophys. J.* **147** 943

Jones T J 1989a *Astron. J.* **98** 2062

Jones T J 1989b *Astrophys. J.* **346** 728

Jones T W and Merrill K M 1976 *Astrophys. J.* **209** 509

Jura M 1980 *Astrophys. J.* **235** 63

Jura M 1986 *Astrophys. J.* **303** 327

Jura M 1989 in *Evolution of Interstellar Dust and Related Topics*, eds A
    Bonetti J M Greenberg and S Aiello (North-Holland Publishing Com-
    pany, Amsterdam) p 143
Jura M and Kleinmann S G 1989 *Astrophys. J.* **341** 359
Kapteyn J C 1909 *Astrophys. J.* **29** 46
Kerr T H Adamson A J and Whittet D C B 1991 *Mon. Not. R. Astron.
    Soc.* **251**, 60P
Kerridge J F 1989 in *IAU Symp. no. 135, Interstellar Dust*, eds L J Alla-
    mandola and A G G M Tielens (Kluwer, Dordrecht) p 383
Kirsten T 1978 in *The Origin of the Solar System*, ed. S F Dermott (John
    Wiley & Sons, New York) p 267
Kissel J and Krueger F R 1987 *Nature* **326** 755
Kitta K and Krätschmer W 1983 *Astron. Astrophys.* **122** 105
Klare G Neckel Th and Schnur G 1972 *Astron. Astrophys. Suppl.* **5** 239
Knacke R F Cudaback D D and Gaustad J E 1969 *Astrophys. J.* **158** 151
Knacke R F Puetter R C Erickson E and McCorkle S 1985 *Astron. J.* **90**
    1828
Knuth D E 1986 *The TEXbook* (Addison-Wesley Publishing Company,
    Reading, Massachusetts)
Koike C Hasegawa H and Hattori T 1987 *Astrophys. Space Sci.* **134** 95
Koornneef J 1982 *Astron. Astrophys.* **107** 247
Kraft R P 1979 *Ann. Rev. Astron. Astrophys.* **17** 309
Krätschmer W and Huffman D R 1979 *Astrophys. Space Sci.* **61** 195
Krätschmer W Lamb L D Fostiropoulos K and Huffman D R 1990 *Nature*
    347 354
Krelowski J and Walker G A H 1987 *Astrophys. J.* **312** 860
Kuijken K and Gilmore G 1989 *Mon. Not. R. Astron. Soc.* **239** 651
Kulkarni S R and Heiles C 1987 in *Interstellar Processes*, eds D J Hollen-
    bach and H A Thronson (Reidel, Dordrecht) p 87
Kutner M L 1987 *Astronomy, a Physical Perspective* (Harper & Row, New
    York)
Kvenvolden K Lawless J Pering K Peterson E Flores J Ponnamperuma C
    Kaplan I R and Moore C 1970 *Nature* **228** 923
Lacy J H Baas F Allamandola L J Persson S E McGregor P J Lonsdale C
    J Geballe T R and van de Bult C E P 1984 *Astrophys. J.* **276** 533
Lada C J 1988 in *Galactic and Extragalactic Star Formation*, eds R E
    Pudritz and M Fich (Kluwer, Dordrecht) p 5
Langevin Y Kissel J Bertaux J L and Chassefiere E 1987 *Astron. Astrophys.*
    187 761
Larson R B 1989 in *Structure and Dynamics of the Interstellar Medium*,
    eds G Tenorio-Tagle M Moles and J Melnick (Springer-Verlag, Berlin)
    p 44
Larson H P Davis D S Black J H and Fink U 1985 *Astrophys. J.* **299** 873
Laureijs R J Mattila K and Schnur G 1987 *Astron. Astrophys.* **184** 269

Leach S 1991 in *Molecular clouds*, eds R A James and T J Millar (Cambridge University Press)

Le Bertre T 1987 *Astron. Astrophys.* **176** 107

Léger A 1983 *Astron. Astrophys.* **123** 271

Léger A and d'Hendecourt L 1985 *Astron. Astrophys.* **146** 81

Léger A and d'Hendecourt L 1988 in *Dust in the Universe*, eds M E Bailey and D A Williams (Cambridge University Press) p 219

Léger A and Puget J L 1984 *Astron. Astrophys.* **137** L5

Léger A Gauthier S Defourneau D and Rouan D 1983 *Astron. Astrophys.* **117** 164

Léger A Jura M and Omont A 1985 *Astron. Astrophys.* **144** 147

Léger A Verstraete L d'Hendecourt L Défourneau D Dutuit O Schmidt W and Lauer J C 1989 in *IAU Symp. no. 135, Interstellar Dust*, eds L J Allamandola and A G G M Tielens (Kluwer, Dordrecht) p 173

Leung C M 1975 *Astrophys. J.* **199** 340

Leung C M Herbst E and Huebner W F 1984 *Astrophys. J. Suppl.* **56** 231

Lewis R S Tang M Wacker J F Anders E and Steel E 1987 *Nature* **326** 160

Lewis R S Anders E and Draine B T 1989 *Nature* **339** 117

Lillie C F and Witt A N 1976 *Astrophys. J.* **208** 64

Lindblad B 1935 *Nature* **135** 133

Little-Marenin I R 1986 *Astrophys. J. Lett.* **307** L15

Longo R Stalio R Polidan R S and Rossi L 1989 *Astrophys. J.* **339** 474

Low F J Beintema D A Gautier T N Gillett F C Beichman C A Neugebauer G Young E Aumann H H Boggess N Emerson J P Habing H J Hauser M G Houck J R Rowan-Robinson M Soifer B T Walker R G and Wesselius P R 1984 *Astrophys. J. Lett.* **278** L19

Lucke P B 1978 *Astron. Astrophys.* **64** 367

MacLean S Duley W W and Miller T J 1982 *Astrophys. J. Lett.* **256** L61

Magalhães A M Coyne G V Piirola V and Rodrigues C V 1989 in *Interstellar Dust*, eds L J Allamandola and A G G M Tielens (NASA Conference Publication 3036) p 347

Mann A P C and Williams D A 1980 *Nature* **283** 721

Manning P G 1976 *Chem. Soc. Rev.* **5** 233

Martin P G 1971 *Mon. Not. R. Astron. Soc.* **152** 279

Martin P G 1972 *Mon. Not. R. Astron. Soc.* **158** 63

Martin P G 1974 *Astrophys. J.* **187** 461

Martin P G 1978 *Cosmic Dust, its Impact on Astronomy* (Oxford University Press)

Martin P G 1989 in *IAU Symp. no. 135, Interstellar Dust*, eds L J Allamandola and A G G M Tielens (Kluwer, Dordrecht) p 55

Martin P G and Angel J R P 1974 *Astrophys. J.* **188** 517

Martin P G and Angel J R P 1975 *Astrophys. J.* **195** 379

Martin P G and Angel J R P 1976 *Astrophys. J.* **207** 126

Martin P G and Rogers C 1987 *Astrophys. J.* **322** 374

Martin P G and Shawl S J 1982 *Astrophys. J.* **253** 86

Martin P G and Whittet D C B 1990 *Astrophys. J.* **357** 113

Martin P G Adamson A J Whittet D C B Hough J H Bailey J A Kim S H Sato S Tamura M and Yamashita T 1992 *Astrophys. J.* in press

Massa D Savage B D and Fitzpatrick E L 1983 *Astrophys. J.* **266** 662

Mathewson D S and Ford V L 1970 *Mem. R. Astron. Soc.* **74** 139

Mathis J S 1979 *Astrophys. J.* **232** 747

Mathis J S 1986 *Astrophys. J.* **308** 281

Mathis J S and Whiffen G 1989 *Astrophys. J.* **341** 808

Mathis J S Rumpl W and Nordseik K H 1977 *Astrophys. J.* **217** 425

Mathis J S Mezger P G and Panagia N 1983 *Astron. Astrophys.* **128** 212

McCarthy J F Forrest W J Briotta D A and Houck J R 1980 *Astrophys. J.* **242** 965

McClure R D and Crawford D L 1971 *Astron. J.* **76** 31

McFadzean A D Whittet D C B Longmore A J Bode M F and Adamson A J 1989 *Mon. Not. R. Astron. Soc.* **241** 873

McKee C F 1989 in *IAU Symp. no. 135, Interstellar Dust*, eds L J Allamandola and A G G M Tielens (Kluwer, Dordrecht) p 431

McKee C F and Ostriker J P 1977 *Astrophys. J.* **218** 148

McLachlan A and Nandy K 1984 *The Observatory* **104** 29

McMillan R S 1976 *Astron. J.* **81** 970

McMillan R S 1978 *Astrophys. J.* **225** 880

McMillan R S and Tapia S 1977 *Astrophys. J.* **212** 714

Merrill P W 1934 *Publ. Astron. Soc. Pacific* **46** 206

Merrill P W 1936 *Astrophys. J.* **83** 126

Mestel L 1985 in *Protostars & Planets II*, eds D C Black and M S Matthews (University of Arizona Press, Tucson) p 320

Meyer D M 1983 *Astrophys. J. Lett.* **266** L51

Meyer D M and Savage B D 1981 *Astrophys. J.* **248** 545

Michalsky J J and Schuster G J 1979 *Astrophys. J.* **231** 73

Mie G 1908 *Ann. Phys., NY* **25** 377

Mihalas D and Binney J 1981 *Galactic Astronomy* (W H Freeman, and Company, San Francisco)

Millar T J 1982 *Mon. Not. R. Astron. Soc.* **200** 527

Miller S L and Urey H C 1959 *Science* **130** 245

Millman P M 1975 in *The Dusty Universe*, eds G B Field and A G W Cameron (Neale Watson Academic Publications, New York) p 185

Mitchell G F Ginsburg J L and Kuntz P J 1978 *Astrophys. J. Suppl.* **38** 39

Mitchell G F Allen M and Maillard J P 1988 *Astrophys. J. Lett.* **333** L55

Mizutani K Suto H and Maihara T 1989 *Astrophys. J.* **346** 675

Moore M H and Donn B 1982 *Astrophys. J. Lett.* **257** L47

Morfill G E and Scholer M (eds) 1987 *Physical Processes in Interstellar Clouds* (Reidel, Dordrecht)

Morgan D H Nandy K and Thompson G I 1978 *Mon. Not. R. Astron. Soc.* **185** 371

Nagata T 1990 *Astrophys. J. Lett.* **348** L13

Nandy K 1984 in *IAU Symp. no. 108, Structure and Evolution of the Magellanic Clouds*, eds S van den Bergh and K S de Boer (Reidel, Dordrecht) p 341

Nandy K and Thompson G I 1975 *Mon. Not. R. Astron. Soc.* **173** 237

Nandy K Thompson G I Jamar C Monfils A and Wilson R 1975 *Astron. Astrophys.* **44** 195

Nandy K Thompson G I Jamar C Monfils A and Wilson R 1976 *Astron. Astrophys.* **51** 63

Natta A and Panagia N 1981 *Astrophys. J.* **248** 189

Norman C A and Paresce F 1989 in *The Formation and Evolution of Planetary Systems*, eds H A Weaver and L Danly (Cambridge University Press) p 151

Nuth J A 1990 *Nature* **345** 207

Nuth J A Moseley S H Silverberg R F Goebel J H and Moore W J 1985 *Astrophys. J. Lett.* **290** L41

Nuth J A Donn B and Nelson R 1986 *Astrophys. J. Lett.* **310** L83

Ogmen M and Duley W W 1988 *Astrophys. J. Lett.* **334** L117

Omont A Moseley S H Forveille T Glaccum W J Harvey P M Likkel L Lowenstein R F and Lisse C M 1990 *Astrophys. J. Lett.* **355** L27

Onaka T de Jong T and Willems F J 1989 *Astron. Astrophys.* **218** 169

Oort J H 1932 *Bull. Astron. Inst. Netherlands* **6** 249

Oort J H and van de Hulst H C 1946 *Bull. Astron. Inst. Netherlands* **10** 187

Ostriker J P and Heisler J 1984 *Astrophys. J.* **278** 1

Pagel B E J 1987 in *The Galaxy*, eds G Gilmore and R Carswell (Reidel, Dordrecht) p 341

Pagel B E J and Edmunds M G 1981 *Ann. Rev. Astron. Astrophys.* **19** 77

Pajot F Gispert R Lamarre J M Peyturaux R Puget J L Serra G Coron N Dambier G Leblanc J Moalic J P Renault J C and Vitry R 1986 *Astron. Astrophys.* **154** 55

Papoular R and Pegourie B 1983 *Astron. Astrophys.* **128** 335

Peimbert M 1975 *Ann. Rev. Astron. Astrophys.* **13** 113

Pendleton Y J Tielens A G G M and Werner M W 1990 *Astrophys. J.* **349** 107

Perrin J M and Sivan J P 1990 *Astron. Astrophys.* **228** 238

Perry C L and Johnston L 1982 *Astrophys. J. Suppl.* **50** 451

Phillips A P Pettini M and Gondhalekar P M 1984 *Mon. Not. R. Astron. Soc.* **206** 337

Pipher J L 1973 in *IAU Symp. no. 52, Interstellar Dust and Related Topics*, eds J M Greenberg and H C van de Hulst (Reidel, Dordrecht) p 559

Platt J R 1956 *Astrophys. J.* **123** 486

Pottasch S R Baud B Beintema D Emerson J Habing H J Harris S Houck J Jennings R and Marsden P 1984 *Astron. Astrophys.* **138** 10

Prasad S S and Huntress W T 1980 *Astrophys. J. Suppl.* **43** 1

Prévot M L Lequeux J Maurice E Prévot L and Rocca-Volmerange B 1984 *Astron. Astrophys.* **132** 389

Purcell E M 1969 *Astrophys. J.* **158** 433

Purcell E M 1975 in *The Dusty Universe*, eds G B Field and A G W Cameron (Neale Watson Academic Publications, New York) p 155

Purcell E M 1976 *Astrophys. J.* **206** 685

Purcell E M 1979 *Astrophys. J.* **231** 404

Purcell E M and Shapiro P R 1977 *Astrophys. J.* **214** 92

Purcell E M and Spitzer L 1971 *Astrophys. J.* **167** 31

Rebane K K 1970 *Impurity Spectra of Solids* (Plenum, New York)

Reid M J 1989 in *IAU Symposium 136, The Center of the Galaxy*, ed. M Morris (Reidel, Dordrecht) p 37

Rice W Boulanger F Viallefond F Soifer B T and Freedman W L 1990 *Astrophys. J.* **358** 418

Robertson J and O'Reilly E P 1987 *Phys. Rev.* B **35** 2946

Roche P F 1988 in *Dust in the Universe*, eds M E Bailey and D A Williams (Cambridge University Press) p 415

Roche P F 1989a in *IAU Symposium 131, Planetary Nebulae*, ed. S Torres-Peimbert (Reidel, Dordrecht) p 117

Roche P F 1989b in *22nd ESLAB Symposium, Infrared Spectroscopy in Astronomy*, ed. B H Kaldeich (ESA publication SP-290) p 79

Roche P F 1989c in *IAU Symp. no. 135, Interstellar Dust*, eds L J Allamandola and A G G M Tielens (Kluwer, Dordrecht) p 303

Roche P F and Aitken D K 1984a *Mon. Not. R. Astron. Soc.* **208** 481

Roche P F and Aitken D K 1984b *Mon. Not. R. Astron. Soc.* **209** 33P

Roche P F and Aitken D K 1985 *Mon. Not. R. Astron. Soc.* **215** 425

Roche P F Aitken D K Smith C H and James S D 1986a *Mon. Not. R. Astron. Soc.* **218** 19P

Roche P F Allen D A and Bailey J A 1986b *Mon. Not. R. Astron. Soc.* **220** 7P

Roche P F Aitken D K and Smith C H 1987 *Mon. Not. R. Astron. Soc.* **228** 269

Roche P F Aitken D K and Smith C H 1989a *Mon. Not. R. Astron. Soc.* **236** 485

Roche P F Aitken D K Smith C H and James S D 1989b *Nature* **337** 533

Roche P F Aitken D K Smith C H and Ward M J 1991 *Mon. Not. R. Astron. Soc.* **248** 606

Rogers C and Martin P G 1979 *Astrophys. J.* **228** 450

Rowan-Robinson M 1986 *Mon. Not. R. Astron. Soc.* **219** 737

Rowan-Robinson M Lock T D Walker D W and Harris S 1986 *Mon. Not. R. Astron. Soc.* **222** 273

Rudkjøbing M 1969 *Astrophys. Space Sci.* **3** 102

Russell R W Soifer B T and Willner S P 1978 *Astrophys. J.* **220** 568

Ryter C Cesarsky C J Audouze J 1975 *Astrophys. J.* **198** 103

Sagan C 1973 in *Molecules in the Galactic Environment*, eds M A Gordon and L E Snyder (John Wiley & Sons, New York) p 451

Sagan C and Khare B N 1979 *Nature* **277** 102

Sakata A Nakagawa N Iguchi T Isobe S Morimoto M Hoyle F and Wickramasinghe N C 1977 *Nature* **266** 241

Sakata A Wada S Okutsu Y Shintani H and Nakata Y 1983 *Nature* **301** 493

Salpeter E E 1955 *Astrophys. J.* **121** 161

Salpeter E E 1974 *Astrophys. J.* **193** 579

Salpeter E E 1977 *Ann. Rev. Astron. Astrophys.* **15** 267

Sandage A 1976 *Astron. J.* **81** 954

Sandford S A 1989 in *IAU Symp. no. 135, Interstellar Dust*, eds L J Allamandola and A G G M Tielens (Kluwer, Dordrecht) p 403

Sandford S A and Allamandola L J 1990 *Astrophys. J.* **355** 357

Sandford S A and Walker R M 1985 *Astrophys. J.* **291** 838

Sandford S A Allamandola L J Tielens A G G M Valero G J 1988 *Astrophys. J.* **329** 498

Sandford S A Allamandola L J Tielens A G G M Sellgren K Tapia M and Pendleton Y 1991 *Astrophys. J.* **371** 607

Sanner F Snell R and Vanden Bout P 1978 *Astrophys. J.* **226** 460

Sargent A I and Beckwith S 1987 *Astrophys. J.* **323** 294

Sato S Nagata T Tanaka M and Yamamoto T 1990 *Astrophys. J.* **359** 192

Savage B D 1975 *Astrophys. J.* **199** 92

Savage B D 1976 *Astrophys. J.* **205** 122

Savage B D and Mathis J S 1979 *Ann. Rev. Astron. Astrophys.* **17** 73

Savage B D and Sitko M L 1984 *Astrophys. Space Sci.* **100** 427

Savage B D Bohlin R C Drake J F and Budich W 1977 *Astrophys. J.* **216** 291

Schalen C 1936 *Medd. Uppsala Astron. Obs.* **64** 1

Schlessinger G and Miller S L 1983 *J. Mol. Evol.* **19** 376

Schmitt B Greenberg J M and Grim R J A 1989 *Astrophys. J. Lett.* **340** L33

Schulz A Lenzen R Schmidt Th and Protel K 1981 *Astron. Astrophys.* **95** 94

Schutte W A and Greenberg J M 1988 in *Dust in the Universe*, eds M E Bailey and D A Williams (Cambridge University Press) p 403

Schutte W A Tielens A G G M Allamandola L J Cohen M and Wooden D H 1990 *Astrophys. J.* **360** 577

Schwartz R D 1983 *Ann. Rev. Astron. Astrophys.* **21** 209

Seab C G 1988 in *Dust in the Universe*, eds M E Bailey and D A Williams (Cambridge University Press) p 303

Seab C G and Shull J M 1983 *Astrophys. J.* **275** 652

Seab C G and Snow T P 1984 *Astrophys. J.* **277** 200

Seab C G and Snow T P 1985 *Astrophys. J.* **295** 485

Seab C G and Snow T P 1989 *Astrophys. J.* **347** 479

Seab C G Snow T P and Joseph C L 1981 *Astrophys. J.* **246** 788

Seaton M J 1979 *Mon. Not. R. Astron. Soc.* **187** 73P

Seddon H 1969 *Nature* **222** 757

Seeley D and Berendzen R 1972a *J. Hist. Astron.* **3** 52

Seeley D and Berendzen R 1972b *J. Hist. Astron.* **3** 75

Sellgren K 1984 *Astrophys. J.* **277** 623

Serkowski K 1973 in *IAU Symp. no. 52, Interstellar Dust and Related Topics*, eds J M Greenberg and H C van de Hulst (Reidel, Dordrecht) p 145

Serkowski K Mathewson D S and Ford V L 1975 *Astrophys. J.* **196** 261

Shapiro P R and Holcomb K A 1986a *Astrophys. J.* **305** 433

Shapiro P R and Holcomb K A 1986b *Astrophys. J.* **310** 872

Shaver P A McGee R X Newton L M Danks A C and Pottasch S R 1983 *Mon. Not. R. Astron. Soc.* **204** 53

Shock E L and Schulte M D 1990 *Nature* **343** 728

Shu F H 1982 *The Physical Universe, an Introduction to Astronomy* (Oxford University Press)

Shu F H Adams F C and Lizano S 1989 in *Evolution of Interstellar Dust and Related Topics*, eds A Bonetti J M Greenberg and S Aiello (North-Holland Publishing Company, Amsterdam) p 213

Simpson J P and Rubin R H 1989 in *Interstellar Dust*, eds L J Allamandola and A G G M Tielens (NASA Conference Publication 3036) p 523

Sitko M L Savage B D and Meade M R 1981 *Astrophys. J.* **246** 161

Skrutskie M F Dutkevitch D Strom S E Edwards S and Strom K M 1990 *Astron. J.* **99** 1187

Slipher V M 1912 *Lowell Obs. Bull.* **11** 26

Smith B A and Terrile R J 1984 *Science* **226** 421

Smith R G Sellgren K and Tokunaga A T 1988 *Astrophys. J.* **334** 209

Smith R G Sellgren K and Tokunaga A T 1989 *Astrophys. J.* **344** 413

Smith W H Snow T P and York D G 1977 *Astrophys. J.* **218** 124

Smith W H Snow T P Jura M and Cochran W D 1981 *Astrophys. J.* **248** 128

Sneden C Gehrz R D Hackwell J A York D G and Snow T P 1978 *Astron. J.* **223** 168

Snell R L and Vanden Bout P A 1981 *Astrophys. J.* **244** 844

Snijders M A J Batt T J Roche P F Seaton M J Morton D C Spoelstra T A T and Blades J C 1987 *Mon. Not. R. Astron. Soc.* **228** 329

Snow T P York D G and Welty D E 1977 *Astron. J.* **82** 113

Snow T P Timothy J G and Saar S 1982 *Astrophys. J.* **262** 611

Snow T P Buss R H Gilra D P and Swings J P 1987 *Astrophys. J.* **321** 921

Sodroski T J Dwek E Hauser M G and Kerr F J 1987 *Astrophys. J.* **322** 101

Sodroski T J Dwek E Hauser M G and Kerr F J 1989 *Astrophys. J.* **336** 762

Sofue Y Fujimoto M and Wielebinski R 1986 *Ann. Rev. Astron. Astrophys.* **24** 459

Soifer B T Willner S P Capps R W and Rudy R J 1981 *Astrophys. J.* **250** 631

Somerville W B 1988 *Mon. Not. R. Astron. Soc.* **234** 655

Somerville W B 1989 in *Interstellar Dust*, eds L J Allamandola and A G G M Tielens (NASA Conference Publication 3036) p 77

Sopka R J Hildebrand R H Jaffe D T Gatley I Roellig T Werner M Jura M and Zuckerman B 1985 *Astrophys. J.* **294** 242

Sorrell W H 1990 *Mon. Not. R. Astron. Soc.* **243** 570

Sorrell W H 1991 *Mon. Not. R. Astron. Soc.* **248** 439

Spitzer L 1978 *Physical Processes in the Interstellar Medium* (John Wiley & Sons, New York)

Spitzer L 1985 *Astrophys. J. Lett.* **290** L21

Stebbins J Huffer C M and Whitford A E 1939 *Astrophys. J.* **90** 209

Stecher T P 1965 *Astrophys. J.* **142** 1683

Stecher T P and Donn B 1965 *Astrophys. J.* **142** 1681

Steel T M and Duley W W 1987 *Astrophys. J.* **315** 337

Stephens J R 1980 *Astrophys. J.* **237** 450

Strom S E Edwards S and Strom K M 1989 in *The Formation and Evolution of Planetary Systems*, eds H A Weaver and L Danly (Cambridge University Press) p 91

Strömgren B 1966 *Ann. Rev. Astron. Astrophys.* **4** 433

Strömgren B 1987 in *The Galaxy*, eds G Gilmore and R Carswell (Reidel, Dordrecht) p 229

Struve F G W 1847 *Etudes d'Astronomie Stellaire*

Tammann G A 1974 in *Supernovae and Supernova Remnants*, ed. C B Cosmovici (Reidel, Dordrecht) p 155

Tanabé T Nakada Y Kamijo F and Sakata A 1983 *Publ. Astron. Soc. Japan* **35** 397

Tanaka M Sato S Nagata T and Yamamoto T 1990 *Astrophys. J.* **352** 724

Tang M and Anders E 1988 *Astrophys. J. Lett.* **335** L31

Tang M Anders E Hoppe P and Zinner E 1989 *Nature* **339** 351

Tapia M Persi P Roth M and Ferrari-Toniolo M 1989 *Astron. Astrophys.* **225** 488

Tayler R J 1975 *The Origin of the Chemical Elements* (Wykeham Publications, London)

Terebey S and Fich M 1986 *Astrophys. J. Lett.* **309** L73

Thronson H A 1988 in *Galactic and Extragalactic Star Formation*, eds R E Pudritz and M Fich (Kluwer, Dordrecht) p 621

Thronson H A Latter W B Black J H Bally J and Hacking P 1987 *Astrophys. J.* **322** 770

Tielens A G G M 1983 *Astron. Astrophys.* **119** 177

Tielens A G G M and Allamandola L J 1987a in *Interstellar Processes*, eds D J Hollenbach and H A Thronson (Reidel, Dordrecht) p 397

Tielens A G G M and Allamandola L J 1987b in *Physical Processes in Interstellar Clouds*, eds G E Morfill and M Scholer (Reidel, Dordrecht) p 333

Tielens A G G M and Hagen W 1982 *Astron. Astrophys.* **114** 245

Tielens A G G M Allamandola L J Bregman J Goebel J d'Hendecourt L B and Witteborn F C 1984 *Astrophys. J.* **287** 697

Tielens A G G M Seab C G Hollenbach D J and McKee C F 1987 *Astrophys. J.* **319** L109

Toller G N 1981 *Thesis* (State University of New York at Stony Brook)

Treffers R R and Cohen M 1974 *Astrophys. J.* **188** 545

Trimble V 1975 *Rev. Mod. Phys.* **47** 877

Trimble V 1991 *Astron. Astrophys. Rev.* **3** 1

Trumpler R J 1930a *Lick Obs. Bull.* **14** 154

Trumpler R J 1930b *Publ. Astron. Soc. Pacific* **42** 214

Trumpler R J 1930c *Publ. Astron. Soc. Pacific* **42** 267

Turon P and Mennessier M O 1975 *Astron. Astrophys.* **44** 209

Turner B E 1990 *Astrophys. J. Lett.* **362** L29

Twarog B A 1980 *Astrophys. J.* **242** 242

Unsöld A 1964 *Publ. R. Obs. Edinburgh* **4** 35

Valentijn E A 1990 *Nature* **346** 153

van Breda I G and Whittet D C B 1981 *Mon. Not. R. Astron. Soc.* **195** 79

van de Bult C E P M Greenberg J M and Whittet D C B 1985 *Mon. Not. R. Astron. Soc.* **214** 289

van de Hulst H C 1946 *Rech. Astron. Obs. Utrecht* **11** 1

van de Hulst H C 1957 *Light Scattering by Small Particles* (John Wiley & Sons, New York)

van de Hulst H C 1989 in *Evolution of Interstellar Dust and Related Topics*, eds A Bonetti J M Greenberg and S Aiello (North-Holland Publishing Company, Amsterdam) p 1

van der Zwet G P and Allamandola L J 1985 *Astron. Astrophys.* **146** 76

van Dishoeck E F and Black J H 1987 in *Physical Processes in Interstellar Clouds*, eds G E Morfill and M Scholer (Reidel, Dordrecht) p 241

van Dishoeck E F Blake G A Draine B T and Lunine J I 1991 in *Protostars & Planets III*, eds E H Levy J I Lunine and M S Matthews (University of Arizona Press, Tucson)

Vrba F J Strom S E and Strom K M 1976 *Astron. J.* **81** 958

Vrba F J Coyne G V and Tapia S 1981 *Astrophys. J.* **243** 489

Wagner J and Lautenschlager P 1986 *J. Appl. Phys.* **59** 2044

Wainscoat R J de Jong T and Wesselius P R 1987 *Astron. Astrophys.* **181** 225

Walker G A H Yang S Fahlman G G and Witt A N 1980 *Publ. Astron. Soc. Pacific* **92** 411

Walker H J and Wolstencroft R D 1988 *Publ. Astron. Soc. Pacific* **100** 1509

Wannier P G 1989 in *IAU Symposium 136, The Center of the Galaxy*, ed. M Morris (Reidel, Dordrecht) p 107

Wannier P G Sahai R Andersson B G and Johnson H R 1990 *Astrophys. J.* **358** 251

Watanabe I Hasegawa S and Kurata Y 1982 *Japanese J. Appl. Phys.* **21** 856

Waters L B F M Lamers H J G L M Snow T P Mathlener E Trams N R van Hoof P A M Waelkens C Seab C G and Stanga R 1989 *Astron. Astrophys.* **211** 208

Watson W D 1976 *Rev. Mod. Phys.* **48** 513

Weidenschilling S J Donn B and Meakin P 1989 in *The Formation and Evolution of Planetary Systems*, eds H A Weaver and L Danly (Cambridge University Press) p 131

Weiland J L Blitz L Dwek E Hauser M G Magnani L and Rickard L J 1986 *Astrophys. J. Lett.* **306** L101

Welter G L and Savage B D 1977 *Astrophys. J.* **215** 788

Westerlund B E and Krelowski J 1988 *Astron. Astrophys.* **189** 221

Wetherill G W 1989 in *The Formation and Evolution of Planetary Systems*, eds H A Weaver and L Danly (Cambridge University Press) p 1

Whipple F L 1987 *Astron. Astrophys.* **187** 852

Whiteoak J B 1966 *Astrophys. J.* **144** 305

Whitford A E 1958 *Astron. J.* **63** 201

Whittet D C B 1977 *Mon. Not. R. Astron. Soc.* **180** 29

Whittet D C B 1979 *Astron. Astrophys.* **72** 370

Whittet D C B 1984a *Mon. Not. R. Astron. Soc.* **210** 479

Whittet D C B 1984b *The Observatory* **104** 159

Whittet D C B 1987 *Quart. J. R. Astron. Soc.* **28** 303

Whittet D C B 1988 in *Dust in the Universe*, eds M E Bailey and D A Williams (Cambridge University Press) p 25

Whittet D C B 1989 in *IAU Symp. no. 135, Interstellar Dust*, eds L J Allamandola and A G G M Tielens (Kluwer, Dordrecht) p 455

Whittet D C B and Blades J C 1980 *Mon. Not. R. Astron. Soc.* **191** 309

Whittet D C B and Duley W W 1991 *Astron. Astrophys. Rev.* **2** 167

Whittet D C B and van Breda I G 1978 *Astron. Astrophys.* **66** 57

Whittet D C B and Walker H J 1991 *Mon. Not. R. Astron. Soc.* **252** 63

Whittet D C B van Breda I G and Glass I S 1976 *Mon. Not. R. Astron. Soc.* **177** 625

Whittet D C B Bode M F Longmore A J Baines D W T and Evans A 1983 *Nature* **303** 218

Whittet D C B McFadzean A D and Geballe T R 1984 *Mon. Not. R. Astron. Soc.* **211** 29P

Whittet D C B Longmore A J and McFadzean A D 1985 *Mon. Not. R. Astron. Soc.* **216** 45P

Whittet D C B Hough J H Bailey J A Rouse M F and Kirrane T M 1986 in *Light on Dark Matter*, ed. F P Israel (Reidel, Dordrecht) p 197

Whittet D C B Bode M F Longmore A J Adamson A J McFadzean A D Aitken D K and Roche P F 1988 *Mon. Not. R. Astron. Soc.* **233** 321

Whittet D C B Adamson A J Duley W W Geballe T R and McFadzean A D 1989 *Mon. Not. R. Astron. Soc.* **241** 707

Whittet D C B Duley W W and Martin P G 1990 *Mon. Not. R. Astron. Soc.* **244** 427

Whittet D C B Prusti T and Wesselius P R 1991 *Mon. Not. R. Astron. Soc.* **249** 319

Whittet D C B Martin P G Hough J H Rouse M F Bailey J A and Axon D J 1992 *Astrophys. J.* in press

Wickramasinghe N C and Nandy K 1971 *Mon. Not. R. Astron. Soc.* **153** 205

Wilking B A 1989 *Publ. Astron. Soc. Pacific* **101** 229

Wilking B A Lebofsky M J Martin P G Rieke G H and Kemp J C 1980 *Astrophys. J.* **235** 905

Wilking B A Lebofsky M J and Rieke G H 1982 *Astron. J.* **87** 695

Wilking B A Lada C J and Young E T 1989 *Astrophys. J.* **340** 823

Williams D A 1987 in *Physical Processes in Interstellar Clouds*, eds G E Morfill and M Scholer (Reidel, Dordrecht) p 377

Willner S P Puetter R C Russell R W and Soifer B T 1979 *Astrophys. Space Sci.* **65** 95

Willner S P Gillett F C Herter T L Jones B Krassner J Merrill K M Pipher J L Puetter R C Rudy R J Russell R W and Soifer B T 1982 *Astrophys. J.* **253** 174

Willstrop R V 1965 *Mem. R. Astron. Soc.* **69** 83

Wilson R 1960 *Mon. Not. R. Astron. Soc.* **120** 51

Witt A N 1988 in *Dust in the Universe*, eds M E Bailey and D A Williams (Cambridge University Press) p 1

Witt A N 1989 in *IAU Symp. no. 135, Interstellar Dust*, eds L J Allamandola and A G G M Tielens (Kluwer, Dordrecht) p 87

Witt A N and Boroson T A 1990 *Astrophys. J.* **355** 182

Witt A N and Schild R E 1988 *Astrophys. J.* **325** 837

Witt A N Bohlin R C and Stecher T P 1981 *Astrophys. J.* **244** 199

Witt A N Bohlin R C and Stecher T P 1983 *Astrophys. J. Lett.* **267** L47

Witt A N Bohlin R C and Stecher T P 1984 *Astrophys. J.* **279** 698

Witt A N Bohlin R C and Stecher T P 1986 *Astrophys. J. Lett.* **305** L23

Wollman E R Smith H A and Larson H P 1982 *Astrophys. J.* **258** 506

Wood J A *et al* 1986 in *Interrelationships among Circumstellar, Interstellar and Interplanetary Dust*, eds J A Nuth and R E Stencel (NASA Conference Publication 2403) p WG–33

Woolf N J 1973 in *IAU Symp. no. 52, Interstellar Dust and Related Topics*, eds J M Greenberg and H C van de Hulst (Reidel, Dordrecht) p 485

Wu C G 1972 *Astrophys. J.* **178** 681

Wu C G Gilra D P and van Duinen R J 1980 *Astrophys. J.* **241** 173

Wu C G York D G and Snow T P 1981 *Astron. J.* **86** 755

Wynn-Williams G C 1982 *Ann. Rev. Astron. Astrophys.* **20** 587

York D G 1971 *Astrophys. J.* **166** 65

York D G Drake J F Jenkins E B Morton D C Rogerson J B and Spitzer L 1973 *Astrophys. J. Lett.* **182** L1

Zaikowski A Knacke R F and Porco C C 1975 *Astrophys. Space Sci.* **35** 97

Index

Absorption, 57–61, 166–168, 244,
    261–262
  efficiency factor for, 59, 166–167,
    211
  in small-particle approximation, 61
Absorption edge, ultraviolet,
  in ices, 27
  in silicates and MgO, 128
Absorption features, 115–164
  infrared, 27–30, 144–164, 203–207,
    225–233
  optical, 129–144
  ultraviolet, 27, 29, 69, 74, 77, 80–82,
    116–129, 207–209
Abundances of the elements, 39–43
  calibration of, 40
  correlation of solar and meteoritic,
    41
  implications for grain composition,
    25–26, 53–56
  in carbonaceous chondrites, 39–41
  in globular clusters, 44
  in H II regions, 44–45, 47
  in OB stars, 44-45
  in red giants, 197
  primordial, 32, 34
  solar, 39–43
  spatial variations in, 46–47
  tabulation of, 42
  temporal variations in, 44–46
Accretion,
  effect on depletion, 54
  in protoplanetary disks, 243, 249
  timescale for, 223–224, 236
Acetylene (C_2H_2), circumstellar,
    199–200
Active galactic nuclei, 189
Adsorption, 216–219
  timescale for, 217

AFGL 961, infrared spectrum of, 163,
    229, 232
Albedo, 13, 61, 68–69, 82, 262
  definition of, 61
  observed spectral dependence of,
    68–69
Alignment,
  efficiency of, 85, 87, 110–111
  in dense molecular clouds, 112–113,
    179–180
  mechanisms for, 106–114, 179–180
  non-magnetic, 113–114
Ambipolar diffusion,
  and grain alignment, 114
  and star formation, 239
Amino acids, 253–255
  interstellar precursors of, 254
  meteoritic, 253–254
  synthesis of, 254–255
Ammonia (NH_3), interstellar,
  gas phase, 220–221
  in ice mantles, 26, 159, 224–225, 263
Andromeda Galaxy,
  polarization law in, 105
Asymmetry factor,
  definition of, 62
  observations of, 68–69

Barnard E E,
  photographic survey of dark
    nebulae, 5
Becklin–Neugebauer (BN) object,
  infrared absorption in, 181, 229
  polarization of IR features in, 95,
    97, 163–164
Big bang,
  element creation in, 32–34
Biological organisms, 255–256
  phosphorus abundance in, 30, 56

ultraviolet absorption of, 30
Bipolar outflow, 240–241
Buckminsterfullerine, 127–128

Carbon, amorphous,
    active surface sites on, 216
    aliphatic surface groups on, 153
    circumstellar origin of, 199–200
    extended red emission from,
        190–192
    graphitization of, 126, 236
    hydrogenation of, 153, 190–191
    infrared 3.4 μm absorption in,
        153–154
    infrared emissivity of, 170, 202–203
    interstellar origin of, 29, 260
    luminescence of, 190–191
    meteoritic, 248
    stochastic heating of, 173
    structure of, 28–29, 173
Carbon, elemental,
    abundance and depletion of, 26, 29,
        42, 49–52, 82, 124, 184, 252
    convective dredge-up of, 38–39
    nucleosynthesis of, 36–37
Carbon, graphitic, see Graphite
Carbon dioxide (CO_2),
    in Earth's atmosphere, 145, 253
    interstellar solid, 226, 231–233, 263
Carbon monoxide (CO),
    circumstellar, 196–199
    interstellar depletion of, 161,
        230–231
    interstellar gas phase, 24, 161, 179,
        220–222, 230–231
    interstellar solid, 160–161, 225–233
Carbonaceous chondrites, 39–41,
        245–256
    amino acids in, 253–254
    diamond in, 247–248
    element abundances in, 39–41, 55
    graphite in, 247–248
    isotopic anomalies in, 246–249
    organic matter in, 247–248, 253–255
    silicon carbide in, 247–249
    structure of, 245–246
Cauchy formula, 59
Centaurus A (NGC 5128),
    polarization law in, 105–107
Chemical elements, origin of, 32–39
Circumstellar dust (see also Dust
        grains, circumstellar),

around evolved stars, 194–215
    around young stars, 242–244
Cirrus, infrared, 13–14, 174
Clouds, interstellar,
    grain alignment in, 112–113,
        179–180
    depletion in, 50–53
    diffuse and dense, 20–21, 145–147
    distribution of, 9–11
    extinction curves in, 73–75, 78–80
    grain destruction in, 234–238
    grain growth in, 222–233
    infrared absorption in, 147–160
    infrared emission from, 168–174
    mass estimation from IR flux, 24,
        169–171, 179
    masses of, 21
    organic molecules in, 254
    phase structure of, 17–21
    polarization in, 104, 106
CNO cycle, 35
Colour-difference technique, 62–65,
        206–207
Colour excess, 7, 63, 80
    (see also Reddening, interstellar)
Comets, 250–252
Condensation temperature, 50–51,
        195–201
C/O ratio, stellar, 28, 46–47, 196–199,
        202–204
Corundum (Al_2O_3), 197, 204, 246, 261
Cosmic cycle, 32–34
Cosmic rays, 220, 230–231
Curve of growth, 48–49
Cygnus OB2 no. 12,
    infrared features in, 148–155
    polarization of, 95–96

Davis–Greenstein mechanism, 8,
        108–111, 171, 261
Debye temperature, 172
Depletion, interstellar elemental, 22,
        47–56, 235–236
    correlation with cloud velocity,
        235–236
    correlation with condensation
        temperature, 50–51
    correlation with density, 52–53,
        235–236
    fractional (definition), 47
    implications of, 26, 53–56
    index of (definition), 47

observations of, 47–53
pattern of, 50–51
Depolarization, 91
Diffuse galactic light (DGL), 13, 66–68
Diffuse interstellar bands, 129–144
  abundance limits on carrier, 138
  catalogue of, 130–131
  correlation with reddening, 133-135
  correlation with λ2175, 135
  families of, 134, 143–144
  gas phase mechanisms for, 142–143
  intercorrelation of, 133–134
  observed properties of, 129–138
  pairs of, 130
  polarization in, 132–133
  production efficiency of, 135
  profile variations in, 136–138
  relation to FUV extinction, 135
  ρ Leonis effect, 134–135
  solid state mechanisms for, 138–142
  standard star (HD 183143) for,
    130–132
Diffuse matter in the Solar System,
  245–256
Diffuse radiation from the galactic
  disk, 13–15, 174–179
Dilution factor, 167–168
Drude profile, 74, 117–119
Dust grains, circumstellar,
  composition of, 196–210, 241
  formation of, 195–201
  infrared emission from, 63, 201–203,
    239–241
  infrared spectra of, 203–207, 242
  in protoplanetary disks, 242–244
  in red giant atmospheres, 194–215
  radiation pressure on, 211–213,
    243–244
  size distribution of, 212–213,
    243–244
  timescale for injection into ISM, 215
  ultraviolet extinction of, 207–210
Dust grains, interplanetary, 55, 245,
  249–251, 254, 260–261
Dust grains, interstellar,
  alignment of, 8, 106–114, 179–180
  angular momentum of, 107–109,
    112, 114
  charge of, 113, 234
  circumstellar origin of, 210–215
  'classical', 14
  coagulation of, 222–223

composition of, 25–30, 258–263
definition of, 1, 7
density contribution of, 50, 72–73
destruction of, 54, 215, 233–238
dielectric properties of, 55, 83–85,
  95–99, 106–107
emissivity of, 24, 167–170
growth of, 222–224
heat capacity of, 172
magnetic susceptibility of, 109
mantles of, 29, 222–233, 259–260
models for, 25–30, 53–56, 80–82,
  257–263
shape of, 8, 83–85, 106–107,
  124–125, 170–171
significance of, 22–25
size of, 7, 14–15, 28, 58–61,
  69, 80–82, 97–99, 125–126,
  171–173, 222–223, 258–259
specific density of, 72–73, 150, 212
spin of, 107–109, 111–112
surface defects on, 216
surface reactions on, 23, 54,
  216–222, 224, 254
temperature of, 109–110, 166–168,
  171–173
Dust to gas ratio, circumstellar,
  213–214
Dust to gas ratio, interstellar,
  abundance limit on, 26, 42
  correlation with metallicity, 46–48
  estimate from depletions, 51
  estimate from extinction, 72–73
  estimate from FIR emission,
    178–179
  variation with mantle growth, 222

Elias 16, infrared spectrum of,
  158–160, 226, 229
Emissivity, far infrared, 167, 244
  spectral index of, 24, 167–170,
    201–203
Emissivity function, silicate, 181, 204
Environments, interstellar, 15–22
Evaporation from grain surfaces, 218,
  233
Extended red emission (ERE),
  190–192
Extinction, 5–13, 57–85, 206–210
  cross-section, 58
  circumstellar, 206–210
  efficiency factor for spheres, 58–60,

efficiency factors for non-spherical
  grains, 84–85
interstellar, 5–13, 65–82
neutral, 60, 71–72, 81
selective, 7 (see also Reddening,
  interstellar)
ratio of visual extinction to
  distance, 11
total, 58
visual, 6–8, 80
Extinction curve, circumstellar,
  206–210
Extinction curve, interstellar, 7,
  26–30, 63–82
broadband structure (VBS) in,
  69–70, 190
infrared, 77–80
mathematical representation of,
  70–71, 74, 76, 78–79
mean, 65–67
models for, 80–82
normalization of, 63, 78–79
spatial variations in, 73–80
tabulation of, 67
ultraviolet, 73–77

Ferromagnetic inclusions, 97, 112–113
Fischer–Tropsch reactions, 255

Galactic Centre,
  distance of, 11, 46
  extinction to, 1, 11
  infrared absorption features in,
    148–155
  infrared sources associated with,
    147–148
  metallicity of, 46
  PAH emission from, 189
Gas–grain collisions, 109, 217
Gas phase reactions, 219–222
Gold mechanism, 113
Gould's Belt, 10, 14, 21, 88, 104–105
Grain–grain collisions, 213, 222,
  234–237
Graphite,
  chemisputtering of, 237
  circumstellar limit on, 203
  contribution to grain models, 27–30,
    80–82, 259–260
  effect of impurities in, 126
  equilibrium temperature of, 168
  FIR emissivity of, 170, 202

meteoritic, 247–248
Mie theory calculations for, 80–82,
  124–127
origin of, 124, 236
plasma oscillations in, 124
structure of, 28–29, 124, 200
ultraviolet λ2175 feature in, 27, 80,
  124–127, 129, 260

Helium,
  abundance of, 32, 34
  nucleosynthesis of, 36
  origin of, 34, 36
Herbig–Haro objects, 242
Hydrocarbons, aliphatic, 153,
  159–160, 188, 260
Hydrocarbons, aromatic, see
  Polycyclic aromatic
  hydrocarbons
Hydrogen, elemental,
  nucleosynthesis of, 35
Hydrogen, interstellar,
  gas to reddening ratio, 15
  mean number density of, 16, 49–50
  molecular, 15, 17–21, 23, 112,
    218–219, 224
  phases of, 17–21
Hydrogen, ionized (H II) regions,
  abundances in, 44–45, 47
  chemisputtering in, 237
  dust temperature in, 168
  extinction in, 74
  FIR emission from, 177
  in starburst galaxies, 189
  PAH features in, 183–189
  polarization in, 104, 106
  silicate emission from, 180–182
  ultraviolet λ2175 absorption feature
    in, 119–120, 122

Ices,
  abundance of CO in, 230–231
  abundance of H_2O in, 157–158,
    224–225
  circumstellar, 198–199, 206–207,
    227, 242
  composition of, 26, 155, 224–225
  destruction of, 232–234
  infrared absorption in, 27, 95, 97,
    144–147, 154–164, 206–207,
    227–229, 233, 242
  interstellar origin of, 222–225

threshold effect for $\tau_{3.0}$ in, 157
ultraviolet absorption in, 27
ultraviolet photolysis of, 153,
    232–233
Impurities in solids, 126, 139, 216, 226
Infrared absorption features, 27–29,
    144–164, 203–207, 225–233
    due to C–H bonds, 29, 153–154,
        159, 161
    due to silicates, 27–28, 95, 97,
        147–153, 162, 164, 181–182,
        203–207
    due to solid CO, 147, 160–161,
        225–233
    due to water-ice, 27, 95, 97, 147,
        154–160, 206–207, 227–229,
        233, 242
    in circumstellar shells, 203–207
    in diffuse clouds, 147–155
    in laboratory ices, 226, 233
    in molecular clouds, 147, 155–164
    list of observed, 147
    polarization in, 95, 97, 163–164
Infrared continuum emission, 4, 13–15,
    165–180, 201–203
Infrared emission features, 180–189
    due to PAHs, 176, 182–189
    due to SiC, 204–205
    due to silicates, 152, 180–182,
        204–205, 242, 250–252
    in comets, 250–252
    in emission and reflection nebulae,
        176, 180–189
    in evolved stars, 185, 203–205
Integrated absorption cross-section,
    158
Intercloud medium, 18
Interstellar grains, see Dust grains,
    interstellar
Interstellar medium (ISM),
    definition of, 1
    environments of, 15–21
    local, 21
    phases of, 15, 17–21
    volume filling factor of, 18
Interstellar molecules,
    abundances of, 221
    deuteration of, 254
    exothermic reactions of, 219–221,
        254
    formation of, 23, 218–222
    organic, 254

Interstellar radiation field (ISRF), 17,
    19, 262
    energy density of, 166–167
Iron (Fe), elemental,
    abundance and depletion of, 43, 52,
        54, 113, 236
    significance in nucleosynthesis, 37,
        43

Johnson (UBV) photometric system,
    intrinsic colours in, 63
    polarization in, 86
    reddening in, 7,

Kirchhoff's law, 167
Kramers–Krönig relationship, 72

Large Magellanic Cloud (LMC),
    dust to gas ratio in, 48, 73
    extinction curves for, 76–78
    metallicity of, 48, 77, 121–122
    polarization in, 104–106
    silicate dust in, 182
    ultraviolet $\lambda 2175$ feature in, 120–122
Life, origin of, 253–256
Lithium, elemental,
    abundance of, 40–43
    nucleosynthesis of, 35

Magnesium oxide (MgO), 128
Magnetic field, interstellar,
    flux density of, 110
    significance for grain alignment,
        90–92, 106–114
    variations in, 91–92
Mass absorption coefficient, 150
    of hydrogenated amorphous carbon
        at 3.4 μm, 154
    of graphite/silicate mixture at
        100 μm, 178
    of organic residues at 3.4 μm, 154
    of SiC at 11.2 μm, 153
    of silicates at 9.7 μm, 150, 164
Mass loss, stellar, 38–39, 210–215
Mass spectroscopy, 252
Metallicity, 44–45
Metal oxides, 55
    absorptions in, 128, 139, 204
    circumstellar formation of, 197–199
Meteorites, 39–43, 245–256 (see also
        Carbonaceous chondrites)

Methanol (CH_3OH), interstellar, 161, 263
Mie theory, 59–60, 84–85, 212
  reliability of, 126–127
Milky Way Galaxy (see also Galactic Centre),
  abundance variations in, 43–47
  distribution of dust in, 8–15
  infrared continuum emission from, 165, 174–179
  supernova detection in, 11–13
  visual appearance of, 2–3
Miller–Urey experiment, 253
Missing mass, 17, 23
Models for interstellar dust, 25–30, 53–56, 80–82, 257–263
  biological, 29–30, 56, 255–256
  core/mantle, 29, 155, 259–260
  ice, 26, 80
  unified, 257–263
  unmantled refractory (MRN), 28–29, 80–82, 97, 260
Mon R2 IRS3, infrared spectrum of, 159, 162, 229

Neutron capture, 37
NGC 7538E, infrared spectrum of, 155–156
Novae, 205, 213–214
Nucleation theory, 195–196
Nucleogenesis, 34
Nucleosynthesis, stellar, 35–37, 43

Occam's razor, 256
OH–IR stars, 206–207, 213–214
Ophiuchus (ρ Oph) dark cloud,
  extinction in, 74–75, 78–79
  ice threshold in, 157
  grain growth in, 222–223
  polarization in, 104–106
  ultraviolet $\lambda2175$ feature in, 119–120
Optical constants, 59, 127
Organic refractory material,
  abundance of, 55
  cometary, 250–252
  delivery to Earth, 255
  infrared 3.4 μm feature in, 153–154, 250–252
  interstellar origin of, 29, 232–233, 260
  meteoritic, 246–248, 253
Orion Bar,

PAH features in, 185–187
Orion Nebula,
  anomalous extinction curve in, 73–75
  anomalous $\lambda2175$ strength in, 119–120
  polarization in, 104, 106
  silicate emission feature in, 181
  star formation in, 239
Oxides, refractory, see Metal oxides
Oxygen, elemental,
  abundance and depletion of, 42–43, 51–56, 158
  nucleosynthesis of, 36–37

Panspermia, 255–256
Paramagnetic relaxation time, 109
Phase diagrams, 198–199
Phosphorus abundance, 42
  significance for biological grain model, 30, 56
Photoelectric effect, 45, 113
Planck function, 167
Planetary nebulae, 183, 201, 208, 213–214
Plasma oscillations, 124, 139
Platt particles, 26
Polarization, interstellar, 8, 83–114, 179–180
  circular, 83, 91, 99–101, 262
  correlation of P_K with τ_K, 91–93
  correlation of P_V with E_{B-V}, 86–87
  correlation of λ_{max} with R_V, 101–102
  degree of, 86
  infrared power-law behaviour of, 95–98
  mathematical representation of, 92–98
  models for spectral dependence of, 95–98
  of IR continuum emission, 171, 179–180
  position angle of, 86, 179–180
  relation to galactic structure, 88–91
  Serkowski law of, 93
  spectral dependence of, 92–108
  Wilking law of, 94
Polycyclic aromatic hydrocarbons,
  abundance limits on, 128, 184
  aliphatic subgroups on, 188

as carriers of diffuse interstellar bands, 143
circumstellar, 203–204, 208–210
formation in C-rich stellar atmospheres, 200
formation in the early Solar System, 254–255
hydrogenation of, 185
infrared emission spectrum of, 176, 182–189
ionization of, 143
molecular structure of, 183, 186
spectral resonances of, 183–186
relation to amorphous carbon, 28–29, 192, 200, 260
relation to graphite, 28–29, 200
ultraviolet absorption in, 127–128
Post-AGB stars, 208–209, 261
Poynting–Robertson effect, 243
Predissociation transitions, 142
Preionization transitions, 142
Pre-main-sequence stars, 146, 189, 227, 239–244
Proteins, ultraviolet absorption in, 123
Protoplanetary disks, 242–244
Protostars, 146, 155, 227–230, 239–240

Quenched carbonaceous composites, 127

Radiation pressure, 25, 211–212, 243–245
efficiency factor for, 211
Rate of visual extinction with respect to distance, 11
Ratio of hydrogen column density to extinction, 15–16
Ratio of polarization to extinction, observational limit, 87
theoretical limit, 85
within spectral feature, 164
Ratio of total to selective extinction, 7–8, 65
evaluation of, 70–72
infrared formulation of, 78–79
mechanism for increases in, 222–223
relation to infrared colours, 71
relation to λ_{max}, 101–103
significance for distance determinations, 8, 22
value in dense clouds, 74, 78
value in diffuse clouds, 8, 71

Rayleigh scattering, 61
Rayleigh–Jeans approximation, 170, 201–202, 240
R Coronae Borealis (RCB) stars, 207–209
Red giants, see Stars, evolved
Red Rectangle (HD 44179), 184–185, 190, 192, 208, 261
Reddening, interstellar, 7
cosecant law of, 10
distribution of, 8–13
Reflection nebulae,
extended red emission from, 190–192
infrared continuum emission from, 175–176, 182
infrared PAH features in, 184–185
scattering in, 13, 66–69
Refractive index, complex, 59

Scattering, 5, 13, 57–69
asymmetry factor for, 61–62
cross-section for, 61
efficiency factor for, 59–60
in small-particle approximation, 60–61
observed properties of, 66–69
Serkowski law, 93
Shocks, interstellar,
grain destruction and processing in, 234–238
timescales for, 215, 238
Silicates,
abundance relative to SiC, 151–153, 262–263
active surface sites on, 216
circumstellar origin of, 197–199, 263
composition and structure of, 28, 164, 197
contribution to dust density, 55, 150–151
contribution to grain models, 28–29, 55, 80–81, 258–260, 262–263
contribution to polarization, 95–99, 163–164
correlation of $\tau_{9.7}$ with A_V, 150–151
crystallinity of, 149, 249–250, 252
destruction of, 234–238, 262
emissivity function of 9.7 μm feature, 181, 204
equilibrium temperature of, 168
FIR emissivity of, 170

in comets and interplanetary dust,
249–252
infrared absorption features in, 129,
147–153, 181, 203–207
infrared emission features in,
180–182, 203–205, 249–250
impurities in, 139
meteoritic, 245–247
ultraviolet absorption in, 128
Silicon carbide (SiC),
abundance relative to silicates,
151–153, 262–263
circumstellar origin of, 199, 201, 248
destruction of, 237, 262–263
infrared 11.2 μm emission feature of,
204–205
limit on interstellar abundance,
151–153
meteoritic, 247–249
Silicon, elemental
abundance and depletion of, 40–43,
51–52, 55, 150–151, 235–236,
262
nucleosynthesis of, 36–37
Size distribution function, 28, 58
effect of grain–grain collisions on,
213, 223, 237
effect of mantle growth on, 223
in stellar winds, 28, 213
of meteoritic particles, 247–248
power-law form of, 28, 80–82, 97,
213, 247, 259–260
Small Magellanic Cloud,
extinction curve of, 76–78
metallicity of, 48, 77, 121, 261
non-detection of $\lambda 2175$ in, 120–122
polarization in, 105
Small-particle approximation, 60–61
Solar System,
age of, 40
diffuse matter in, 245–252
element abundances in, 39–43
interstellar dust within, 10, 245–256
interstellar environment of, 21–22
Soot, formation of, 184, 200
Spectral energy distribution, infrared,
of evolved stars, 201–203
of young stars, 239–241
Spectral index of FIR emissivity, 24,
167–170, 201–203
Spectral lines, equivalent width of,
48–49

Spiral galaxies (see also Milky Way
Galaxy),
Andromeda (M31), 105
M33, 46, 176–177
NGC 891, 2, 4, 14
NGC 4565, 90, 155
Sputtering, 234, 237
Starburst galaxies, 189
Star clusters, globular,
element abundances in, 44
Star clusters, open,
extinction of, 5–6, 71–72
polarization of, 104, 106
Star formation, 24–25, 232–233,
238–242
Stars, evolved
as sources of interstellar grains,
210–215
C/O ratio in, 46–47, 196–199,
202–204
dust condensation in atmospheres
of, 195–201
hydrogen deficiency in, 197, 200
mass loss from, 38–39, 210–215
metallicity of, 46–47, 192, 197
observations of circumstellar dust
in, 201–210
Stefan–Boltzmann law, 166
Stellar distance equation, 5–6
Sticking coefficient, 217–218, 223–224
Stokes parameters, 88
Strömgren ($uvby$) photometric system,
metallicity parameter, 44–45
reddening in, 7
Sun, element abundances in, 39–43
Supernovae,
and phase structure of the ISM,
17–18
element production in, 37–38
dust in the ejecta of, 213–214
shocks and grain destruction,
234–238
visibility of, 11–13
Superparamagnetic alignment,
112–113, 179–180
Suprathermal spin, 111–112

Taurus dark cloud,
anomalous $\lambda 2175$ feature in,
118–119
CO depletion in, 230–231

ice mantles in, 156–160, 224, 227–231
star formation in, 239
Titanium depletion, 53–54, 236
Trapezium emissivity function, 162, 181
Triple-α process, 36
Trumpler cluster technique, 5–6, 71–72
T Tauri stars, 240, 242–243

Ultraviolet absorption feature (λ2175), 27–30, 74–82, 116–129,
  absence in circumstellar dust, 207–209
  abundance limits on carrier, 123–124
  albedo observations of, 68–69
  anomalous strength in dense clouds, 118–120
  correlation with diffuse interstellar bands, 135
  correlation with FUV extinction, 119–120
  correlation with reddening, 117
  graphite identification of, 124–127, 129, 259–261
  mathematical representation of, 117–118
  observed properties of, 116–123
  oxygen-rich carriers for, 128–129
  polarization in, 122
  scattering in, 122

Ultraviolet absorption features (other than λ2175), 27, 30, 120, 122–124, 127–128, 207–209

Vega phenomenon, 25, 243–244
Very small grains (VSGs) (see also Polycyclic aromatic hydrocarbons)
  carbon-rich nature of, 182
  FUV extinction of, 69, 80–82
  infrared continuum emission from, 14–15, 174–178
  stochastic heating of, 171–173
Vibrational resonances,
  in polycyclic aromatic hydrocarbons, 183–184
  in solids, 144–146

Water, gas phase, 220–221, 224
Wavelength of maximum polarization, 92–106
Wien displacement law, 169
Wilking law, 94
Wolf–Rayet stars, 213–214
W33A, infrared spectrum of, 146–147, 161

X-rays, absorption of, 45–46

Young stellar objects, 240 (see also Protostars)

Zero-age main sequence, 239